Computer Modelling of
Electrical Power Systems

Computer Modelling of Electrical Power Systems

J. ARRILLAGA and C. P. ARNOLD

Department of Electrical Engineering,
University of Canterbury, New Zealand

and

B. J. HARKER

New Zealand Electricity

A Wiley–Interscience Publication

JOHN WILEY & SONS

Chichester · New York · Brisbane · Toronto · Singapore

Library of Congress Cataloging in Publication Data:
Arrillaga, J.
  Computer modelling of electrical power systems.
  'A Wiley–Interscience publication.'
  Includes bibliographies and index.
  1. Electric power systems—Mathematical models.
    2. Electric power systems—Data processing.
    I. Arnold, C. P. II. Harker, B. J. III. Title.
    TK1005.A76   621.319′1′0724   82–2664

    ISBN 0 471 10406 X   AACR2

British Library Cataloguing in Publication Data:
Arrillaga, J.
  Computer modelling of electrical power systems.
  1. Electric power systems—Simulation methods
  2. Electric power systems—Data processing
  I. Title   II. Arnold, C. P.   III. Harker, B. J.
  621.31′0724   TK1005

    ISBN 0 471 10406 X

Photosetting by Thomson Press (India) Ltd., New Delhi

Printed in Great Britain by
Page Bros (Norwich) Ltd

Contents

v

851251

Preface

This book describes the use of power-system component models and efficient computational techniques in the development of a new generation of computer program representing the steady and dynamic states of Electrical Power Systems.

The content, although directed to practicing engineers, should also be appropriate for Advanced Power System courses at final year degree and Masters levels. Much of the material included was developed by the authors and their colleagues at the University of Manchester Institute of Science and Technology and used as part of a postgraduate course in Electrical Power Systems Analysis and Control. The integration of h.v.d.c. transmission and the final development of programs were carried out by the authors since 1975 at the University of Canterbury (New Zealand), in collaboration with New Zealand Electricity.

Some basic knowledge of Power-System theory, matrix analysis and numerical techniques is presumed, and specific references are given to help the uninitiated to pick up the relevant material.

An introductory chapter describes the main computational and transmission system developments justifying the purpose of the book. Steady state models of a.c. and d.c. power-system plant components with varying degrees of complexity are derived in Chapters 2 and 3 respectively.

A general purpose single-phase a.c. load-flow program is described in Chapter 4 with particular reference to the Newton Fast-Decoupled algorithm. In Chapter 5, the decoupling technique is used to develop a three-phase load-flow program and Chapters 6 and 7 extend the previous algorithms to incorporate static converters and h.v.d.c. transmission. Chapter 8 is devoted to the modelling of conventional a.c. faults.

The remaining chapters consider the power system in the dynamic state. Chapter 9 describes basic dynamic models of a.c. power-system plant and their use in multi-machine transient stability studies. More advanced dynamic models and a quasi-steady-state representation of large converter plant and h.v.d.c. transmission are developed in Chapter 10.

A detailed Transient Converter Simulation model and its application to the analysis of a.c. and d.c. short-circuit faults are described in Chapters 11 and 12 respectively.

Finally, Chapter 13 combines the Transient Converter Simulation and multi-machine transient stability programs for the assessment of transient stability in a.c. systems containing h.v.d.c. transmission links.

The authors should like to express their gratitude to P. W. Blakeley (ex-Manager) and K. McCool present General Manager of New Zealand Electricity for their continuous support in this work. They also wish to acknowledge the considerable help received over a decade from many of their colleagues and in particular from A. Brameller, B. Stott, M. D. Heffernan, K. S. Turner, P. Bodger, J. G. Campos Barros, H. Al-Khashali and D. B. Giesner. Finally, the authors are very grateful to Mrs. A. Haughan for her active cooperation in the preparation of the manuscript.

1
Introduction

1.1. GENERAL BACKGROUND

In parallel with the development of larger and faster digital computers in the 1950's, considerable effort was devoted to the computer modelling of large Electrical Power Systems. Several good books have collected such experience emphasizing the potential of matrix analysis and network theory for the solution of Power-System problems.

In the early stages, however, the lack of coordination regarding computer hardware and languages made it difficult to develop techniques for general use and the tendency was for various supply authorities to develop their own programs. Later, with the general acceptance of FORTRAN in the Power-System field, some success was achieved by computer manufacturers in producing program packages which included load flow, short-circuit and stability studies.

The dramatic expansion of the subject in the '70's can no longer be recorded in a single textbook and various specialist treatises are regularly compiled to review and compare the developments.

While the academic search for new methods continues, the general feeling is that no major contribution is expected over the basic techniques developed in the last decade for the modelling of a.c. power systems.

The primary subject of computer modelling is the load-flow problem, which finds application in all phases of power-system analysis. Due to space limitations, only the solution of the basic load-flow equations is normally considered. It is acknowledged, however, that the load-flow problem is not restricted to the solution of the basic continuously differentiable equations. There is probably not a single routine program in use anywhere that does not model other features. Such features often have more influence on convergence than the performance of the basic algorithm.

The most successful contribution to the load-flow problem has been the application of Newton–Raphson and derived algorithms. These were finally established with the development of programming techniques for the efficient handling of large matrices and in particular the sparsity-oriented ordered-elimination methods. The Newton algorithm was first enhanced by taking advantage of the decoupling characteristics of power flow and finally by the use of reasonable approximations directed towards the use of constant Jacobian matrices.

In dynamic studies the most significant modelling development has probably been the application of implicit integration techniques which allow the differential equations to be algebraized and then incorporated with the network algebraic equations to be solved simultaneously. The use of implicit trapezoidal integration has proved to be very stable, permitting step lengths greater than the smallest time constant of the system. This technique allows detailed representation of synchronous machines with their voltage regulators and governors, induction motors and nonimpedance loads.

1.2. TRANSMISSION SYSTEM DEVELOPMENT

Large increases in transmission distances and voltage levels have resulted in more sophisticated means of active and reactive power control and the use of d.c. transmission.

Long-distance transmission presents voltage and power-balancing problems which can only be accurately assessed by using correspondingly accurate models of power transmission plant in the phase frame of reference.

Following the first commercial scheme in 1954, h.v.d.c. transmission has been gradually accepted as a viable alternative to a.c. for many applications and is now well established.

An all-purpose package of power-system programs can no longer ignore such developments and the basic algorithms used in conventional a.c. systems need to be reconsidered for their efficient incorporation.

At first glance the book may appear to overemphasize the role of h.v.d.c. transmission, while the number of such schemes in existence and under consideration is still relatively small. However, most of the world's large power systems contain some large power converters and d.c. transmission links. Moreover, considering the large power ratings of such schemes, their presence influences the behaviour of the rest of the system and must be represented properly.

Whenever possible, any equivalent models used to simulate the converter behaviour should involve traditional power-system concepts, for easy incorporation within existing power-system programs.

However, the number of degrees of freedom of d.c. power transmission is higher and any attempt to model its behaviour in the more restricted a.c. framework will have limited application.

The integration of h.v.d.c. transmission with conventional a.c. load-flow models has been given sufficient coverage in recent years and is now well understood. Steady-state unbalanced studies, on the other hand, require for more elaborate modelling in the presence of h.v.d.c. converters and have so far been given very little consideration.

With reference to power-system disturbances, the behaviour of h.v.d.c. converters cannot be modelled by conventional fault study or stability programs.

As a result, a.c.–d.c. system planning is normally carried out with the help of elaborate scaled-down models and simulators. There is considerable reluctance to the introduction of computer modelling for this purpose on the basis of inadequate control representation and prohibitive computer runs. However, it is reasonable to expect that this is only a temporary problem and that in the long run, computer modelling will become the only practical alternative for the analysis of a.c.–d.c. system disturbances.

1.3. THEORETICAL MODELS AND COMPUTER PROGRAMS

The dissemination of information on Power System computation takes the form of technical publications and program instruction manuals.

Technical papers are by necessity concise, their information is mostly theoretical and their structure schematic. Textbooks, although less restricted, follow a similar pattern and the extent of practical information normally included is insufficient to provide adequate understanding of the programs.

Instruction manuals on the other hand, provide full information on the practical structure of the programs but lack the technical background necessary for the user to perform the inevitable modifications required in the long run.

It is almost expected that a specialist book of this type should provide a comprehensive survey and comparison of the various conventional alternatives. Such approach, although academically satisfying, would detract from the main object of the book and would occupy invaluable space. Instead, generally recognized as efficient up-to-date modelling techniques will be described from theoretical and practical considerations.

Finally, the authors have tried to combine theoretical and practical considerations to introduce in the most direct way the application of such techniques to the development of a new generation of power-system programs.

Modelling of Power Transmission Plant

2.1. INTRODUCTION

Transmission plant components are modelled by their equivalent circuits in terms of inductance, capacitance and resistance. Each unit constitutes an electric network in its own right and their interconnection constitutes the transmission system.

Among the many alternative ways of describing transmission systems to comply with Kirchhoff's laws, two methods, mesh and nodal analysis, are normally used. The latter has been found to be particularly suitable for digital computer work, and is almost exclusively used for routine network calculations.

The nodal approach has the following advantages:

— The numbering of nodes, performed directly from a system diagram, is very simple.
— Data preparation is easy.
— The number of variables and equations is usually less than with the mesh method for power networks.
— Network crossover branches present no difficulty.
— Parallel branches do not increase the number of variables or equations.
— Node voltages are available directly from the solution, and branch currents are easily calculated.
— Off-nominal transformer taps can easily be represented.

This chapter deals with the derivation of equivalent circuits of transmission plant components and with the formation of the system admittance matrix relating the current and voltage at every node of the transmission system. This constitutes the basic framework for the algorithms developed in following chapters.

2.2. LINEAR TRANSFORMATION TECHNIQUES

Linear transformation techniques are used to enable the admittance matrix of any network to be found in a systematic manner. Consider, for the purposes of illustration, the network drawn in Fig. 2.1.

Five steps are necessary to form the network admittance matrix by linear transformation, i.e.

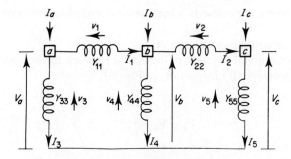

Fig. 2.1. Actual connected network

(i) Label the nodes in the original network.
(ii) Number, in any order, the branches and branch admittances.
(iii) Form the primitive network admittance matrix by inspection.

This matrix relates the nodal injected currents to the node voltages of the primitive network. The primitive network is also drawn by inspection of the actual network. It consists of the unconnected branches of the original network with a current equal to the original branch current injected into the corresponding node of the primitive network. The voltages across the primitive network branches then equal those across the same branch in the actual network.

The primitive network for Fig. 2.1. is shown in Fig. 2.2.

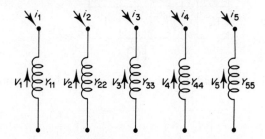

Fig. 2.2. Primitive or unconnected network

The primitive admittance matrix relationship is:

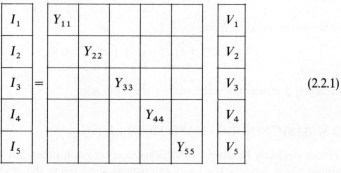

$$[Y_{PRIM}] \tag{2.2.1}$$

Off-diagonal terms are present where mutual coupling between branches is present.

(iv) Form the connection matrix $[C]$.

This relates the nodal voltages of the actual network to the nodal voltages of the primitive network. By inspection of Fig. 2.1,

$$V_1 = V_a - V_b$$
$$V_2 = V_b - V_c$$
$$V_3 = V_a \qquad (2.2.2)$$
$$V_4 = V_b$$
$$V_5 = V_c$$

or in matrix form

V_1	1	-1			V_a
V_2		1	-1		V_b
V_3 $=$	1				V_c
V_4		1			
V_5			1		

$$[C] \qquad (2.2.3)$$

(v) The actual network admittance matrix which relates the nodal currents to the voltages by,

I_a		V_a
I_b $=$	$[Y_{abc}]$	V_b
I_c		V_c

$$(2.2.4)$$

can now be derived from,

$$[Y_{abc}] = [C]^T [Y_{\text{PRIM}}][C] \qquad (2.2.5)$$
$$3 \times 3 \quad 3 \times 5 \quad 5 \times 5 \quad 5 \times 3$$

which is a straightforward matrix multiplication.

2.3. BASIC SINGLE-PHASE MODELLING

Under perfectly balanced conditions, transmission plant can be represented by single-phase models, the most extensively used being the equivalent-π circuit.

Transmission lines

In the case of a transmission line the total resistance and inductive reactance of the line is included in the series arm of the π-equivalent and the total capacitance to neutral is divided between its shunt arms.

Transformer on nominal ratio

The equivalent-π model of a transformer is illustrated in Fig. 2.3, where y_{oc} is the reciprocal of z_{oc} (magnetizing impedance) and y_{sc} is the reciprocal of z_{sc} (leakage impedance). z_{sc} and z_{oc} are obtained from the standard short-circuit and open-circuit tests.

This yields the following matrix equation:

$$
\begin{bmatrix} I_p \\ I_s \end{bmatrix} = \begin{bmatrix} y_{sc} & -y_{sc} + y_{oc/2} \\ -y_{sc} + y_{oc/2} & y_{sc} \end{bmatrix} \begin{bmatrix} V_p \\ V_s \end{bmatrix}
\tag{2.3.1}
$$

where

y_{sc} is the short-circuit or leakage admittance,
y_{oc} is the open-circuit or magnetizing admittance.

The use of a three-terminal network is restricted to the single-phase representation and cannot be used as a building block for modelling three-phase transformer banks.

The magnetizing admittances are usually removed from the transformer model and added later as small shunt-connected admittances at the transformer terminals. In the per unit system the model of the single-phase transformer can then be reduced to a lumped leakage admittance between the primary and secondary busbars.

Off-nominal transformer tap settings

A transformer with turns ratio 'a' interconnecting two nodes i, k can be represented by an ideal transformer in series with the nominal transformer leakage admittance as shown in Fig. 2.4(a).

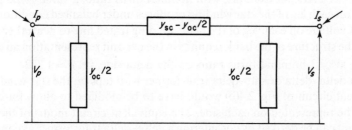

Fig. 2.3. Transformer equivalent circuit

If the transformer is on nominal tap ($a = 1$), the nodal equations for the network branch in the per unit system are:

$$I_{ik} = y_{ik}V_i - y_{ik}V_k \qquad (2.3.2)$$

$$I_{ki} = y_{ik}V_k - y_{ik}V_i \qquad (2.3.3)$$

In this case $I_{ik} = -I_{ki}$.

For an off-nominal tap setting and letting the voltage on the k side of the ideal transformer be V_t we can write

$$V_t = \frac{V_i}{a} \qquad (2.3.4)$$

$$I_{ki} = y_{ik}(V_k - V_t) \qquad (2.3.5)$$

$$I_{ik} = -\frac{I_{ki}}{a} \qquad (2.3.6)$$

Eliminating V_t between equations (2.3.4) and (2.3.5) we obtain

$$I_{ki} = y_{ik}V_k - \frac{y_{ik}}{a}V_i \qquad (2.3.7)$$

$$I_{ik} = -\frac{y_{ik}}{a}V_k + \frac{y_{ik}}{a^2}V_i \qquad (2.3.8)$$

A simple equivalent π circuit can be deduced from equations (2.3.7) and (2.3.8) the elements of which can be incorporated into the admittance matrix. This circuit is illustrated in Fig. 2.4(b)

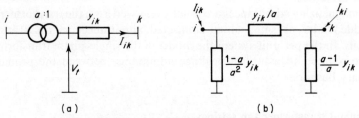

(a)   (b)

Fig. 2.4. Transformer with off-nominal tap setting

The equivalent circuit of Fig. 2.4(b) has to be used with care in banks containing delta-connected windings. In a star–delta bank of single-phase transformer units, for example, with nominal turns ratio, a value of 1.0 per unit voltage on each leg of the star winding produces under balanced conditions 1.732 per unit voltage on each leg of the delta winding (rated line to neutral voltage as base). The structure of the bank requires in the per unit representation an effective tapping at $\sqrt{3}$ nominal turns ratio on the delta side, i.e. $a = 1.732$.

For a delta–delta or star–delta transformer with taps on the star winding, the equivalent circuit of Fig. 2.4(b) would have to be modified to allow for effective taps to be represented on each side. The equivalent-circuit model of the single-phase unit can be derived by considering a delta–delta transformer as comprising a delta–star transformer connected in series (back to back) via a zero-impedance link to a star–delta transformer, i.e. star windings in series. Both neutrals are solidly earthed. The leakage impedance of each transformer would be half the impedance of the equivalent delta–delta transformer. An equivalent per unit representation of this coupling is shown in Fig. 2.5. Solving this circuit for terminal currents:

$$I_p = \frac{I'}{\alpha} = \frac{(V' - V'')y}{\alpha}$$

$$= \frac{(V_p/\alpha - V_s/\beta)y}{\alpha} = \frac{y}{\alpha^2}V_p - \frac{y}{\alpha\beta}V_s \qquad (2.3.9)$$

$$-I_s = \frac{I'}{\beta} = \frac{y}{\alpha\beta}V_p - \frac{y}{\beta^2}V_s \qquad (2.3.10)$$

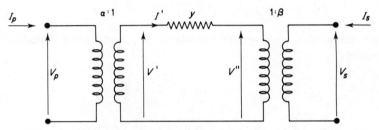

Fig. 2.5. Basic equivalent circuit in p.u. for coupling between primary and secondary coils with both primary and secondary off-nominal tap ratios of α and β

or in matrix form:

$$\begin{vmatrix} I_p \\ I_s \end{vmatrix} = \begin{vmatrix} y/\alpha^2 & -y/\alpha\beta \\ -y/\alpha\beta & y/\beta^2 \end{vmatrix} \begin{vmatrix} V_p \\ V_s \end{vmatrix} \qquad (2.3.11)$$

These admittance parameters form the primitive network for the coupling between a primary and secondary coil.

Phase-shifting transformers

To cope with phase-shifting, the transformer of Fig. 2.5 has to be provided with a complex turns ratio. Moreover, the invariance of the product VI* across the ideal transformer requires a distinction to be made between the turns ratios for current and voltage, i.e.

$$V_p I_p^* = -V'I'^*$$

or

$$V_p = (a + jb)V' = \alpha V'$$

$$I_p^* = -\frac{I'^*}{a + jb}$$

$$I_p = -\frac{I'}{a - jb} = -\frac{I'}{\alpha^*}$$

Thus the circuit of Fig. 2.5 has two different turns ratios, i.e.

$$\alpha_v = a + jb \quad \text{for the voltages}$$

and

$$\alpha_i = a - jb \quad \text{for the currents}$$

Solving the modified circuit for terminal currents:

$$I_p = \frac{I'}{\alpha_i} = \frac{(V' - V'')y}{\alpha_i}$$

$$= \frac{(V_p/\alpha_v - V_s/\beta)y}{\alpha_i} = \frac{y}{\alpha_v \alpha_i} V_p - \frac{y}{\alpha_i \beta} V_s \tag{2.3.12}$$

$$-I_s = \frac{I'}{\beta} = \frac{y}{\alpha_v \beta} V_p - \frac{y}{\beta^2} V_s \tag{2.3.13}$$

Thus, the general single-phase admittance of a transformer including phase shifting is:

$$[y] = \begin{array}{|c|c|} \hline \dfrac{y}{\alpha_i \alpha_v} & -\dfrac{y}{\alpha_i \beta} \\ \hline -\dfrac{y}{\alpha_v \beta} & \dfrac{y}{\beta^2} \\ \hline \end{array} \tag{2.3.14}$$

It should be noted that although an equivalent lattice network similar to that in Fig. 2.5 could be constructed, it is no longer a bilinear network as can be seen from the asymmetry of y in equation (2.3.14). The equivalent circuit of a single-phase phase-shifting transformer is thus of limited value and the transformer is best represented analytically by its admittance matrix.

2.4. THREE-PHASE SYSTEM ANALYSIS

Discussion of the frame of reference

Sequence components have long been used to enable convenient examination of the balanced power system under both balanced and unbalanced loading conditions.

The symmetrical component transformation is a general mathematical technique developed by Fortescue whereby any 'system of n vectors or quantities may be resolved, when n is prime, into n different symmetrical n phase systems'.[1] Any set of three-phase voltages or currents may therefore be transformed into three symmetrical systems of three vectors each. This, in itself, would not commend the method and the assumptions, which lead to the simplifying nature of symmetrical components, must be examined carefully.

Consider, as an example, the series admittance of a three-phase transmission line, shown in Fig. 2.6, i.e. three mutually coupled coils. The admittance matrix relates the illustrated currents and voltages by

$$[I]_{abc} = [Y]_{abc}[V]_{abc} \tag{2.4.1}$$

where

$$[I]_{abc} = [I_a \ I_b \ I_c]^T$$
$$[V]_{abc} = [V_a \ V_b \ V_c]^T$$

and

$$[Y]_{abc} = \begin{array}{|c|c|c|} \hline y_{aa} & y_{ab} & y_{ac} \\ \hline y_{ba} & y_{bb} & y_{bc} \\ \hline y_{ca} & y_{cb} & y_{cc} \\ \hline \end{array}$$

(2.4.2)

By the use of the symmetrical component transformation the three coils of Fig. 2.6 can be replaced by three uncoupled coils. This enables each coil to be treated separately with a great simplification of the mathematics involved in the analysis.

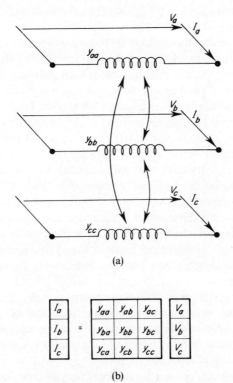

(a)

(b)

Fig. 2.6. Admittance representation of a three-phase series element. (a) Series admittance element; (b) Admittance matrix representation

The transformed quantities (indicated by subscripts 012 for the zero, positive and negative sequences respectively) are related to the phase quantities by:

$$[V_{012}] = [T_s]^{-1}[V_{abc}] \tag{2.4.3}$$

$$[I_{012}] = [T_s]^{-1}[I_{abc}] \tag{2.4.4}$$

$$= [T_s]^{-1}[Y_{abc}][T_s][V_{012}] \tag{2.4.5}$$

where $[T_s]$ is the transformation matrix.

The transformed voltages and currents are thus related by the transformed admittance matrix,

$$[Y_{012}] = [T_s]^{-1}[Y_{abc}][T_s] \tag{2.4.6}$$

Assuming that the element is balanced, we have

$$
\begin{aligned}
y_{aa} &= y_{bb} = y_{cc} \\
y_{ab} &= y_{bc} = y_{ca} \\
y_{ba} &= y_{cb} = y_{ac}
\end{aligned}
\tag{2.4.7}
$$

and a set of invariant matrices $[T]$ exist. Transformation (2.4.6) will then yield a diagonal matrix $[y]_{012}$.

In this case the mutually coupled three-phase system has been replaced by three uncoupled symmetrical systems. In addition, if the generation and loading are balanced, or may be assumed balanced, then only one system, the positive sequence system, has any current flow and the other two sequences may be ignored. This is essentially the situation with the single-phase load flow.

If the original phase admittance matrix $[y_{abc}]$ is in its natural unbalanced state then the transformed admittance matrix $[y_{012}]$ is full. Therefore, current flow of one sequence will give rise to voltages of all sequences, i.e. the equivalent circuits for the sequence networks are mutually coupled. In this case the problem of analysis is no simpler in sequence components than in the original phase components and symmetrical components should not be used.

From the above considerations it is clear that the asymmetry inherent in all power systems cannot be studied with any simplification by using the symmetrical component frame of reference. Data in the symmetrical component frame should only be used when the network element is balanced, for example, synchronous generators.

In general, however, such an assumption is not valid. Unsymmetrical interphase coupling exists in transmission lines and to a lesser extent in transformers and this results in coupling between the sequence networks. Furthermore, the phase shift introduced by transformer connections is difficult to represent in sequence component models.

With the use of phase coordinates the following advantages become apparent:

— Any system element maintains its identity.
— Features such as asymmetric impedances, mutual couplings between phases

and between different system elements, and line transpositions are all readily considered.

— Transformer phase shifts present no problem.

The use of compound admittances

When analysing three-phase networks, where the three nodes at a busbar are always associated together in their interconnections, the graphical representation of the network is greatly simplified by means of 'compound admittances', a concept which is based on the use of matrix quantities to represent the admittances of the network.

The laws and equations of ordinary networks are all valid for compound networks by simply replacing single quantities by appropriate matrices.[2]

Consider six mutually coupled single admittances, the primitive network of which is illustrated below in Fig. 2.7.

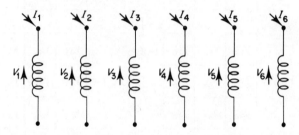

Fig. 2.7. Primitive network of six coupled admittances

The primitive admittance matrix relates the nodal injected currents to the branch voltages as follows:

$$
\begin{bmatrix} I_1 \\ I_2 \\ I_3 \\ I_4 \\ I_5 \\ I_6 \end{bmatrix}
=
\begin{bmatrix}
y_{11} & y_{12} & y_{13} & y_{14} & y_{15} & y_{16} \\
y_{21} & y_{22} & y_{23} & y_{24} & y_{25} & y_{26} \\
y_{31} & y_{32} & y_{33} & y_{34} & y_{35} & y_{36} \\
y_{41} & y_{42} & y_{43} & y_{44} & y_{45} & y_{46} \\
y_{51} & y_{52} & y_{53} & y_{54} & y_{55} & y_{56} \\
y_{61} & y_{62} & y_{63} & y_{64} & y_{65} & y_{66}
\end{bmatrix}
\begin{bmatrix} V_1 \\ V_2 \\ V_3 \\ V_4 \\ V_5 \\ V_6 \end{bmatrix}
\qquad (2.4.8)
$$

$6 \times 1 \qquad\qquad 6 \times 6 \qquad\qquad 6 \times 1$

Partitioning equation (2.4.8) into 3×3 matrices and 3×1 vectors, the equation becomes,

$$
\left[\begin{array}{c} [I_a] \\ \hline [I_b] \end{array}\right] = \left[\begin{array}{c|c} [Y_{aa}] & [Y_{ab}] \\ \hline [Y_{ba}] & [Y_{bb}] \end{array}\right] \left[\begin{array}{c} [V_a] \\ \hline [V_b] \end{array}\right] \qquad (2.4.9)
$$

where

$$[I_a] = [I_1 \ I_2 \ I_3]^T$$
$$[I_b] = [I_4 \ I_5 \ I_6]^T$$

$$
[Y_{aa}] = \begin{array}{|c|c|c|}
\hline
y_{11} & y_{12} & y_{13} \\
\hline
y_{21} & y_{22} & y_{23} \\
\hline
y_{31} & y_{32} & y_{33} \\
\hline
\end{array}
\qquad
[Y_{bb}] = \begin{array}{|c|c|c|}
\hline
y_{44} & y_{45} & y_{46} \\
\hline
y_{54} & y_{55} & y_{56} \\
\hline
y_{64} & y_{65} & y_{66} \\
\hline
\end{array}
$$

$$(2.4.10)$$

$$
[Y_{ab}] = \begin{array}{|c|c|c|}
\hline
y_{14} & y_{15} & y_{16} \\
\hline
y_{24} & y_{25} & y_{26} \\
\hline
y_{34} & y_{35} & y_{36} \\
\hline
\end{array}
\qquad
[Y_{ba}] = \begin{array}{|c|c|c|}
\hline
y_{41} & y_{42} & y_{43} \\
\hline
y_{51} & y_{52} & y_{53} \\
\hline
y_{61} & y_{62} & y_{63} \\
\hline
\end{array}
$$

Graphically we represent this partitioning as grouping the six coils into two compound coils (a) and (b), each composed of three individual admittances. This is illustrated in Fig. 2.8.

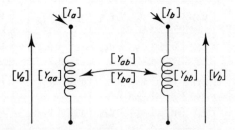

Fig. 2.8. Two coupled compound admittances

On examination of $[Y_{ab}]$ and $[Y_{ba}]$ it can be seen that,

$$[Y_{ba}] = [Y_{ab}]^T$$

if, and only if $y_{ik} = y_{ki}$ for $i = 1$ to 3 and $k = 4$ to 6. That is, if and only if the couplings between the two groups of admittances are bilateral.

In this case equation (2.4.9) may be written

$$
\begin{bmatrix} [I_a] \\ [I_b] \end{bmatrix} = \begin{bmatrix} [Y_{aa}] & [Y_{ab}] \\ [Y_{ab}]^T & [Y_{bb}] \end{bmatrix} \begin{bmatrix} [V_a] \\ [V_b] \end{bmatrix}
$$

(2.4.11)

The primitive network for any number of compound admittances is formed in exactly the same manner as for single admittances, except in that all quantities are matrices of the same order as the compound admittances.

The actual admittance matrix of any network composed of the compound admittances can be formed by the usual method of linear transformation; the elements of the connection matrix are now $n \times n$ identity matrices where n is the dimension of the compound admittances.

If the connection matrix of any network can be partitioned into identity elements of equal dimensions greater than one, the use of compound admittances is advantageous.

As an example, consider the network shown in Figs. 2.9 and 2.10, this represents a simple line section. The admittance matrix will be derived using single and compound admittances to show the simple correspondence. The primitive networks and associated admittance matrices are drawn in Fig. 2.11. The connection matrices for the single and compound networks are illustrated by equations (2.4.12) and (2.4.13) respectively.

V_1		-1			1				V_a
V_2			-1			1			V_b
V_3				-1			1		V_c
V_4		1							V_d
V_5	$=$		1						V_e
V_6				1					V_f
V_7					1				
V_8						1			
V_9							1		

(2.4.12)

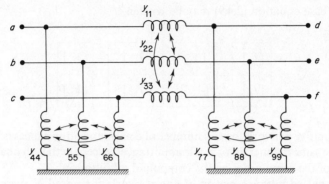

Fig. 2.9. Sample network represented by single admittances

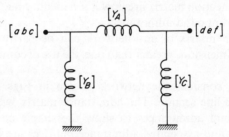

Fig. 2.10. Sample network represented by compound admittances

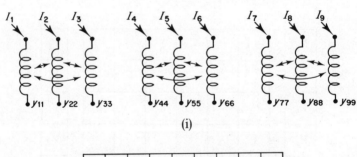

(i)

y_{11}	y_{12}	y_{13}						
y_{21}	y_{22}	y_{23}						
y_{31}	y_{32}	y_{33}						
			y_{44}	y_{45}	y_{46}			
			y_{54}	y_{55}	y_{56}			
			y_{64}	y_{65}	y_{66}			
						y_{77}	y_{78}	y_{79}
						y_{87}	y_{88}	y_{89}
						y_{97}	y_{98}	y_{99}

(ii)

17

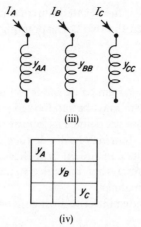

(iii)

<table>
<tr><td>y_A</td><td></td><td></td></tr>
<tr><td></td><td>y_B</td><td></td></tr>
<tr><td></td><td></td><td>y_C</td></tr>
</table>

(iv)

Fig. 2.11. Primitive networks and corresponding admittance matrices. (i) Primitive network using single admittances; (ii) Primitive admittance matrix; (iii) Primitive network using compound admittances; (iv) Primitive admittance matrix

$$
\begin{array}{|c|c c|c|}
\hline
[V_A] & -I & I & [V_{abc}] \\
\hline
[V_B] & I & & [V_{def}] \\
\hline
[V_c] & & I & \\
\hline
\end{array}
\qquad (2.4.13)
$$

The exact equivalence, with appropriate matrix partitioning, is clear.

The network admittance matrix is given by the linear transformation equation,

$$[Y_{\text{NODE}}] = [C]^T [Y_{\text{PRIM}}][C]$$

This matrix multiplication can be executed using the full matrices or in partitioned form. The result in partitioned form is,

$$
[Y_{\text{NODE}}] =
\begin{array}{|c|c|}
\hline
[Y_A] + [Y_B] & -[Y_A] \\
\hline
-[Y_A] & [Y_A] + [Y_C] \\
\hline
\end{array}
$$

Rules for forming the admittance matrix of simple networks

The method of linear transformation may be used to obtain the admittance matrix of any network. For the special case of networks where there is no mutual coupling, simple rules may be used to form the admittance matrix by inspection. These rules, which apply to compound networks with no mutual coupling between the compound admittances, may be stated as follows:

(a) Any diagonal term is the sum of the individual branch admittances connected to the node corresponding to that term.

(b) Any off-diagonal term is the negated sum of the branch admittances which are connected between the two corresponding nodes.

Network subdivision

To enable the transmission system to be modelled in a systematic, logical and convenient manner the system must be subdivided into more manageable units. These units, called subsystems, are defined as follows: A SUBSYSTEM is the unit into which any part of the system may be divided such that no subsystem has any mutual couplings between its constituent branches and those of the rest of the system. This definition ensures that the subsystems may be combined in an extremely straightforward manner.

The system is first subdivided into the most convenient subsystems consistent with the definition above.

The most convenient unit for a subsystem is a single network element. In previous sections the nodal admittance matrix representation of all common elements has been derived.

The subsystem unit is retained for input data organization. The data for any subsystem is input as a complete unit, the subsystem admittance matrix is formulated and stored and then all subsystems are combined to form the total system admittance matrix.

2.5. THREE-PHASE MODELS OF TRANSMISSION LINES

Transmission line parameters are calculated from the line geometrical characteristics. The calculated parameters are expressed as a series impedance and shunt admittance per unit length of line. The effect of ground currents and earth wires are included in the calculation of these parameters[3][4].

SERIES IMPEDANCE

A three-phase transmission line with a ground wire is illustrated in Fig. 2.12(a). The following equations can be written for phase 'a':

$$V_a - V_a' = I_a(R_a + j\omega L_a) + I_b(j\omega L_{ab}) + I_c(j\omega L_{ac})$$

$$+ j\omega L_{ag} \cdot I_g - j\omega L_{an} \cdot I_n + V_n$$

$$V_n = I_n(R_n + j\omega L_n) - I_a j\omega L_{na} - I_b j\omega L_{nb} - I_c j\omega L_{nc} - I_g \cdot j\omega L_{ng}$$

and substituting

$$I_n = I_a + I_b + I_c + I_g$$

$$V_a - V_a' = I_a(R_a + j\omega L_a) + I_b j\omega L_{ab} + I_c j\omega L_{ac}$$

$$+ j\omega L_{ag} \cdot I_g - j\omega L_{an}(I_a + I_b + I_c + I_g) + V_n$$

Regrouping and substituting for V_n, i.e.

$$\Delta V_a = V_a - V_a'$$

$$= I_a(R_a + j\omega L_a - j\omega L_{an} + R_n + j\omega L_n - j\omega L_{na})$$

$$+ I_b(j\omega L_{ab} - j\omega L_{an} + R_n + j\omega L_n - j\omega L_{nb})$$
$$+ I_c(j\omega L_{ac} - j\omega L_{an} + R_n + j\omega L_n - j\omega L_{nc})$$
$$+ I_g(j\omega L_{ag} - j\omega L_{an} + R_n + j\omega L_n - j\omega L_{ng})$$
$$\Delta V_a = I_a(R_a + j\omega L_a - 2j\omega L_{an} + R_n + j\omega L_n)$$
$$+ I_b(j\omega L_{ab} - j\omega L_{bn} - j\omega L_{an} + R_n + j\omega L_n)$$
$$+ I_c(j\omega L_{ac} - j\omega L_{cn} - j\omega L_{an} + R_n + j\omega L_n)$$
$$+ I_g(j\omega L_{ag} - j\omega L_{gn} - j\omega L_{an} + R_n + j\omega L_n)$$

or

$$\Delta V_a = Z_{aa-n}I_a + Z_{ab-n}I_b + Z_{ac-n}I_c + Z_{ag-n}I_g \tag{2.5.1}$$

and writting similar equations for the other phases the following matrix equation results:

$$
\begin{vmatrix} \Delta V_a \\ \Delta V_b \\ \Delta V_c \\ \hline \Delta V_g \end{vmatrix} =
\begin{vmatrix} Z_{aa\text{-}n} & Z_{ab\text{-}n} & Z_{ac\text{-}n} & Z_{ag\text{-}n} \\ Z_{ba\text{-}n} & Z_{bb\text{-}n} & Z_{bc\text{-}n} & Z_{bg\text{-}n} \\ Z_{ca\text{-}n} & Z_{cb\text{-}n} & Z_{cc\text{-}n} & Z_{cg\text{-}n} \\ \hline Z_{ga\text{-}n} & Z_{gb\text{-}n} & Z_{gc\text{-}n} & Z_{gg\text{-}n} \end{vmatrix}
\begin{vmatrix} I_a \\ I_b \\ I_c \\ \hline I_g \end{vmatrix}
\tag{2.5.2}
$$

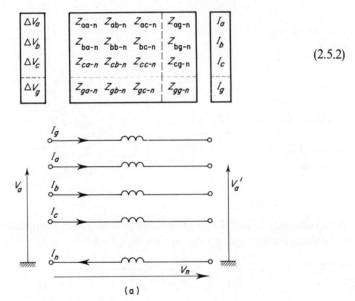

(a)

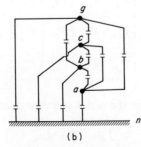

(b)

Fig. 2.12. (a) Three-phase transmission series impedance equivalent; (b) Three-phase transmission shunt impedance equivalent

Since we are interested only in the performance of the phase conductors it is more convenient to use a three-conductor equivalent for the transmission line. This is achieved by writing Matrix equation (2.5.2) in partitioned form as follows:

$$
\left[\begin{array}{c} \Delta V_{abc} \\ \hline \Delta V_g \end{array}\right] = \left[\begin{array}{c|c} Z_A & Z_B \\ \hline Z_C & Z_D \end{array}\right] \left[\begin{array}{c} I_{abc} \\ \hline I_g \end{array}\right] \tag{2.5.3}
$$

From (2.5.3).

$$
\Delta V_{abc} = Z_A I_{abc} + Z_B I_g \tag{2.5.4}
$$

$$
\Delta V_g = Z_C I_{abc} + Z_D I_g \tag{2.5.5}
$$

From equations (2.5.3) and (2.5.4) and assuming that the ground wire is at zero potential:

$$
\Delta V_{abc} = Z_{abc} I_{abc} \tag{2.5.6}
$$

where:

$$
Z_{abc} = Z_A - Z_B Z_D^{-1} Z_C = \left[\begin{array}{c|c|c} Z'_{aa-n} & Z'_{ab-n} & Z'_{ac-n} \\ \hline Z'_{ba-n} & Z'_{bb-n} & Z'_{bc-n} \\ \hline Z'_{ca-n} & Z'_{cb-n} & Z'_{cc-n} \end{array}\right]
$$

SHUNT ADMITTANCE

With reference to Fig. 2.12(b) the potentials of the line conductors are related to the conductor charges by the matrix equation[3]:

$$
\left[\begin{array}{c} V_a \\ \hline V_b \\ \hline V_c \\ \hline V_g \end{array}\right] = \left[\begin{array}{c|c|c|c} P_{aa} & P_{ab} & P_{ac} & P_{ag} \\ \hline P_{ba} & P_{bb} & P_{bc} & P_{bg} \\ \hline P_{ca} & P_{cb} & P_{cc} & P_{cg} \\ \hline P_{ga} & P_{gb} & P_{gc} & P_{gg} \end{array}\right] \times \left[\begin{array}{c} Q_a \\ \hline Q_b \\ \hline Q_c \\ \hline Q_g \end{array}\right] \tag{2.5.7}
$$

Similar considerations as for the series impedance matrix lead to

$$
V_{abc} = P'_{abc} Q_{abc} \tag{2.5.8}
$$

where P'_{abc} is a 3×3 matrix which includes the effects of the ground wire. The capacitance matrix of the transmission line of Fig. 2.12 is given by

$$C'_{abc} = P'^{-1}_{abc} = \begin{array}{|c|c|c|} \hline C_{aa} & -C_{ab} & -C_{ac} \\ \hline -C_{ba} & C_{bb} & -C_{bc} \\ \hline -C_{ca} & -C_{cb} & C_{cc} \\ \hline \end{array}$$

The series impedance and shunt admittance lumped-π model representation of the three-phase line is shown in Fig. 2.13(a) and its matrix equivalent is illustrated in Fig. 2.13(b). These two matrices can be represented by compound admittances, (Fig. 2.13(c)) as described earlier.

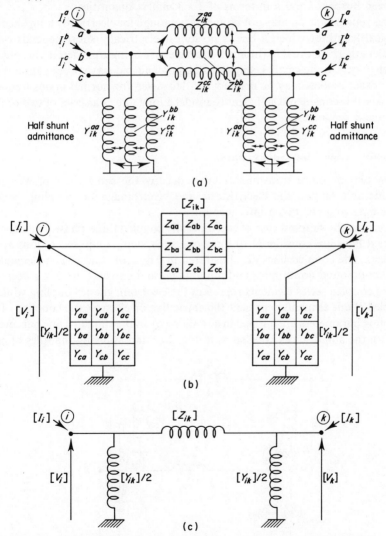

Fig. 2.13. Lumped-π model of a short three-phase line series impedance. (a) Full circuit representation; (b) Matrix equivalent; (c) Using three-phase compound admittances

Following the rules developed for the formation of the admittance matrix using the compound concept, the nodal injected currents of Fig. 2.13(c) can be related to the nodal voltages by the equation,

$$
\begin{array}{|c|}
\hline
[I_i] \\
\hline
[I_k] \\
\hline
\end{array}
=
\begin{array}{|c|c|}
\hline
[Z]^{-1} + [Y]/2 & -[Z]^{-1} \\
\hline
-[Z]^{-1} & [Z]^{-1} + [Y]/2 \\
\hline
\end{array}
\begin{array}{|c|}
\hline
[V_i] \\
\hline
[V_k] \\
\hline
\end{array}
\qquad (2.5.9)
$$

$$6 \times 1 \qquad\qquad 6 \times 6 \qquad\qquad 6 \times 1$$

This forms the element admittance matrix representation for the short line between busbars i and k in terms of 3×3 matrix quantities.

This representation may not be accurate enough for electrically long lines. The physical length at which a line is no longer electrically short depends on the wavelength, therefore if harmonic frequencies are being considered, this physical length may be quite small. Using transmission line and wave propagation theory more exact models may be derived[5][6]. However, for normal mains frequency analysis, it is considered sufficient to model a long line as a series of two or three nominal-π sections.

Mutually coupled three-phase lines

When two or more transmission lines occupy the same right of way for a considerable length, the electrostatic and electromagnetic coupling between those lines must be taken into account.

Consider the simplest case of two mutually coupled three-phase lines. The two coupled lines are considered to form one subsystem composed of four system busbars. The coupled lines are illustrated in Fig. 2.14, where each element is a 3×3 compound admittance and all voltages and currents are 3×1 vectors.

The coupled series elements represent the electromagnetic coupling while the coupled shunt elements represent the capacitive or electrostatic coupling. These coupling parameters are lumped in a similar way to the standard line parameters.

With the admittances labelled as in Fig. 2.14 and applying the rules of linear

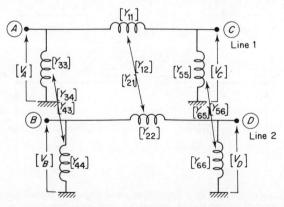

Fig. 2.14. Two coupled three-phase lines

transformation for compound networks the admittance matrix for the subsystem is defined as follows:

$$
\begin{array}{|c|}
I_A \\ I_B \\ I_C \\ I_D
\end{array}
=
\begin{array}{|cccc|}
Y_{11}+Y_{33} & Y_{12}+Y_{34} & -Y_{11} & -Y_{12} \\
Y_{12}^T+Y_{34}^T & Y_{22}+Y_{44} & -Y_{12}^T & -Y_{22} \\
-Y_{11} & -Y_{12} & Y_{11}+Y_{55} & Y_{12}+Y_{56} \\
-Y_{12}^T & -Y_{22} & Y_{12}^T+Y_{56}^T & Y_{22}+Y_{66}
\end{array}
\begin{array}{|c|}
V_A \\ V_B \\ V_c \\ V_D
\end{array}
\qquad (2.5.10)
$$

$12 \times 1 \qquad\qquad\qquad 12 \times 12 \qquad\qquad\qquad 12 \times 1$

It is assumed here that the mutual coupling is bilateral. Therefore. $Y_{21} = Y_{12}^T$ etc.

The subsystem may be redrawn as Fig. 2.15. The pairs of coupled 3×3 compound admittances are now represented as a 6×6 compound admittance. The matrix representation is also shown. Following this representation and the labelling of the admittance blocks in the figure, the admittance matrix may be written in terms of the 6×6 compound coils as,

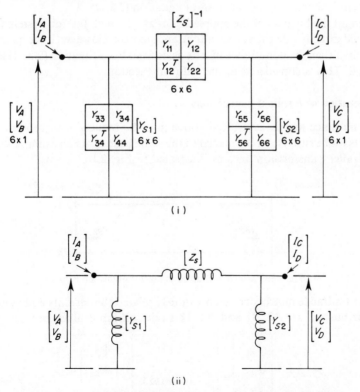

Fig. 2.15. 6×6 compound admittance representation of two coupled three-phase lines. (i) 6×6 Matrix representation; (ii) 6×6 Compound admittance representation

$$
\begin{bmatrix} \begin{bmatrix} I_A \\ I_B \end{bmatrix} \\ \begin{bmatrix} I_C \\ I_D \end{bmatrix} \end{bmatrix} = \begin{bmatrix} [Z_s]^{-1} + [Y_{s1}] & -[Z_s]^{-1} \\ -[Z_s]^{-1} & [Z_s]^{-1} + [Y_{s2}] \end{bmatrix} \begin{bmatrix} \begin{bmatrix} V_A \\ V_B \end{bmatrix} \\ \begin{bmatrix} V_C \\ V_D \end{bmatrix} \end{bmatrix} \qquad (2.5.11)
$$

$12 \times 1 \qquad\qquad 12 \times 12 \qquad\qquad 12 \times 1$

This is clearly identical to equation (2.5.10) with the appropriate matrix partitioning.

The representation of Fig. 2.15 is more concise and the formation of equation (2.5.11) from this representation is straightforward, being exactly similar to that which results from the use of 3×3 compound admittances for the normal single three-phase line.

The data which must be available, to enable coupled lines to be treated in a similar manner to single lines, is the series impedance and shunt admittance matrices. These matrices are of order 3×3 for a single line, 6×6 for two coupled lines, 9×9 for three and 12×12 for four coupled lines.

Once the matrices $[Z_s]$ and $[Y_s]$ are available, the admittance matrix for the subsystem is formed by application of equation (2.5.11).

When all the busbars of the coupled lines are distinct, the sub-system may be combined directly into the system admittance matrix. However, if the busbars are not distinct then the admittance matrix as derived from equation (2.5.11) must be modified. This is considered in the following section.

Consideration of terminal connections

The admittance matrix as derived above must be reduced if there are different elements in the subsystem connected to the same busbar. As an example consider two parallel transmission lines as illustrated in Fig. 2.16.

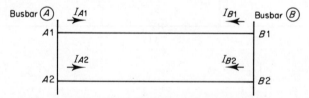

Fig. 2.16. Mutually coupled parallel transmission lines

The admittance matrix derived previously related the currents and voltages at the four busbar $A1$, $A2$, $B1$ and $B2$. This relationship is given by:

$$
\begin{bmatrix} I_{A1} \\ I_{A2} \\ I_{B1} \\ I_{B2} \end{bmatrix} = [Y_{A1A2B1B2}] \begin{bmatrix} V_{A1} \\ V_{A2} \\ V_{B1} \\ V_{B2} \end{bmatrix} \qquad (2.5.12)
$$

The nodal injected current at busbar $A, (I_A)$, is given by,

$$I_A = I_{A1} + I_{A2}$$

similarly

$$I_B = I_{B1} + I_{B2}$$

Also from inspection of Fig. 2.16,

$$V_A = V_{A1} = V_{A2}$$
$$V_B = V_{B1} = V_{B1}$$

The required matrix equation relates the nodal injected currents, I_A and I_B, to the voltages at these busbar. This is readily derived from equation (2.5.12) and the conditions specified above. This is simply a matter of adding appropriate rows and columns and yields,

$$
\begin{bmatrix} I_A \\ I_B \end{bmatrix} = [Y_{AB}] \begin{bmatrix} V_A \\ V_B \end{bmatrix}
\qquad (2.5.13)
$$

This matrix $[Y_{AB}]$ is the required nodal admittance matrix for the subsystem.

It should be noted that the matrix in equation (2.5.12) must be retained as it is required in the calculation of the individual line power flows.

Shunt elements

Shunt reactors and capacitors are used in a power system for reactive power control. The data for these elements are usually given in terms of their rated MVA and rated kV; the equivalent phase admittance in p.u. is calculated from these data.

Consider, as an example, a three-phase capacitor bank shown in Fig. 2.17. A similar triple representation as that for a line section is illustrated. The final two forms are the most compact and will be used exclusively from this point on.

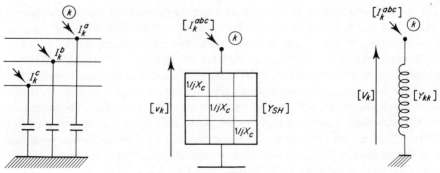

Fig. 2.17. Representation of a shunt capacitor bank

26

The admittance matrix for shunt elements is usually diagonal as there is normally no coupling between the components of each phase. This matrix is then incorporated directly into the system admittance matrix, contributing only to the self admittance of the particular bus.

Series elements

Any element connected directly between two buses may be considered a series element. Series elements are often taken as being a section in a line sectionalization which is described later in the chapter.

A typical example is the series capacitor bank which is usually taken as uncoupled, i.e. the admittance matrix is diagonal.

This can be represented graphically as in Fig. 2.18.

The admittance matrix for the subsystem can be written by inspection as:

$$[Y] = \begin{array}{|c|c|} \hline [Y_{SE}] & -[Y_{SE}] \\ \hline -[Y_{SE}] & [Y_{SE}] \\ \hline \end{array} \qquad (2.5.14)$$

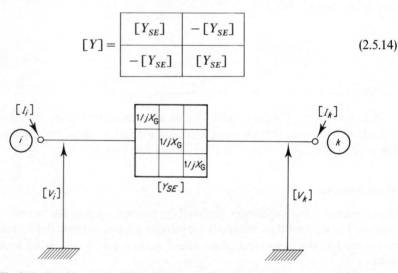

Fig. 2.18. Graphic representation of series capacitor bank between nodes i and k

2.6. THREE-PHASE MODELS OF TRANSFORMERS

The inherent assumption, that the transformer is a balanced three-phase device, is justified in the majority of practical situations, and traditionally, three-phase transformers are represented by their equivalent sequence networks.

More recently, however, methods have been developed[3],[4] to enable all three-phase transformer connections to be accurately modelled in phase coordinates. In phase coordinates no assumptions are necessary although physically justifiable assumptions are still used in order to simplify the model. The primitive admittance matrix, used as a basis for the phase coordinate transformer model is derived from the primitive or unconnected network for the transformer windings and the method of linear transformation enables the admittance matrix of the actual connected network to be found.

Primitive admittance model of three-phase transformers

Many three-phase transformers are wound on a common core and all windings are therefore coupled to all other windings. Therefore, in general, a basic two-winding three-phase transformer has a primitive or unconnected network consisting of six coupled coils. If a tertiary winding is also present the primitive network consists of nine coupled coils. The basic two-winding transformer shown in Fig. 2.19 is now considered, the addition of further windings being a simple but cumbersome extension of the method.

The primitive network, Fig. 2.20, can be represented by the primitive admittance matrix which has the following general form:

$$
\begin{bmatrix} I_1 \\ I_2 \\ I_3 \\ I_4 \\ I_5 \\ I_6 \end{bmatrix}
=
\begin{bmatrix}
y_{11} & y_{12} & y_{13} & y_{14} & y_{15} & y_{16} \\
y_{21} & y_{22} & y_{23} & y_{24} & y_{25} & y_{26} \\
y_{31} & y_{32} & y_{33} & y_{34} & y_{35} & y_{36} \\
y_{41} & y_{42} & y_{43} & y_{44} & y_{45} & y_{46} \\
y_{51} & y_{52} & y_{53} & y_{54} & y_{55} & y_{56} \\
y_{61} & y_{62} & y_{63} & y_{64} & y_{65} & y_{66}
\end{bmatrix}
\begin{bmatrix} V_1 \\ V_2 \\ V_3 \\ V_4 \\ V_5 \\ V_6 \end{bmatrix}
\qquad (2.6.1)
$$

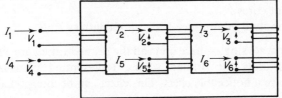

Fig. 2.19. Diagrammatic representation of two-winding transformer

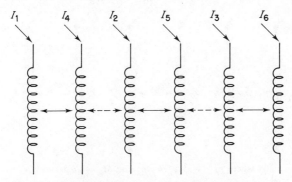

Fig. 2.20. Primitive network of two-winding transformer. Six coupled coil primitive network. (Note the dotted coupling represents parasitic coupling between phases)

The elements of matrix $[Y]$ can be measured directly, i.e. by energizing coil i and short-circuiting all other coils, column i of $[Y]$ can be calculated from $y_{ki} = I_k/V_i$.

Considering the reciprocal nature of the mutual couplings in equation (2.6.1) twenty-one short circuit measurements would be necessary to complete the admittance matrix. Such a detailed representation is seldom required.

By assuming that the flux paths are symmetrically distributed between all windings equation (2.6.1) may be simplified to equation (2.6.2).

$$
\begin{bmatrix} I_1 \\ I_2 \\ I_3 \\ I_4 \\ I_5 \\ I_6 \end{bmatrix}
=
\begin{bmatrix}
y_p & y'_m & y'_m & -y_m & y''_m & y''_m \\
y'_m & y_p & y'_m & y''_m & -y_m & y''_m \\
y'_m & y'_m & y_p & y''_m & y''_m & -y_m \\
-y_m & y''_m & y''_m & y_s & y'''_m & y'''_m \\
y''_m & -y_m & y''_m & y'''_m & y_s & y'''_m \\
y''_m & y''_m & -y_m & y'''_m & y'''_m & y_s
\end{bmatrix}
\begin{bmatrix} V_1 \\ V_2 \\ V_3 \\ V_4 \\ V_5 \\ V_6 \end{bmatrix}
\qquad (2.6.2)
$$

where

y'_m is the mutual admittance between primary coils;

y''_m is the mutual admittance between primary and secondary coils on different cores;

y'''_m is the mutual admittance between secondary coils.

For three separate single-phase units all the primed values are effectively zero. In three-phase units the primed values, representing parasitic interphase coupling, do have a noticeable effect. This effect can be interpreted through the symmetrical component equivalent circuits.

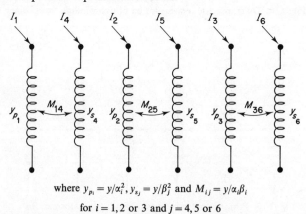

where $y_{p_i} = y/\alpha_i^2$, $y_{s_j} = y/\beta_j^2$ and $M_{ij} = y/\alpha_i\beta_i$

for $i = 1, 2$ or 3 and $j = 4, 5$ or 6

Fig. 2.21. Primitive network

If the values in equation (2.6.2) are available then this representation of the primitive network should be used. If interphase coupling can be ignored, the coupling between a primary and a secondary coil is modelled as for the single-phase unit, giving rise to the primitive network of Fig. 2.21.

The new admittance matrix equation is

$$
\begin{bmatrix} I_1 \\ I_2 \\ I_3 \\ I_4 \\ I_5 \\ I_6 \end{bmatrix}
=
\begin{bmatrix}
y_{p_1} & & & M_{14} & & \\
& y_{p_2} & & & M_{25} & \\
& & y_{p_3} & & & M_{36} \\
M_{41} & & & y_{s_4} & & \\
& M_{52} & & & y_{s_5} & \\
& & M_{63} & & & y_{s_6}
\end{bmatrix}
\begin{bmatrix} V_1 \\ V_2 \\ V_3 \\ V_4 \\ V_5 \\ V_6 \end{bmatrix}
$$

$$(2.6.3)$$

Models for common transformer connections

The network admittance matrix for any two-winding three-phase transformer can now be formed by the method of linear transformation.

As a simple example, consider the formation of the admittance matrix for a star–star connection with both neutrals solidly earthed in the absence of interphase mutuals. This example is chosen as it is the simplest computationally.

The connection matrix is derived from consideration of the actual connected network. For the star–star transformer illustrated in Fig. 2.22, the connection matrix $[C]$ relating the branch voltages (i.e. voltages of the primitive network) to the node voltages (i.e. voltages of the actual network) is a 6×6 identity matrix, i.e.

$$
\begin{bmatrix} V_1 \\ V_2 \\ V_3 \\ V_4 \\ V_5 \\ V_6 \end{bmatrix}
=
\begin{bmatrix}
1 & & & & & \\
& 1 & & & & \\
& & 1 & & & \\
& & & 1 & & \\
& & & & 1 & \\
& & & & & 1
\end{bmatrix}
\begin{bmatrix} v_a \\ v_b \\ v_c \\ V_a \\ V_b \\ V_c \end{bmatrix}
$$

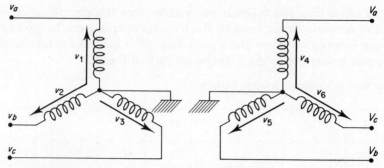

Fig. 2.22. Network connection diagram for three-phase star–star transformer

The nodal admittance matrix $[Y]_{NODE}$ is given by,

$$[Y]_{NODE} = [C]^t [Y]_{PRIM} [C] \qquad (2.6.5)$$

Substituting for $[C]$ yields,

$$[Y]_{NODE} = [Y]_{PRIM} \qquad (2.6.6)$$

Let us now consider the Wye G–Delta connection illustrated in Fig. 2.23.

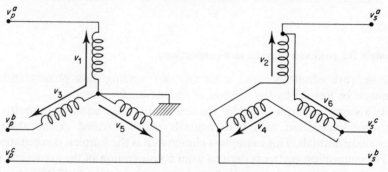

Fig. 2.23. Network connection diagram for Wyê G–Delta transformer

The following connection can be written by inspection between the primitive branch voltages and the node voltages

V_1	1	0	0	0	0	0	V_p^a
V_2	0	0	0	1	-1	0	V_p^b
V_3	0	1	0	0	0	0	V_p^c
V_4	0	0	0	0	1	-1	V_s^a
V_5	0	0	1	0	0	0	V_s^b
V_6	0	0	0	-1	0	1	V_s^c

$$(2.6.7)$$

or

$$[V]_{\text{branch}} = [C][V]_{\text{node}} \qquad (2.6.8)$$

we can also write

$$[Y]_{\text{NODE}} = [C]^t [Y]_{\text{PRIM}}[C] \qquad (2.6.9)$$

and using $[Y]_{\text{PRIM}}$ from equation (2.6.2).

$[Y]_{\text{NODE}} =$

y_p	y'_m	y'_m	$-(y_m + y''_m)$	$(y_m + y''_m)$	0	a
y'_m	y_p	y'_m	0	$-(y_m + y''_m)$	$(y_m + y''_m)$	b
y'_m	y'_m	y_p	$(y_m + y''_m)$	0	$-(y_m + y''_m)$	c
$-(y_m + y''_m)$	0	$(y_m + y''_m)$	$2(y_s - y'''_m)$	$-(y_s - y'''_m)$	$-(y_s - y'''_m)$	A
$(y_m + y''_m)$	$-(y_m + y''_m)$	0	$-(y_s - y'''_m)$	$2(y_s - y'''_m)$	$-(y_s - y'''_m)$	B
0	$(y_m + y''_m)$	$-(y_m + y''_m)$	$-(y_s - y'''_m)$	$-(y_s - y'''_m)$	$2(y_s - y'''_m)$	C

$$(2.6.10)$$

Moreover, if the primitive admittances are expressed in per unit, with both the primary and secondary voltages being one per unit, the Wye–Delta transformer model must include an effective turns ratio of $\sqrt{3}$. The upper right and lower left quadrants of matrix (2.6.10) must be divided by $\sqrt{3}$ and the lower right quadrant by 3.

In the particular case of three-single phase transformer units connected in Wye G–Delta all the y' and y'' terms will disappear. Ignoring off nominal taps (but keeping in mind the effective $\sqrt{3}$ turns ratio in per unit) the nodal admittance matrix equation relating the nodal currents to the nodal voltages is

I_p^a							V_p^a
	y			$-y/\sqrt{3}$	$y/\sqrt{3}$		
I_p^b		y			$-y/\sqrt{3}$	$y/\sqrt{3}$	V_p^b
I_p^c			y	$y/\sqrt{3}$		$-y/\sqrt{3}$	V_p^c
I_s^A	$-y/\sqrt{3}$		$y/\sqrt{3}$	$\frac{2}{3}y$	$-\frac{1}{3}y$	$-\frac{1}{3}y$	V_s^A
I_s^B	$y/\sqrt{3}$	$-y/\sqrt{3}$		$-\frac{1}{3}y$	$\frac{2}{3}y$	$-\frac{1}{3}y$	V_s^B
I_s^C		$y/\sqrt{3}$	$-y/\sqrt{3}$	$-\frac{1}{3}y$	$-\frac{1}{3}y$	$\frac{2}{3}y$	V_s^C

$$(2.6.11)$$

where Y is the transformer leakage admittance in p.u.

An equivalent circuit can be drawn, corresponding to this admittance model of the transformer, as illustrated in Fig. 2.24.

The large shunt admittances to earth from the nodes of the star connection are apparent in the equivalent circuit. These shunts are typically around 10 p.u. (for a 10 percent leakage reactance transformer).

32

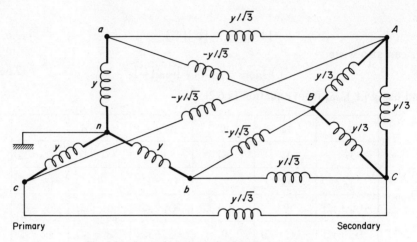

Fig. 2.24. Equivalent circuit for Star–Delta transformer

The models for the other common connections can be derived following a similar procedure.

In general, any two-winding three-phase transformer may be represented using two coupled compound coils. The network and admittance matrix for this representation is illustrated in Fig. 2.25.

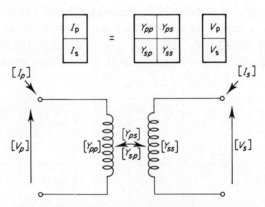

Fig. 2.25. Two winding three-phase transformer as two coupled compound coils

It should be noted that,

$$[Y_{sp}] = [Y_{ps}]^T$$

as the coupling between the two compound coils is bilateral.

Often, because more detailed information is not required, the parameters of all three phases are assumed balanced. In this case the common three-phase connections are found to be modelled by three basic submatrices.

The submatrices, $[Y_{pp}], [Y_{ps}]$ etc., are given in Table 2.1 for the common connections.

Table 2.1. Characteristic submatrices used in forming the transformer admittance matrices

Trans. connection		Self admittance		Mutual admittance
Bus P	Bus S	Y_{PP}	Y_{SS}	Y_{PS}, Y_{SP}
Wye $- G$	Wye $- G$	Y_I	Y_I	$- Y_I$
Wye $- G$	Wye	$Y_{II/3}$	$Y_{II/3}$	$- Y_{II/3}$
Wye $- G$	Delta	Y_I	Y_{II}	$+ Y_{III}$
Wye	Wye	$Y_{II/3}$	$Y_{II/3}$	$- Y_{II/3}$
Wye	Wye	$Y_{II/3}$	Y_{II}	Y_{III}
Delta	Delta	Y_{II}	Y_{II}	$- Y_{II}$

Basic submatrices used in node admittance formulation of common three-phase transformer connections, where:

$$Y_I = \begin{bmatrix} y_t & & \\ & y_t & \\ & & y_t \end{bmatrix}$$

$$Y_{II} = \begin{bmatrix} 2y_t & -y_t & -y_t \\ -y_t & 2y_t & -y_t \\ -y_t & -y_t & 2y_t \end{bmatrix}$$

$$Y_{III} = \begin{bmatrix} -y_t & y_t & \\ & -y_t & y_t \\ y_t & & -y_t \end{bmatrix}$$

Finally, these submatrices must be modified to accounts for off-nominal tap ratio as follows:

(i) Divide the self admittance of the primary by α^2.
(ii) Divide the self admittance of the secondary by β^2.
(iii) Divide the mutual admittance matrices by $(\alpha\beta)$.

It should be noted that in the p.u. system a delta winding has an off-nominal tap of $\sqrt{3}$.

For transformers with ungrounded Wye connections, or with neutrals connected through an impedance, an extra coil is added to the primitive network for each unearthed neutral and the primitive admittance matrix increases in dimension. By noting that the injected current in the neutral is zero, these extra terms can be eliminated from the connected network admittance matrix.[7]

Once the admittance matrix has been formed for a particular connection it represents a simple subsystem composed of the two busbars interconnected by the transformer.

Three-phase transformer models with independent phase tap control

Disregarding interphase mutual couplings the per unit primitive admittance matrix in terms of the transformer leakage admittance (y_{ti}) is:

$$[Y_{\text{prim}}] =$$

$\dfrac{y_{t1}}{a_1^2}$			$-\dfrac{y_{t1}}{a_1}$		
	$\dfrac{y_{t2}}{a_2^2}$			$-\dfrac{y_{t2}}{a_2}$	
		$-\dfrac{y_{t3}}{a_3^2}$			$-\dfrac{y_{t3}}{a_3}$
$-\dfrac{y_{t1}}{a_1}$			y_{t1}		
	$-\dfrac{y_{t2}}{a_2}$			y_{t2}	
		$-\dfrac{y_{t3}}{a_3}$			y_{t3}

where a_1, a_2 and a_3 are the off-nominal taps on windings 1, 2, and 3 respectively. In addition any windings connected in delta will, because of the per unit system, have an effective tap of $\sqrt{3}$.

The nodal admittance matrix for the transformer windings is:

$$[Y_{\text{node}}] = [C]^T [Y_{\text{prim}}][C]$$

where $[C]$ is the connection (windings to nodes) matrix.

As an example $[Y_{\text{node}}]$ for a star–delta transformer with earthed neutral is as follows:

$$[Y_{node}] = \begin{bmatrix} \dfrac{y_{t1}}{a_1^2} & & & \dfrac{-y_{t1}}{\sqrt{3}a_1} & \dfrac{y_{t1}}{\sqrt{3}a_1} & \\[2ex] & \dfrac{y_{t2}}{a_2^2} & & & \dfrac{-y_{t2}}{\sqrt{3}a_2} & \dfrac{y_{t2}}{\sqrt{3}a_2} \\[2ex] & & \dfrac{y_{t3}}{a_3^2} & \dfrac{y_{t3}}{\sqrt{3}a_3} & & \dfrac{-y_{t3}}{\sqrt{3}a_3} \\[2ex] \dfrac{-y_{t1}}{\sqrt{3}a_1} & & \dfrac{y_{t3}}{\sqrt{3}a_3} & \dfrac{y_{t1}+y_{t3}}{3} & \dfrac{-y_{t1}}{3} & \dfrac{-y_{t3}}{3} \\[2ex] \dfrac{y_{t1}}{\sqrt{3}a_1} & \dfrac{-y_{t2}}{\sqrt{3}a_2} & & \dfrac{-y_{t1}}{3} & \dfrac{y_{t2}+y_{t3}}{3} & \dfrac{-y_{t2}}{3} \\[2ex] & \dfrac{y_{t2}}{\sqrt{3}a_2} & \dfrac{-y_{t3}}{\sqrt{3}a_3} & \dfrac{-y_{t3}}{3} & \dfrac{-y_{t2}}{3} & \dfrac{y_{t2}+y_{t3}}{3} \end{bmatrix}$$

Sequence components modelling of three-phase transformers

In most cases lack of data will prevent the use of the general model based on the primitive admittance matrix and will justify the conventional approach in terms of symmetrical components. Let us now derive the general sequence components equivalent circuits and the assumptions introduced in order to arrive at the conventional models.

With reference to the Wye G–Delta common-core transformer of Fig. 2.23, represented by equation (2.6.10), and partioning this matrix to separate self and mutual elements the following transformations apply:

Primary side:

$$y^p_{120} = T_s^{-1} \begin{bmatrix} y_p & y'_m & y'_m \\ y'_m & y_p & y'_m \\ y'_m & y'_m & y_p \end{bmatrix} T_s$$

where

$$[T_s] = \begin{bmatrix} 1 & 1 & 1 \\ 1 & a^2 & a \\ 1 & a & a^2 \end{bmatrix} \quad \text{and} \quad a = e^{j2\pi/3}$$

Therefore

$$
y^p_{120} =
\begin{bmatrix}
y_p - y'_m & 0 & 0 \\
0 & y_p - y'_m & 0 \\
0 & 0 & y_p + 2y'_m
\end{bmatrix}
\tag{2.6.12}
$$

Secondary side

The delta connection on the secondary side introduces an effective $\sqrt{3}$ turns ratio and the sequence components admittance matrix is

$$
y^s_{120} = \tfrac{1}{3} T_s^{-1}
\begin{bmatrix}
2(y_s - y'''_m) & -(y_s - y'''_m) & -(y_s - y'''_m) \\
-(y_s - y'''_m) & 2(y_s - y'''_m) & -(y_s - y'''_m) \\
-(y_s - y'''_m) & -(y_s - y'''_m) & 2(y_s - y'''_m)
\end{bmatrix}
T_s
$$

$$
=
\begin{bmatrix}
y_s - y'''_m & 0 & 0 \\
0 & y_s - y'''_m & 0 \\
0 & 0 & 0
\end{bmatrix}
\tag{2.6.13}
$$

Mutual terms

The mutual admittance submatrix of equation (2.6.10), modified for effective turns ratio is transformed as follows:

$$
y^M_{120} = \frac{\sqrt{3}}{3} T_s^{-1}
\begin{bmatrix}
-(y_m + y''_m) & (y_m + y''_m) & 0 \\
0 & -(y_m + y''_m) & (y_m + y''_m) \\
(y_m + y''_m) & 0 & -(y_m + y''_m)
\end{bmatrix}
T_s
$$

$$
=
\begin{bmatrix}
-(y_m + y''_m)\underline{/30^\circ} & 0 & 0 \\
0 & -(y_m + y''_m)\underline{/-30^\circ} & 0 \\
0 & 0 & 0
\end{bmatrix}
\tag{2.6.14}
$$

Recombining the sequence components submatrices yields

$$
\begin{bmatrix} I_1^p \\ I_2^p \\ I_0^p \\ I_1^s \\ I_2^s \\ I_0^s \end{bmatrix}
=
\begin{bmatrix}
y_p - y_m' & & & -(y_m + y_m'')\angle 30 & & 0 \\
& y_p - y_m' & & & -(y_m + y_m'')\angle{-30} & 0 \\
& & y_p + 2y_m' & & & 0 \\
-(y_m + y_m'')\angle 30 & & & y_s - y_m''' & & 0 \\
& -(y_m + y_m'')\angle{-30} & & & y_s - y_m''' & 0 \\
& & & & & 0
\end{bmatrix}
\begin{bmatrix} V_1^p \\ V_2^p \\ V_0^p \\ V_1^s \\ V_2^s \\ V_0^s \end{bmatrix}
$$

$$(2.6.15)$$

Equation (2.6.15) can be represented by the three sequence networks of Figs. 2.26, 2.27 and 2.28 respectively.

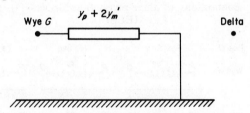

Fig. 2.26. Zero-sequence node admittance model for a common-core grounded Wye–Delta transformer (3). (© 1982 IEEE)

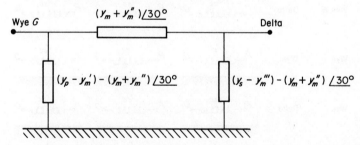

Fig. 2.27. Positive-sequence node admittance model for a common-core grounded Wye–Delta Transformer (3). (© 1982 IEEE)

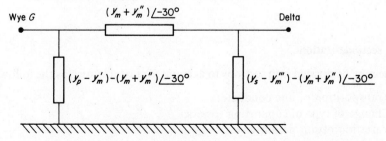

Fig. 2.28. Negative-sequence node admittance model for a common-core grounded Wye–Delta transformer (3). (© 1982 IEEE)

In general, therefore, the three sequence impedances are different on a common-core transformer.

The complexity of these equivalent models is normally eliminated by the following simplifications:

— The 30° phase shifts of Wye–Delta connections are ignored.
— The interphase mutuals admittances are assumed equal, i.e. $y'_m = y''_m = y'''_m$. These are all zero with uncoupled single-phase units.
— The differences $(y_p - y_m)$ and $(y_s - y_m)$ are very small and are therefore ignored.

With these simplifications, Table 2.2. illustrates the sequence impedance models of three-phase transformers in conventional steady-state balanced transmission system studies.

Table 2.2. Typical symmetrical-component models for the six most common connections of three-phase transformers (3). (© 1982 IEEE)

Line sectionalization

A line may be divided into sections to account for features such as the following:

— Transposition of line conductors.
— Change of type of supporting towers.
— Variation of soil permitivity.

— Improvement of line representation. (Series of two or more equivalent π networks.)
— Series capacitors for line compensation.
— Lumping of series elements not central to a particular study.

An example of a line divided into a number of sections is shown in Fig. 2.29.

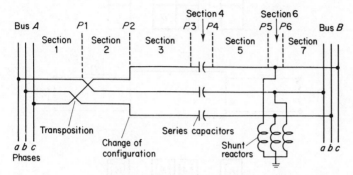

Fig. 2.29. Example of a transmission line divided into sections

The network of Fig. 2.29 is considered to form a single subsystem. The resultant admittance matrix between bus A and bus B may be derived by finding, for each section, the $ABCD$ or transmission parameters, then combining these by matrix multiplications to give the resultant transmission parameters. These are then converted to the required admittance parameters.

This procedure involves an extension of the usual two port network theory to multi-two-port networks. Currents and voltages are new matrix quantities and are defined in Fig. 2.30. The $ABCD$ matrix parameters are also shown.

The dimensions of the parameters matrices correspond to those of the section being considered, i.e. 3, 6, 9, or 12 for 1, 2, 3 or 4 mutually coupled three-phase elements respectively. All sections must contain the same number of mutually coupled three-phase elements, ensuring that all the parameter matrices are of the same order and that the matrix multiplications are executable. To illustrate this feature, consider the example of Fig. 2.31.

FEATURES OF INTEREST

(a) As a matter of programming convenience an ideal transformer is created and included in Section 1.
(b) The dotted coupling represents coupling which is zero. It is included to ensure correct dimensionality of all matrices.
(c) In the p.u. system the mutual coupling between the 220-kV and 66-kV lines is expressed to a voltage base given by the geometric mean of the base line-neutral voltages of the two parallel circuits.

40

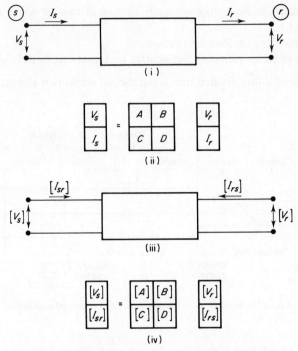

$$\begin{array}{|c|} V_s \\ \hline I_s \end{array} = \begin{array}{|c|c|} A & B \\ \hline C & D \end{array} \begin{array}{|c|} V_r \\ \hline I_r \end{array}$$

(ii)

$$\begin{array}{|c|} [V_s] \\ \hline [I_{sr}] \end{array} = \begin{array}{|c|c|} [A] & [B] \\ \hline [C] & [D] \end{array} \begin{array}{|c|} [V_r] \\ \hline [I_{rs}] \end{array}$$

(iv)

Fig. 2.30. Two port network transmission parameters.
(i) Normal two port network; (ii) Transmission parameters;
(iii) Multi-two port network; (iv) Matrix transmission
parameters

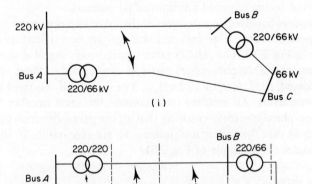

Fig. 2.31. Sample system to illustrate line sectionalization.
(i) System single line diagram; (ii) system redrawn to illustrate line
sectionalization

Table 2.3. *ABCD* Parameter matrices for the common section types

Transmission line	$[u] + [Z][Y]/2$	$-[Z]$
	$[Y]\{[u] + [Z][Y]/4\}$	$-\{[u] + [Y][Z]/2\}$

Transformer	$-[Y_{SP}]^{-1}[Y_{SS}]$	$[Y_{SP}]^{-1}$
	$[Y_{PS}] - [Y_{PP}][Y_{SP}]^{-1}[Y_{SS}]$	$[Y_{PP}][Y_{SP}]^{-1}$

Shunt element	$[u]$	$[0]$
	$[Y_{SH}]$	$-[u]$

Series element	$[u]$	$-[Y_{SE}]^{-1}$
	$[0]$	$-[u]$

In Table 2.3 $[u]$ is the unit matrix, $[0]$ is a matrix of zeros, and all other matrices have been defined in their respective sections.

Note, all the above matrices have dimensions corresponding to the number of coupled three-phase elements in the section.

Once the resultant *ABCD* parameters have been found the equivalent nodal admittance matrix for the subsystem can be calculated from the following equation.

$$[Y] = \begin{array}{|c|c|} \hline [D][B]^{-1} & [C] - [D][B]^{-1}[A] \\ \hline [B]^{-1} & -[B]^{-1}[A] \\ \hline \end{array} \qquad (2.6.16)$$

2.7. FORMATION OF THE SYSTEM ADMITTANCE MATRIX

It has been shown that the element (and subsystem) admittance matrices can be manipulated efficiently if the three nodes at the busbar are associated together. This association proves equally helpful when forming the admittance matrix for the total system.

The subsystem, as defined in Section 2.4., may have common busbars with other subsystems, but may not have mutual coupling terms to the branches of other subsystems. Therefore the subsystem admittance matrices can be combined to form the overall system admittance matrix as follows:

42

— The self-admittance of any busbar is the sum of all the individual self-admittance matrices at that busbar.
— The mutual-admittance between any two busbars is the sum of the individual mutual-admittance matrices from all the subsystems containing those two nodes.

2.8. REFERENCES

1. E. Clarke, 1943. *Circuit Analysis of A.C. Power Systems*, Vol. I. Wiley and Sons Ltd., New York.
2. G. Kron, republished in 1965. *Tensor Analysis of Networks*. MacDonald, London.
3. M. S. Chen and W. E. Dillon, July 1974. 'Power-system modelling', *Proc. IEEE*, **62** (7), 901.
4. M. A. Laughton, 1968. 'Analysis of unbalanced polyphase networks by the method of phase coordinates, Part I. System representation in phase frame of reference', *Proc. IEE*, **115** (8), 1163–1172.
5. K. J. Bowman and J. M. McNamee, 1964. 'Development of equivalent *pi* and *T* matrix circuits for long untransposed transmission lines', *IEEE*, **PAS-84**, 625–632.
6. L. M. Wedepohl and R. G. Wasley, 1966. 'Wave propagation in multiconductor overhead lines', *Proc. IEE*, **113** (4), 627–632.
7. W. E. Dillon and M. S. Chen, 1972. 'Transformer modelling in unbalanced three-phase networks.' IEEE Summer Power Meeting, Vancouver.

Modelling of Static A.C.–D.C. Conversion Plant

3.1. INTRODUCTION

Although the power electronic device is basically a switch, it is only explicitly represented as such in dynamic studies (Chapters 11 and 12). The periodicity of switching sequences can be used in steady-state studies to model the active and reactive power loading conditions of a.c.–d.c. converters at the relevant busbars. Such modelling is discussed here with reference to the most common configuration used in power systems, i.e. the three-phase bridge rectifier shown in Fig. 3.1.

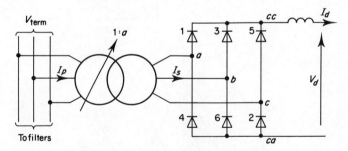

Fig. 3.1. Basic three-phase rectifier bridge

For large power ratings static converter units generally consist of a number of series and/or parallel connected bridges, some or all bridges being phase-shifted relative to the others. With these configurations twelve-pulse and higher pulse numbers can be achieved to reduce the distortion of the supply current with limited or no filtering. A multiple bridge rectifier can therefore be modelled as a single equivalent bridge with a sinusoidal supply voltage at the terminals.

The following basic assumptions are normally made in the development of the steady-state model[1],[2],[3],[4]:

(i) The forward voltage drop in a conducting valve is neglected so that the valve may be considered as a switch. This is justified by the fact that the voltage drop is very small in comparison with the normal operating voltage. It is, further, quite independent of the current and should, therefore, play an insignificant part in the commutation process since all valves commutating on the same side of the bridge suffer similar drops. Such a voltage drop is taken into account by adding it to the d.c. line resistance. The transformer windings resistance is also ignored in the development of the equations, though it should also be included to calculate the power loss.

(ii) The converter transformer leakage reactances as viewed from the secondary terminals are identical for the three phases, and variations of leakage reactance caused by on-load tap-changing are ignored.

(iii) The direct current ripple is ignored, i.e. sufficient smoothing inductance is assumed on the d.c. side.

3.2. RECTIFICATION

Rectifier loads can use diode and thyristor elements in full or half-bridge configurations. In some cases the diode bridges are complemented by transformer on-load tap-changer and saturable reactor control. Saturable reactors produce the same effect as thyristor control over a limited range of delay angles

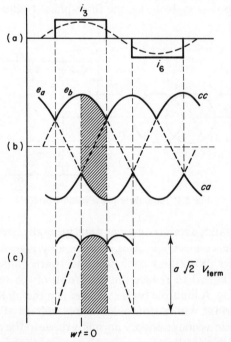

Fig. 3.2. Diode rectifier waveforms. (a) Alternating current in phase '*b*'; (b) Common anode (*ca*) and cathode (*cc*) voltage waveform; (c) Rectified voltage

Referring to the voltage waveforms in Fig. 3.2. and using as time reference the instant when the phase to neutral voltage in phase 'b' is a maximum, the commutating voltage of valve 3 can be expressed as:

$$e_b - e_a = \sqrt{2}aV_{\text{term}}\sin\left(\omega t + \frac{\pi}{3}\right)$$

where 'a' is the off-nominal tap-change position of the converter transformer. The shaded area in Fig. 3.2(b) indicates the potential difference between the common cathode (cc) and common anode (ca) bridge poles for the case of uncontrolled rectification. The maximum average rectified voltage is therefore:

$$V_0 = \frac{1}{\pi/3}\int_0^{\pi/3}\sqrt{2}a\cdot V_{\text{term}}\sin\left(\omega t + \frac{\pi}{3}\right)d(\omega t) = \frac{3\sqrt{2}}{\pi}aV_{\text{term}} \qquad (3.2.1)$$

However, uncontrolled rectification is rarely used in large power conversion. Controlled rectification is achieved by phase-shifting the valve conducting periods with respect to their corresponding phase voltage waveforms.

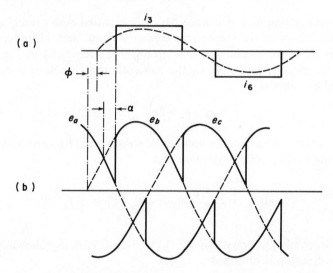

Fig. 3.3. Thyristor-controlled waveform. (a) Alternating current in phase 'b'; (b) Rectified d.c. voltage wafeforms

With delay angle control the average rectified voltage (shown in Fig. 3.3) is thus

$$V_d = \frac{1}{\pi/3}\int_\alpha^{\pi/3+\alpha}\sqrt{2}a\cdot V_{\text{term}}\sin\left(\omega t + \frac{\pi}{3}\right)d(\omega t) = V_0\cos\alpha \qquad (3.2.2)$$

In practice the voltage waveform is that of Fig. 3.4, where a voltage area (δA) is lost due to the reactance (X_c) of the a.c. system (as seen from the converter),

46

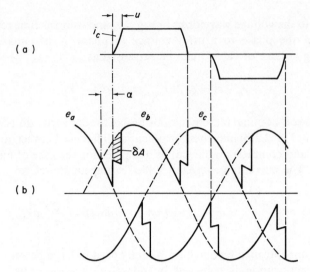

Fig. 3.4. Effect of commutation reactance. (a) Alternating current; (b) D.c. voltage waveforms

referred to as commutation reactance. The energy stored in this reactance has to be transferred from the outgoing to the incoming phase, and this process results in a commutation or conduction overlap angle (u). Referring to Fig. 3.4, and ignoring the effect of resistance in the commutation circuit, area δA can be determined as follows:

$$e_b - e_a = 2\frac{X_c}{\omega}\frac{di_c}{dt} \tag{3.2.3}$$

where e_a, e_b are the instantaneous voltages of phases a and b respectively, and i_c is the incoming valve (commutating) current.

$$\delta A = \int_{\alpha}^{\alpha+\gamma} \frac{e_b - e_a}{2}\, d(\omega t) = X_c \int_{0}^{I_d} di_c = X_c I_d \tag{3.2.4}$$

Finally, by combining equations (3.2.1), (3.2.2) and (3.2.4) the following a.c.–d.c. voltage relationship is obtained

$$V_d = V_0 \cos\alpha - \frac{\delta A}{\pi/3} = \frac{3\sqrt{2}}{\pi} a V_{\text{term}} \cos\alpha - \frac{3}{\pi} X_c I_d \tag{3.2.5}$$

It must be emphasized that the commutating voltage (V_{term}) is the a.c. voltage at the closest point to the converter bridge where sinusoidal waveforms can be assumed. The commutation reactance (X_c) is the reactance between the point at which V_{term} exists and the bridge. Where filters are installed the filter busbar voltage can be used as V_{term}. In the absence of filters, V_{term} must be established at some remote point and X_c must be modified to include the system impedance from the remote point to the converter.

With perfect filtering, only the fundamental component of the current waveform will appear in the a.c. system. This component is obtained from the Fourier analysis of the current waveform in Fig. 3.4, and requires information of i_c and u.

Taking as a reference the instant when the line voltage $(e_b - e_a)$ is zero, equation (3.2.3) can be written as

$$\sqrt{2}a \cdot V_{\text{term}} \sin \omega t = 2\frac{X_c}{\omega}\frac{di_c}{dt}$$

and integrating with respect to (ωt):

$$\int \frac{a \cdot V_{\text{term}}}{\sqrt{2}} \sin \omega t \, d(\omega t) = X_c \int di_c$$

or

$$-\frac{1}{\sqrt{2}}a \cdot V_{\text{term}} \cos \omega t + K = X_c i_c$$

From the initial condition: $i_c = 0$ at $\omega t = \alpha$ the following expressions for K and i_c are obtained.

$$K = \frac{1}{\sqrt{2}}a \cdot V_{\text{term}} \cos \alpha$$

$$i_c = \frac{a \cdot V_{\text{term}}}{\sqrt{2}X_c}[\cos \alpha - \cos \omega t] \qquad (3.2.6)$$

From the final condition: $i_c = I_d$ at $\omega t = \alpha + u$ the following expressions for I_d and u are obtained.

$$I_d = \frac{a \cdot V_{\text{term}}}{\sqrt{2}X_c}[\cos \alpha - \cos(\alpha + u)] \qquad (3.2.7)$$

$$u = \cos^{-1}\left[\cos \alpha - \frac{\sqrt{2}X_c I_d}{a \cdot V_{\text{term}}}\right] - \alpha \qquad (3.2.8)$$

Equation (3.2.6) provides the time-varying commutating current and equation (3.2.8) the limits for the Fourier analysis.

Fourier analysis of the a.c. current waveform, including the effect of commutation (Fig. 3.4) leads to the following relationship between the rms of the fundamental component and the direct current[1]:

$$I_s = k\frac{\sqrt{6}}{\pi}I_d \qquad (3.2.9)$$

where

$$k = \frac{\sqrt{[\cos 2\alpha - \cos 2(\alpha + u)]^2 + [2u + \sin 2\alpha - \sin 2(\alpha + u)]^2}}{4[\cos \alpha - \cos(\alpha + u)]}$$

for values of u not exceeding $60°$.

48

The values of k are very close to unity under normal operating conditions, i.e., when the voltage and currents are close to their nominal values and the a.c. voltage waveforms are symmetrical and undistorted. Alternative steady state models for operating conditions deviating from the above are described in Chapters 7 and 10.

Taking into account the transformer tap position the current on the primary side becomes

$$I_p = k \cdot \frac{\sqrt{6}}{\pi} \cdot a \cdot I_d \qquad (3.2.10)$$

When using per unit values based on a common power and voltage base on both sides of the converter, the direct current base has to be $\sqrt{3}$ times larger than the a.c. current base (as explained in Section 6.3) and equation (3.2.10) becomes

$$I_p = k \cdot \frac{3\sqrt{2}}{\pi} \cdot a \cdot I_d \qquad (3.2.11)$$

Using the fundamental components of voltage and current and assuming perfect filtering at the converter terminals the power factor angle at the converter terminals is ϕ (the displacement between fundamental voltage and current waveforms) and we may write

$$P = \sqrt{3}\,V_{\text{term}}I_p \cos\phi = V_d I_d \qquad (3.2.12)$$

or

$$\cos\phi = \frac{1}{2k}(\cos\alpha + \cos\delta) \qquad (3.2.13)$$

and

$$Q = \sqrt{3}\,V_{\text{term}}I_p \sin\phi \qquad (3.2.14)$$

3.3. INVERSION

Owing to the unidirectional nature of current flow through the converter valves, power reversal (i.e. power flow from the d.c. to the a.c. side) requires direct voltage polarity reversal. This is achieved by delay angle control, which, in the absence of commutation overlap produces rectification between $0° < \alpha < 90°$ and inversion between $90° < \alpha < 180°$. In the presence of overlap, the value of 'α' at which inversion begins is always less than 90°. Moreover, unlike with rectification, full inversion (i.e. $\alpha = 180°$) can not be achieved in practice. This is due to the existence of a certain deionization angle γ at the end of the conducting period, before the voltage across the commutating valve reverses, i.e.

$$\alpha + u \leq 180 - \gamma_0$$

If the above condition is not met (γ_0 being the minimum required extinction angle) a commutation failure occurs; this event would upset the normal conducting sequence and preclude the use of the steady-state model derived in this chapter.

The inverter voltage, although of opposite polarity with respect to the rectifier, is usually expressed as positive when considered in isolation.

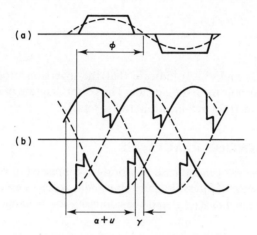

Fig. 3.5. Invertor waveforms. (a) Alternating current; (b) D.c. voltage waveforms

Typical inverter voltage and current waveforms are illustrated in Fig. 3.5. By similarity with the waveforms of Fig. 3.4, the following expression can be written for the inverter voltage in terms of the extinction angle:

$$V_d = \frac{3\sqrt{2}}{\pi} a V_{\text{term}} \cos \gamma - \frac{3X_c}{\pi} I_d \qquad (3.3.1)$$

which is the same as equation (3.2.5) substituting γ for α.

It should by now be obvious that inverter operation requires the existence of three conditions as follows:

(i) An active a.c. system which provides the commutating voltages.
(ii) A d.c. power supply of opposite polarity to provide continuity for the unidirectional current flow (i.e. from anode to cathode through the switching devices).
(iii) Fully controlled rectification to provide firing delays beyond 90°.

When these three conditions are met, a negative voltage of a magnitude given by equation (3.3.1), is impressed across the converter bridge and power $(-V_d I_d)$ is inverted. Note that the power factor angle (ϕ) is now larger than 90°, i.e.

$$P = \sqrt{3} V_{\text{term}} I_p \cos \phi = -\sqrt{3} V_{\text{term}} I_p \cos (\pi - \phi) \qquad (3.3.2)$$

$$Q = \sqrt{3} V_{\text{term}} I_p \sin \phi = \sqrt{3} V_{\text{term}} I_p \sin (\pi - \phi) \qquad (3.3.3)$$

50

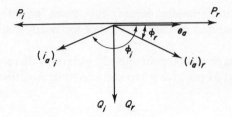

Fig. 3.6. P and Q vector diagram

Equations (3.3.2) and (3.3.3) indicate that the inversion process still requires reactive power supply from the a.c. side. The vector diagram of Fig. 3.6 illustrates the sign of P and Q for rectification and inversion.

3.4. COMMUTATION REACTANCE

Figure 3.7 shows the general case of n bridges connected in parallel on the a.c. side. In the absence of filters the pure sinusoidal voltages exist only behind the system source impedance (X_{ss}) and the commutation reactance (X_{c_j}) for the jth bridge is thus

$$X_{cj} = X_{ss} + X_{tj} \tag{3.4.1}$$

However, if the bridges are under the same controller or under identical controllers then it is preferable to create a single equivalent bridge. The commutation reactance of such an equivalent bridge depends upon the d.c. connections and also the phase shifting between bridges.

If there are k bridges with the same phase shift then they will commutate at the same time and the equivalent reactance must reflect this. For a series connection of bridges the commutation reactance of the equivalent bridge is:

$$X_{c_{series}} = kX_{ss} + X_{tj} \tag{3.4.2}$$

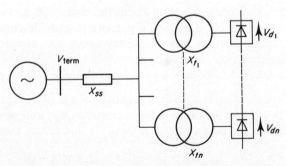

Fig. 3.7. 'n' bridges connected in series on the d.c. side and in parallel on the a.c. side

where j represents any of the n bridges. For bridges connected in parallel on the d.c. side the equivalent bridge commutation reactance is:

$$X_{c_{\text{parallel}}} = X_{ss} + \frac{1}{k}X_{tj} \qquad (3.4.3)$$

It should be noted that with perfect filtering or when many bridges are used with different transformer phase shifts the voltage on the a.c. side of the converter transformers may be assumed to be sinusoidal and hence X_{ss} has no influence on the commutation.

Moreover, the presence of local plant components at the converter terminals may affect the commutation reactance. By way of example, let us consider the two ends of the New Zealand h.v.d.c. link (with reference to Fig. 3.8). It must be noted that h.v.d.c. schemes are normally designed for twelve-pulse operation and that filters are always provided (i.e. the system impedance can be ignored).

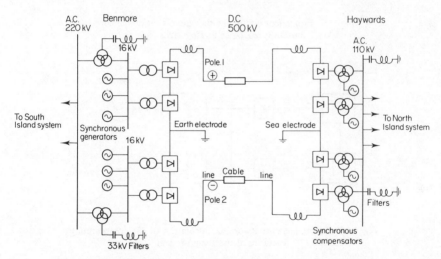

Fig. 3.8. Simplified diagram of the New Zealand h.v.d.c. interconnection

(i) At Haywards the effect of the subtransient reactance of the synchronous compensators on the tertiaries of the converter transformers must be taken into account. The approximate equivalent circuit is illustrated in Fig. 3.9 and the commutation reactance is

$$X_c = X_s + \frac{X_p \cdot (X_t + X_d'')}{X_p + X_t + X_d''} \qquad (3.4.4)$$

where

X_s—transformer secondary leakage reactance
X_p—transformer primary leakage reactance
X_t—transformer tertiary leakage reactance
X_d'' is the subtransient reactance of the synchronous condenser unit

52

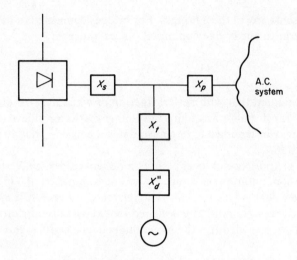

Fig. 3.9. Equivalent circuit for the calculation of the commutation reactance at Haywards end

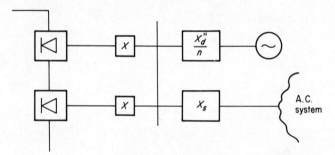

Fig. 3.10. Equivalent circuit for the calculation of the commutation reactance at the Benmore end

(ii) At the Benmore end the subtransient reactance of the generators is combined in parallel with the secondary reactance of the interconnecting transformer. (The primary reactance is beyond the filters and can thus be neglected.) The approximate equivalent circuit is illustrated in Fig. 3.10. Although there are two converter groups commutating on this reactance, the commutations are not simultaneous due to the 30-degree phase-shift of their respective transformers. Thus the effective commutation reactance per group is

$$X_c = X + \frac{X_d'' \cdot X_s}{X_d'' + n \cdot X_s} \tag{3.4.5}$$

where

X—two-winding transformer leakage reactance

X_s—interconnecting transformer secondary leakage reactance (note filters connected to tertiary winding)
X_d''—generator subtransient reactance
n—number of generators connected

3.5. D.C. TRANSMISSION

The sending and receiving ends of a two-terminal d.c. transmission link such as that illustrated in Fig. 3.8 can be modelled as single equivalent bridges with terminal voltages V_{d_r} and V_{d_i} respectively. The direct current is thus given by:

$$I_d = I_{d_r} = I_{d_i} = \frac{V_{d_r} - V_{d_i}}{R_d} \qquad (3.5.1)$$

where R_d is the resistance of the link and includes the loop transmission resistance (if any), the resistance of the smoothing reactors and the converter valves.

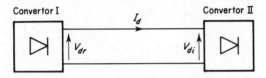

Fig. 3.11. Two-terminal d.c. link

The prime considerations in the operation of a d.c. transmission system are to minimize the need for reactive power at the terminals and reduce system losses. These objectives require maintaining the highest possible transmission voltage and this is achieved by minimizing the inverter end extinction angle, i.e. operating the inverter on constant extinction angle (E.A.) control while controlling the direct current at the rectifier end by means of temporary delay angle backed by transformer tap-change control.

E.A. control applied to the inverter automatically varies the firing angle of advance to maintain the extinction angle γ at a constant value. Deionization imposes a definite minimum limit on γ, and the E.A. control usually maintains it at this limit.

Constant-current (C.C.) control applied to the rectifier regulates the firing angle α to maintain a prespecified link current I_d^{sp}, within the range of α. If the value of α required to maintain I_d^{sp} falls below its minimum limit, current control is transferred to the inverter, i.e. α is fixed on its minimum limit, and the inverter firing angle is advanced to control the current.

The converter-transformer tap-change is a composite part of this control. The rectifier transformer attempts to maintain α within its permitted range. The inverter transformer attempts to regulate the d.c. voltage at some point along the line to a specified level. For minimum loss and minimum-reactive-power absorption, this voltage is required to be as high as possible, and the firing angle of the rectifier should be as low as possible.

54

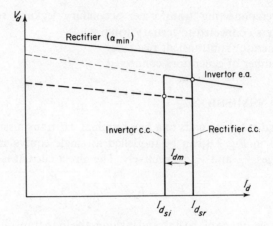

Fig. 3.12. Normal control characteristics

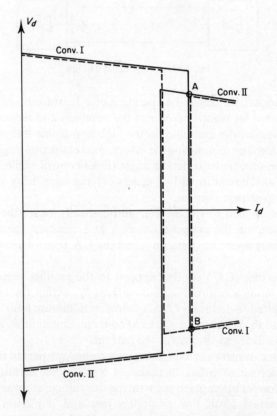

Fig. 3.13. Control characteristics and power flow
reversal

Fig. 3.12 shows the d.c. voltage/current characteristics at the rectifier and inverter ends (the latter have been drawn with reverse polarity in order to illustrate the operating point). The current controller gains are very large and for all practical purposes the slopes of the constant current characteristics can be ignored. Consequently the operating current is equal to the relevant current setting, i.e. $I_{d_{sr}}$ and $I_{d_{si}}$ for rectifier and inverter constant current control respectively.

The direction of power flow is determined by the current settings, the rectifier end always having the larger setting. The difference between the settings is the current margin I_{d_m} and is given by

$$I_{d_m} = I_{d_{sr}} - I_{d_{si}} > 0 \qquad (3.5.4)$$

Many d.c. transmission schemes are bidirectional, i.e. each converter operates sometimes as a rectifier and sometimes as an inverter. Moreover, during d.c. line faults, both converters are forced into the inverter mode in order to de-energize the line faster. In such cases each converter is provided with a combined characteristic as shown in Fig. 3.13 which includes natural rectification, constant current control and constant extinction angle control.

With the characteristics shown by solid lines (i.e. operating at point A), power is transmitted from Converter I to Converter II. Both stations are given the same current command but the current margin setting is subtracted at the inverter end. When power reversal is to be implemented the current settings are reversed and the broken line characteristics apply. This results in operating point B, with direct voltage reversed and no change in direct current.

Alternative forms of control

A commonly used operating mode is constant power (C.P.) control. As with constant current control either converter can control power. The power setting at the rectifier terminal $P_{d_{sr}}$ must be larger than that at the invertor terminal $P_{d_{si}}$ by a suitable power margin P_{d_m}, that is:

$$P_{d_m} = P_{d_{sr}} - P_{d_{si}} > 0 \qquad (3.5.5)$$

The C.P. controller adjusts the C.C. control setting I_d^{sp} to maintain a specified power flow P_d^{sp} through the link, which is usually more practical than C.C. control from a system-operation point of view. The voltage/current loci now become non-linear, as shown in Fig. 3.14.

Several limits are added to the C.P. characteristics as shown in Fig. 3.15. These are:

— A maximum current limit with the purpose of preventing thermal damage to the converter valves; normally between 1 and 1.2 times the nominal current.
— A minimum current limit (about 10 per cent of the nominal value) in order to avoid possible current discontinuities which can cause overvoltages.
— Voltage dependent current limit (line OA in the figure) in order to reduce the power loss and reactive power demand.

56

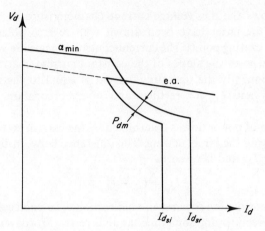

Fig. 3.14. Constant power characteristics

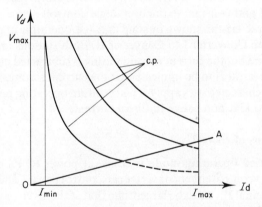

Fig. 3.15. Voltage and current limits

In cases where the power rating of the d.c. link is comparable with the rating of either the sending or receiving a.c. systems interconnected by the link, the frequency of the smaller a.c. system is often controlled to a large extent by the d.c. link. With power-frequency (P.F.) control if the frequency goes out of pre-specified limits, the output power is made proportional to the deviation of frequency from its nominal value. Frequency control is analogous to the current control described earlier, i.e. the converter with lower voltage determines the direct voltage of the line and the one with higher voltage determines the frequency. Again, current limits have to be imposed, which override the frequency error signal.

C.P./E.A. and C.C./E.A. controls were evolved principally for bulk point-to-point power transmission over long distances or submarine crossings and are still the main control modes in present use.

Multiterminal d.c. schemes are also being considered, based on the basic controls already described. Two alternatives are possible, i.e. constant voltage parallel[5] and constant current series[6] schemes.

3.6. REFERENCES

1. C. Adamson and N. G. Hingorani, 1960. *H.V.D.C. Power Transmission*. Garraway Ltd., London.
2. E. W. Kimbark, 1971. *Direct Current Transmission*. Vol. I. Wiley Interscience, London.
3. B. J. Cory (ed.), 1965. *High-Voltage Direct-Current Converters and Systems*. Macdonald, London.
4. E. Uhlmann, 1975. *Power Transmission by Direct Current*. Springer-Verlag, Berlin/Heidelberg.
5. U. Lamm and E. Uhlmann and P. Danfors, 1963. 'Some aspects of tapping of h.v.d.c. transmission systems', *Direct Current*, **8**, 124–129.
6. C. Adamson and J. Arrillaga, 1968. 'behaviour of multiterminal a.c.–d.c. interconnections with series-connected stations', *Proc. IEE*, **115** (11), 1685–1692.

4

Load Flow

4.1. INTRODUCTION

Under normal conditions electrical transmission systems operate in their steady-state mode and the basic calculation required to determine the characteristics of this state is termed load flow (or power flow).

The object of load-flow calculations is to determine the steady-state operating characteristics of the power generation/transmission system for a given set of busbar loads. Active power generation is normally specified according to economic-dispatching practice and the generator voltage magnitude is normally maintained at a specified level by the automatic voltage regulator acting on the machine excitation. Loads are normally specified by their constant active and reactive power requirement, assumed unaffected by the small variations of voltage and frequency expected during normal steady-state operation.

The solution is expected to provide information of voltage magnitudes and angles, active and reactive power flows in the individual transmission units, losses and the reactive power generated or absorbed at voltage-controlled buses.

The load flow problem is formulated in its basic analytical form in this chapter with the network represented by linear, bilateral and balanced lumped parameters. However the power and voltage constraints make the problem non-linear and the numerical solution must therefore be iterative in nature.

The currents problem faced in the development of load flow are: an ever increasing size of systems to be solved, on-line applications for automatic control, and system optimization. Hundreds of contributions have been offered in the literature to overcome these problems.[1]

Five main properties are required of a load-flow solution method.

(i) High computational speed. This is especially important when dealing with large systems, real time applications (on-line), multiple case load-flow such as in system security assessment, and also in inter-active applications.

(ii) Low computer storage. This is important for large systems and in the use of computers with small core storage availability, e.g. mini-computers for on-line application.

(iii) Reliability of solution. It is necessary that a solution be obtained for

illconditioned problems, in outage studies and for real time applications.
(iv) Versatility. An ability on the part of load-flow to handle conventional and special features (e.g. the adjustment of tap ratios on transformers; different representations of power-system apparatus), and its suitability for incorporation into more complicated processes.
(v) Simplicity. The ease of coding a computer program of the load-flow algorithm.

The type of solution required from a load-flow also determines the method used:

accurate	or	approximate
unadjusted	or	adjusted
off-line	or	on-line
single case	or	multiple cases

The first column are requirements needed for considering optimal load-flow and stability studies, and the second column those needed for assessing security of a system. Obviously, solutions may have a mixture of the properties from either column.

The first practical digital solution methods for load flow were the Y matrix-iterative methods.[2] These were suitable because of the low storage requirements, but had the disadvantage of converging slowly or not at all. Z matrix methods[3] were developed which overcame the reliability problem but a sacrifice was made of storage and speed with large systems.

The Newton–Raphson method[4],[5] was developed at this time and was found to have very strong convergence. It was not, however, made competitive until sparsity programming and optimally ordered Gaussian-elimination[6],[7],[8] were introduced, which reduced both storage and solution time.

Nonlinear programming and hybrid methods have also been developed, but these have created only academic interest and have not been accepted by industrial users of load flow. The Newton–Raphson method and techniques derived from this algorithm, satisfy the requirements of solution type and programming properties better than previously used techniques and are gradually replacing them.

4.2. BASIC NODAL METHOD

In the nodal method as applied to power-system networks, the variables are the complex node (busbar) voltages and currents, for which some reference must be designated. In fact, two different references are normally chosen: for voltage magnitudes the reference is ground, and for voltage angles the reference is chosen as one of the busbar voltage angles, which is fixed at the value zero (usually). A nodal current is the nett current entering (injected into) the network at a given node, from a source and/or load external to the network. From this definition, a current entering the network (from a source) is positive in sign, while a current leaving the network (to a load) is negative, and the nett nodal injected current is

60

the algebraic sum of these. One may also speak in the same way of nodal injected powers $S = P + jQ$.

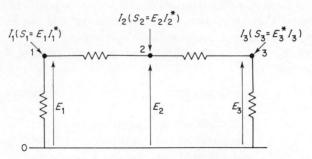

Fig. 4.1. Simple network showing nodal quantities

Figure 4.1 gives a simple network showing the nodal currents, voltages and powers.

In the nodal method, it is convenient to use branch admittances rather than impedances. Denoting the voltages of nodes k and i as E_k and E_i respectively, and the admittance of the branch between them as y_{ki}, then the current flowing in this branch from node k to node i is given by:

$$I_{ki} = y_{ki}(E_k - E_i) \qquad (4.2.1)$$

Let the nodes in the network be numbered 0, 1 ... n, where 0 designates the reference node (ground). By Kirchhoff's current law, the injected current I_k must be equal to the sum of the currents leaving node k, hence:

$$I_k = \sum_{i=0}^{n} I_{ki} = \sum_{i=0}^{n} y_{ki}(E_k - E_i) \qquad (4.2.2)$$

Since $E_0 = 0$, and if the system is linear,

$$I_k = \sum_{i=0 \neq k}^{n} y_{ki} E_k - \sum_{i=1 \neq k}^{n} y_{ki} E_i \qquad (4.2.3)$$

If this equation is written for all the nodes except the reference, i.e. for all busbar in the case of a power-system network, then a complete set of equations defining the network is obtained in matrix form as:

I_1		Y_{11}	Y_{12}		Y_{1n}		E_1
I_2	=	Y_{21}	Y_{22}		Y_{2n}		E_2
I_n		Y_{n1}	Y_{n2}				E_n

(4.2.4)

where

$$Y_{kk} = \sum_{i=0 \neq k}^{n} y_{ki} = \text{self-admittance of node } k$$

$$Y_{ki} = -y_{ki} = \text{mutual-admittance between nodes } k \text{ and } i$$

In shorthand matrix notation, equation (4.2.4) is simply

$$I = Y \cdot E \tag{4.2.5}$$

or in summation notation, we have

$$I_k = \sum_{i=1}^{n} Y_{ki} E_i \quad \text{for } i = 1 \ldots n \tag{4.2.6}$$

The nodal admittance matrix in equations (4.2.4) or (4.2.5) has a well-defined structure, which makes it easy to construct automatically. Its properties are as follows:

  (i) Square of order $n \times n$
 (ii) Symmetrical, since $y_{ki} = y_{ik}$
(iii) Complex
 (iv) Each off-diagonal element y_{ki} is the negative of the branch admittance between nodes k and i, and is frequently of value zero.
  (v) Each diagonal element y_{kk} is the sum of the admittance of the branches which terminate on node k, including branches to ground.
 (vi) Because in all but the smallest practical networks very few nonzero mutual admittances exist, matrix Y is highly sparse.

4.3. CONDITIONING OF Y MATRIX

The set of equations $I = Y \cdot E$ may or may not have a solution. If not, a simple physical explanation exists, concerning the formulation of the network problem. Any numerical attempt to solve such equations is found to break down at some stage of the process (what happens in practice is usually that a finite number is divided by zero).

The commonest case of this is illustrated in the example of Fig. 4.2.

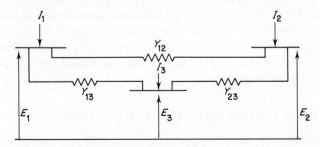

Fig. 4.2. Example of singular network

The nodal equations are constructed in the usual way as:

$$
\begin{vmatrix} I_1 \\ I_2 \\ I_3 \end{vmatrix} =
\begin{vmatrix}
y_{12} + y_{13} & -y_{12} & -y_{13} \\
-y_{12} & y_{12} + y_{23} & -y_{23} \\
-y_{13} & -y_{23} & y_{13} + y_{23}
\end{vmatrix}
\begin{vmatrix} E_1 \\ E_2 \\ E_3 \end{vmatrix}
\qquad (4.3.1)
$$

Suppose that the injected currents are known, and nodal voltages are unknown. In this case no solution for the latter is possible. The Y matrix is described as being singular, i.e. it has no inverse, and this is easily detected in this example by noting that the sum of the elements in each row and column is zero, which is a sufficient condition for singularity, mathematically speaking. Hence, if it is not possible to express the voltages in the form $E = Y^{-1}$. I, it is clearly impossible to solve equation (4.3.1.) by any method, whether involving inversion of Y or otherwise.

The reason for this is obvious: we are attempting to solve a network whose reference node is disconnected from the remainder, i.e. there is no effective reference node, and an infinite number of voltage solutions will satisfy the given injected current values.

When, however, a shunt admittance from at least one of the busbar in the network of Fig. 4.2. is present, the problem of insolubility immediately vanishes in theory, but not necessarily in practice. Practical computation cannot be performed with absolute accuracy, and during a sequence of arithmetic operations, rounding errors due to working with a finite number of decimal places accumulate. If the problem is well-conditioned and the numerical solution technique is suitable, these errors remain small and do not mask the eventual results. If the problem is ill-conditioned, and this usually depends upon the properties of the system being analysed, any computational errors introduced are likely to become large with respect to the true solution.

It is easy to see intuitively that if a network having zero shunt admittances cannot be solved even when working with absolute computational accuracy, then a network having very small shunt admittances may well present difficulties when working with limited computational accuracy. This reasoning provides a key to the practical problems of network, i.e. Y matrix, conditioning. A network with shunt admittances which are small with respect to the other branch admittances is likely to be ill-conditioned, and the conditioning tends to improve with the size of the shunt admittances, i.e. with the electrical connection between the network busbars and the reference node.

4.4. THE CASE WHERE ONE VOLTAGE IS KNOWN

In load-flow studies, it usually happens that one of the voltages in the network is specified, and instead the current at that busbar is unknown. This immediately alleviates the problem of needing at least one good connection with ground, because the fixed busbar voltage can be interpreted as an infinitely strong ground tie. If it is represented as a voltage source with a series impedance of zero value, and then converted to the Norton equivalent, the fictitious shunt admittance is infinite, as is the injected current. This approach is not computationally feasible, however.

The usual way to deal with a voltage which is fixed is to eliminate it as a variable from the nodal equations. Purely for the sake of analytical convenience, let this busbar be numbered 1 in an n-busbar network. The nodal equations are then

$$
\begin{aligned}
I_1 &= Y_{11}E_1 + Y_{12}E_2 + \cdots\cdots\cdots\cdots Y_{1n}E_n \\
I_2 &= Y_{21}E_1 + Y_{22}E_2 + \cdots\cdots\cdots\cdots Y_{2n}E_n \\
&\vdots \\
I_n &= Y_{n1}E_1 + Y_{n2}E_2 + \cdots\cdots\cdots\cdots Y_{nn}E_n
\end{aligned}
\tag{4.4.1}
$$

The terms in E_1 on the right-hand side of equations (4.4.1) are known quantities, and as such are transferred to the left-hand side.

$$
\begin{aligned}
I_1 - Y_{11}E_1 &= Y_{12}E_2 + \cdots\cdots\cdots\cdots Y_{1n}E_n \\
I_2 - Y_{21}E_1 &= Y_{22}E_2 + \cdots\cdots\cdots\cdots Y_{2n}E_n \\
&\vdots \\
I_n - Y_{n1}E_1 &= Y_{n2}E_2 + \cdots\cdots\cdots\cdots Y_{nn}E_n
\end{aligned}
\tag{4.4.2}
$$

The first row of this set may now be eliminated, leaving $(n - 1)$ equations in $(n - 1)$ unknowns, $E_2 \ldots E_n$. In matrix form, this becomes:

$$
\begin{bmatrix}
I_2 - Y_{21}E_1 \\
\\
I_n - Y_{n1}E_1
\end{bmatrix}
=
\begin{bmatrix}
Y_{22} & & Y_{2n} \\
& & \\
Y_{n2} & & Y_{nn}
\end{bmatrix}
\begin{bmatrix}
E_2 \\
\\
E_n
\end{bmatrix}
\tag{4.4.3}
$$

or

$$
I = Y \cdot E
\tag{4.4.4}
$$

The new matrix Y is obtained from the full admittance matrix Y merely by removing the row and column corresponding to the fixed-voltage busbar, both in the present case where it is numbered 1 or in general.

In summation notation, the new equations are

$$
I_k - Y_{k1}E_1 = \sum_{i=2}^{n} Y_{ki}E_i \quad \text{for } k = 2 \ldots n
\tag{4.4.5}
$$

which is an $(n - 1)$ set in $(n - 1)$ unknowns. The equations are then solved by any of the available techniques for the unknown voltages. It is noted that the problem of singularity when there are no ground ties disappears if one row and column are removed from the original Y matrix.

Eliminating the unknown current I_1 and the equation in which it appears is the simplest way of dealing with the problem, and reduces the order of the equations by one. I_1 is evaluated after the solution from the first equation in equation (4.4.1).

4.5. ANALYTICAL DEFINITION OF THE PROBLEM

The complete definition of power flow requires knowledge of four variables at each bus k in the system.

P_k—real or active power
Q_k—reactive or quadrature power
V_k—voltage magnitude
θ_k—voltage phase angle

only two are known *a priori* to solve the problem, and the aim of the load flow is to solve the remaining two variables at a bus.

We define three different bus conditions based on the steady-state assumptions of constant system frequency and constant voltages, where these are controlled.

(i) Voltage controlled bus. The total injected active power P_k is specified, and the voltage magnitude V_k is maintained at a specified value by reactive power injection. This type of bus generally corresponds to a generator where P_k is fixed by turbine governor setting and V_k is fixed by automatic voltage regulators acting on the machine excitation; or a bus where the voltage is fixed by supplying reactive power from static shunt capacitors or rotating synchronous compensators, e.g. at substations.

(ii) Nonvoltage-controlled bus. The total injected power, $P_k + jQ_k$, is specified at this bus. In the physical power system this corresponds to a load centre such as a city or an industry, where the consumer demands his power requirements. Both P_k and Q_k are assumed to be unaffected by small variations in bus voltage.

(iii) Slack (swing) bus. This bus arises because the system losses are not known precisely in advance of the load flow calculation. Therefore the total injected power cannot be specified at every single bus. It is usual to choose one of the available voltage controlled buses as slack, and to regard its active power as unknown. The slack bus voltage is usually assigned as the system phase reference, and its complex voltage

$$E_s = V_s \underline{/\theta_s}$$

is therefore specified. The analogy in a practical power system is the generating station which has the responsibility of system frequency control.

Load flow solves a set of simultaneous nonlinear algebraic power equations for

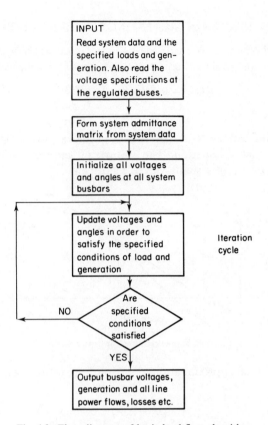

Fig. 4.3. Flow diagram of basic load-flow algorithm

the two unknown variables at each node in a system. A second set of variable equations, which are linear, are derived from the first set, and an iteration method is applied to this second set.

The basic algorithm which load-flow programs use is depicted in Fig. 4.3. System data, such as busbar power conditions, network connections and impedance, are read in and the admittance matrix formed. Initial voltages are specified to all buses; for base case load flows, P, Q buses are set to $1 + j0$ while P, V busbars are set to $V + j0$.

The iteration cycle is terminated when the busbar voltages and angles are such that the specified conditions of load and generation are satisfied. This condition is accepted when power mismatches for all buses are less than a small tolerance, η_1, or voltage increments less than η_2. Typical figures for η_1 and η_2 are 0.01 p.u. and 0.001 p.u. respectively. The sum of the square of the absolute values of power mismatches is a further criterion sometimes used.

When a solution has been reached, complete terminal conditions for all buses are computed. Line power flows and losses and system totals can then be calculated.

4.6. NEWTON–RAPHSON METHOD OF SOLVING LOAD FLOWS

The generalized Newton–Raphson method is an iterative algorithm for solving a set of simultaneous nonlinear equations in an equal number of unknowns.

$$f_k(x_m) = 0 \quad \text{for } k = 1 \rightarrow N \tag{4.6.1}$$

$$m = 1 \rightarrow N$$

At each iteration of the $N-R$ method, the nonlinear problem is approximated by the linear matrix equation. The linearizing approximation can best be visualized in the case of a single-variable problem.

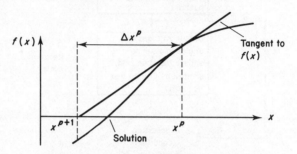

Fig. 4.4. Single-variable linear approximately

In Fig. 4.4. x^p is an approximation to the solution, with error Δx^p at iteration p. Then

$$f(x^p + \Delta x^p) = 0 \tag{4.6.2}$$

This equation can be expanded by Taylor's theorem,

$$f(x^p + \Delta x^p) = 0$$

$$= f(x^p) + \Delta x^p f'(x^p) + \frac{(\Delta x^p)^2}{2!} f''(x^p) + \dots \tag{4.6.3}$$

If the initial estimates of the variable x^p is near the solution value, Δx^p will be relatively small and all terms of higher powers can be neglected. Hence,

$$f(x^p) + \Delta x^p f'(x^p) = 0 \tag{4.6.4}$$

or

$$\Delta x^p = \frac{-f(x^p)}{f'(x^p)} \tag{4.6.5}$$

The new value of the variable is then obtained from

$$x^{p+1} = x^p + \Delta x^p \tag{4.6.6}$$

Equation (4.6.4) may be rewritten as

$$f(x^p) = -J\Delta x^p \tag{4.6.7}$$

The method is readily extended to the set of N equations in N unknowns. J becomes the square Jacobian matrix of first order partial differentials of the functions $f_k(x_m)$. Elements of $[J]$ are defined by

$$J_{km} = \frac{\partial f_k}{\partial x_m} \qquad (4.6.8)$$

and represents the slopes of the tangent hyperplanes which approximate the functions $f_k(x_m)$ at each iteration point.

The Newton–Raphson algorithm will converge quadratically if the functions have continuous first derivates in the neighbourhood of the solution, the Jacobian matrix is nonsingular, and the initial approximations of x are close to the actual solutions. However the method is sensitive to the behaviours of the functions $f_k(x_m)$ and hence to their formulation. The more linear they are, the more rapidly and reliably Newton's method converges. Nonsmoothness, i.e. humps, in any one of the functions in the region of interest, can cause convergence delays, total failure or misdirection to a nonuseful solution.

Equations relating to power-system load flow

The nonlinear network governing equations are

$$I_k = \sum_{m \in k} y_{km} E_m \quad \text{for all } k \qquad (4.6.9)$$

where I_k is the current injected into a bus k. The power at a bus is then given by

$$S_k = P_k + jQ_k = E_k I_k^*$$
$$= E_k \sum_{m \in k} y_{km}^* E_m^* \qquad (4.6.10)$$

Mathematically speaking, the complex load-flow equations are nonanalytic, and cannot be differentiated in complex form. In order to apply Newton's method, the problem is separated into real equations and variables. Polar or rectangular coordinates may be used for the bus voltages. Hence we obtain two equations

$$P_k = P(V, \theta) \quad \text{or} \quad P(e, f)$$

and

$$Q_k = Q(V, \theta) \quad \text{or} \quad Q(e, f)$$

In polar coordinates the real and imaginary parts of equation (4.6.10) are

$$P_k = \sum_{m \in k} V_k V_m (G_{km} \cos \theta_{km} + B_{km} \sin \theta_{km}) \qquad (4.6.11)$$

$$Q_k = \sum_{m \in k} V_k V_m (G_{km} \sin \theta_{km} - B_{km} \cos \theta_{km}) \qquad (4.6.12)$$

68

where

$$\theta_{km} = \theta_k - \theta_m$$

Linear relationships are obtained for small variations in the variables θ and V by forming the total differentials, the resulting equations being:
For a PQ busbar:

$$\Delta P_k = \sum_{m \in k} \frac{\partial P_k}{\partial \theta_m} \Delta \theta_m + \sum_{m \in k} \frac{\partial P_k}{\partial V_m} \Delta V_m \qquad (4.6.13)$$

and

$$\Delta Q = \sum_{m \in k} \frac{\partial Q_k}{\partial \theta_m} \Delta \theta_m + \sum_{m \in k} \frac{\partial Q_k}{\partial V_m} \Delta V_m \qquad (4.6.14)$$

For a PV busbar:
Only equation (4.6.13) is used, since Q_k is not specified.
For a slack busbar:
No equations.

The voltage magnitudes appearing in equations (4.6.13) and (4.6.14) for PV and slack busbars are not variables, but are fixed at their specified values. Similarly θ at the slack busbar is fixed.

The complete set of defining equations is made up of two for each PQ busbar and one for each PV busbar. The problem variables are V and θ for each PQ busbar and θ for each PV busbar. The number of variables is therefore equal to the number of equations. Algorithm (4.6.7) then becomes:

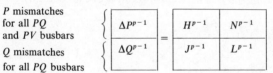

P mismatches for all PQ and PV busbars — θ corrections for all PQ and PV busbars

Q mismatches for all PQ busbars — V corrections for all PQ busbars

$$\begin{Bmatrix} \Delta P^{p-1} \\ \Delta Q^{p-1} \end{Bmatrix} = \begin{bmatrix} H^{p-1} & N^{p-1} \\ J^{p-1} & L^{p-1} \end{bmatrix} \begin{Bmatrix} \Delta \theta^p \\ \dfrac{\Delta V^p}{V^{p-1}} \end{Bmatrix}$$

Jacobian matrix

$$(4.6.15)$$

The division of each ΔV_i^p by V_i^{p-1} does not numerically affect the algorithm, but simplifies some of the Jacobian-matrix terms. For busbars k and m (not row k and column m in the matrix)

$$H_{km} = \frac{\partial P_k}{\partial \theta_m} = V_k V_m (G_{km} \sin \theta_{km} - B_{km} \cos \theta_{km})$$

$$N_{km} = V_m \frac{\partial P_k}{\partial V_m} = V_k V_m (G_{km} \cos \theta_{km} + B_{km} \sin \theta_{km})$$

$$J_{km} = \frac{\partial Q_k}{\partial \theta_m} = - V_k V_m (G_{km} \cos \theta_{km} + B_{km} \sin \theta_{km})$$

$$L_{km} = V_m \frac{\partial Q_k}{\partial V_m} = V_k V_m (G_{km} \sin \theta_{km} - B_{km} \cos \theta_{km})$$

and for $m = k$:

$$H_{kk} = \frac{\partial P_k}{\partial \theta_k} = -Q_k - B_{kk}V_k^2$$

$$N_{kk} = V_k \frac{\partial P_k}{\partial V_k} = P_k + G_{kk}V_k^2$$

$$J_{kk} = \frac{\partial Q_k}{\partial \theta_k} = P_k - G_{kk}V_k^2$$

$$L_{kk} = V_k \frac{\partial Q_k}{\partial V_k} = Q_k - B_{kk}V_k^2$$

In practice, some programs express these coefficients using voltages in rectangular form, i.e. $e_i + jf_i$. This only effects the speed of calculation of the mismatches and the matrix elements, by eliminating the time-consuming trigonometrical functions.

In rectangular coordinates the complex power equations are given as

$$P_k + jQ_k = E_k \sum_{m \in k} Y_{km}^* E_m^*$$

$$= (e_k + jf_k) \sum_{m \in k} (G_{km} - jB_{km})(e_m - jf_m)$$

and these are divided into real and imaginary parts

$$P_k = e_k \sum_{m \in k} (G_{km}e_m - B_{km}f_m) + f_k \sum_{m \in k} (G_{km}f_m + B_{km}e_m)$$

$$Q_k = f_k \sum_{m \in k} (G_{km}e_m - B_{km}f_m) - e_k \sum_{m \in k} (G_{km}f_m + B_{km}e_m)$$

At a voltage controlled bus the voltage magnitude is fixed but not the phase angle. Hence both e_k and f_k vary at each iteration. It is necessary to provide another equation

$$V_k^2 = e_k^2 + f_k^2$$

to be solved with the real power equation for these buses.

Linear relationships are obtained for small variations in e and f, by forming the total differentials,

$$\Delta P_k = \sum_{m \in k} \frac{\partial P_k}{\partial e_m} \Delta e_m + \sum_{m \in k} \frac{\partial P_k}{\partial f_m} \Delta f_m$$

$$= \sum_{m \in k} S_{km} \Delta e_m + \sum_{m \in k} T_{km} \Delta f_m$$

for all buses except the slack bus;

$$\Delta Q_k = \sum_{m \in k} \frac{\partial Q_k}{\partial e_m} \Delta e_m + \sum_{m \in k} \frac{\partial Q_k}{\partial f_m} \Delta f_m$$

$$= \sum_{m \in k} U_{km} \Delta e_m + \sum_{m \in k} W_{km} \Delta f_m$$

for all nonvoltage controlled buses;

$$\Delta V_k^2 = \frac{\partial V_k^2}{\partial e_k} \Delta e_k + \frac{\partial V_k^2}{\partial f_k} \Delta f_k$$

$$= EE_k \Delta e_k + FF_k \Delta f_k$$

for voltage controlled buses.

The Jacobian matrix has the form

ΔP		S	T		Δe
ΔQ	$=$	U	W	$\cdot$	Δf
ΔV^2		EE	FF		

(4.6.16)

and the values of the partial differentials, which are the Jacobian elements, are given by

$$S_{km} = -W_{km} = G_{km}e_k + B_{km}f_k \quad \text{for } m \neq k$$

$$T_{km} = U_{km} = G_{km}f_k - B_{km}e_k \quad \text{for } m \neq k$$

$$S_{kk} = a_k + G_{kk}e_k + B_{kk}f_k$$

$$W_{kk} = a_k - G_{kk}e_k - B_{kk}f_k$$

$$T_{kk} = b_k - B_{kk}e_k + G_{kk}f_k$$

$$U_{kk} = -b_k - B_{kk}e_k + G_{kk}f_k$$

$$EE_k = 2e_k$$

$$FF_k = 2f_k$$

For voltage controlled buses, V is specified, but not the real and imaginary components of voltage, e and f. Approximations can be made, for example, by ignoring the off-diagonal elements in the Jacobian matrix, as the diagonal elements are the largest. Alternatively for the calculation of the elements the voltages can be considered as $E = 1 + j0$. The off-diagonal elements then become constant.

The polar coordinate representation appears to have computational advantages over rectangular coordinates. Real power mismatch equations are present for all buses except the slack bus, while reactive power mismatch equations are needed for nonvoltage controlled buses only.

The Jacobian matrix has the sparsity of the admittance matrix $[Y]$ and has positional but not numerical symmetry. To gain in computation, the form of $[\Delta\theta, \Delta V/V]^T$ is normally used for the variable voltage vector. Both increments are dimensionless and the Jacobian coefficients are now symmetric in structure though not in value. The values of $[J]$ are all functions of the voltage variables V and θ and must be recalculated each iteration.

As an example, the Jacobian matrix equation for the four-busbar system of Fig. 4.5 is given as equation (4.6.17)

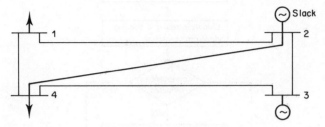

Fig. 4.5. Sample system

$$
\begin{bmatrix} \Delta P_1 \\ \Delta P_3 \\ \Delta P_4 \\ \Delta Q_1 \\ \Delta Q_4 \end{bmatrix} =
\begin{bmatrix}
H_{11} & 0 & H_{14} & N_{11} & N_{14} \\
0 & H_{33} & H_{34} & 0 & N_{34} \\
H_{41} & H_{43} & H_{44} & N_{41} & N_{44} \\
J_{11} & 0 & J_{14} & L_{11} & L_{14} \\
J_{41} & J_{43} & J_{44} & L_{41} & L_{44}
\end{bmatrix}
\cdot
\begin{bmatrix} \Delta\theta_1 \\ \Delta\theta_3 \\ \Delta\theta_4 \\ \Delta V_1/V_1 \\ \Delta V_4/V_4 \end{bmatrix}
\qquad (4.6.17)
$$

The differences in bus powers are obtained from

$$\Delta P_k = P_k^{SP} - P_k \qquad (4.6.18)$$

$$\Delta Q_k = Q_k^{SP} - Q_k \qquad (4.6.19)$$

A further improvement is to replace the reactive power residual ΔQ in the Jacobian matrix equations by $\Delta Q/V$. The performance of the Newton–Raphson method is closely associated with the degree of problem non-linearity; the best left-hand defining functions are the most linear ones. If the system power equation (4.6.19) is divided throughout by V_k, only one term Q_k^{sp}/V_k on the right-hand side of this equation is nonlinear in V_k. For practical values of Q_k^{sp} and V_k, this nonlinear term is numerically relatively small. Hence it is preferable to use $\Delta Q/V$ instead of ΔQ in the Jacobian matrix equation.

Dividing ΔP by V is also helpful, but is less effective since the real power component of the problem is not strongly coupled with voltage magnitudes. A further alternative is to formulate current residuals at a bus. While computationally simple, this method shows poor convergence in the same way as Y matrix iterative methods.

A flow diagram of the basic Newton–Raphson algorithm is given in Fig. 4.6.

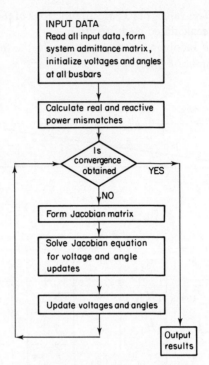

Fig. 4.6. Flow diagram of the basic
Newton–Raphson load-flow algorithm

4.7. TECHNIQUES WHICH MAKE THE NEWTON–RAPHSON METHOD COMPETITIVE IN LOAD FLOW

The efficient solution of (4.6.15) at each iteration is crucial to the success of the N–R method. If conventional matrix techniques were to be used, the storage (αn^2) and computing time (αn^3) would be prohibitive for large systems.

For most power-system networks the admittance matrix is relatively sparse, and in the Newton–Raphson method of load-flow the Jacobian matrix has this same sparsity.

The techniques which have been used to make the Newton–Raphson competitive with other load flow methods involve the solution of the Jacobian matrix equation and the preservation of the sparsity of the matrix by ordered triangular factorization.

Sparsity programming

In conventional matrix programming, double subscript arrays are used for the location of elements. With sparsity programming,[6] only the nonzero elements are stored, in one or more vectors, plus integer vectors for identification.

For the admittance matrix of order n the conventional storage requirements are n^2 words, but by sparsity programming $6b + 3n$ words are required, where b is the number of branches in the system. Typically $b = 1.5n$, and the total storage is $12n$ words. For a large system (say 500 buses) the ratio of storage requirements of conventional and sparse techniques is about 40:1.

Triangular factorization

To solve the Jacobian matrix equation (4.6.15), represented here as

$$[\Delta S] = [J][\Delta E]$$

for increments in voltage, the direct method is to find the inverse of $[J]$ and solve for $[\Delta E]$ from

$$[\Delta E] = [J]^{-1}[\Delta S] \tag{4.7.1}$$

In power systems $[J]$ is usually sparse but $[J]^{-1}$ is a full matrix.

The method of triangular factorization solves for the vector $[\Delta E]$ by eliminating $[J]$ to an upper triangular matrix with a leading diagonal, and then back-substituting for $[\Delta E]$, i.e. eliminate to

$$[\Delta S'] = [U][\Delta E]$$

and back-substitute

$$[U]^{-1}[\Delta S'] = [\Delta E]$$

The triangulation of the Jacobian is best done by rows. Those rows below the one being operated on need not be entered until required. This means that the maximum storage is that of the resultant upper triangle and diagonal. The lower triangle can then be used to record operations.

The number of multiplications and additions to triangulate a full matrix is $\frac{1}{3}N^3$, compared to N^3 to find the inverse. With sparsity programming the number of operations varies as a factor of N. If rows are normalized N further operations are saved.

Optimal ordering

In power-system load flow, the Jacobian matrix is usually diagonally dominant which implies small round-off errors in computation. When a sparse matrix is triangulated, nonzero terms are added in the upper triangle. The number added is affected by the order of the row eliminations, and total computation time increases with more terms.

The pivot element is selected to minimize the accumulation of nonzero terms, and hence conserve sparsity, rather than minimizing round off error. The diagonals are used as pivots.

Optimal ordering of row eliminations to conserve sparsity is a practical impossibility due to the complexity of programming and time involved. However, semioptimal schemes are used and these can be divided into two sections.

(a) *Preordering*.[7] Nodes are renumbered before triangulation. No complicated programming or storage is required to keep track of row and column interchanges.

  (i) Nodes are numbered in sequence of increasing number of connected lines.
  (ii) Diagonal banding—nonzero elements are arranged about either the major or minor diagonals of the matrix.

(b) *Dynamic ordering*.[8] Ordering is effected at each row during the elimination.

  (i) At each step in the elimination, the next row to be operated on is that with the fewest nonzero terms.
  (ii) At each step in the elimination, the next row to be operated on is that which introduces the fewest new nonzero terms, one step ahead.
  (iii) At each step in the elimination, the next row to be operated on is that which introduces the fewest new nonzero terms, two steps ahead. This may be extended to the fully optimal case of looking at the effect in the final step.
  (iv) With cluster ordering, the network is subdivided into groups which are then optimally ordered. This is most efficient if the groups have a minimum of physical intertie. The matrix is then anchor banded.

The best method arises from a trade-off between a processing sequence which requires the least number of operations, and time and memory requirements.

The dynamic ordering scheme of choosing the next row to be eliminated as that with the fewest nonzero terms, appears to be better than all other schemes in sparsity conservation, number of arithmetic operations required, ordering times and total solution time.

However, there are conditions under which other ordering would be preferable, e.g. with system changes affecting only a few rows these rows should be numbered last; when the subnetworks have relatively few interconnections it is better to use cluster ordering.

Aids to convergence

The N–R method can diverge very rapidly or converge to the wrong solution if the equations are not well behaved or if the starting voltages are badly chosen. Such problems can often be overcome by a variety of techniques. The simplest device is to impose a limit on the size of each $\Delta\theta$ and ΔV correction at each iteration. Figure 4.7. illustrates a case which would diverge without this device.

Another more complicated method is to calculate good starting values for the θ's and V's, which also reduces the number of iterations required.

In power-system load flow, setting voltage controlled buses to $V + j0$ and nonvoltage controlled buses to $1 + j0$ may give a poor starting point for the Newton–Raphson method.

If previously stored solutions for a network are available these should be used. One or two iterations of a Y matrix iterative method[2] can be applied before

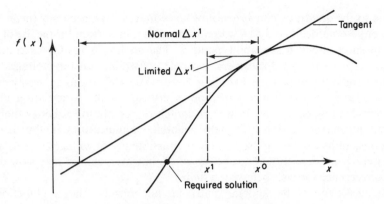

Fig. 4.7. Example of diverging solution

commencing the Newton method. This shows fast initial convergence unless the problem is ill-conditioned, in which case divergence occurs.

A more reliable method is the use of one iteration of a d.c. load flow (i.e. neglecting losses and reactive power conditions) to provide estimates of voltage angles; followed by one iteration of a similar type of direct solution to obtain voltage magnitudes. The total computing time for both sets of equations is about 50 percent of one Newton–Raphson iteration and the extra storage required is only in the programming statements. The resulting combined algorithm is faster and more reliable than the formal Newton method and can be used to monitor diverging or difficult cases, before commencing the Newton–Raphson algorithm.

4.8. CHARACTERISTICS OF THE NEWTON–RAPHSON LOAD FLOW

With sparse programming techniques and optimally ordered triangular factorization, the Newton method for solving load flow has become faster than other methods for large systems. The number of iterations is virtually independent of system size (from a flat voltage start and with no automatic adjustments) due to the quadratic characteristic of convergence. Most systems are solved in $2 \rightarrow 5$ iterations with no acceleration factors being necessary.

With good programming the time per iteration rises nearly linearly with the number of system buses N, so that the overall solution time varies as N. One Newton iteration is equivalent to about seven Gauss–Seidel iterations. For a 500-bus system, the conventional Gauss–Seidel method takes about 500 iterations and the speed advantage of the Newton method is then 15:1. Storage requirements of the Newton method are greater, however, but increase linearly with system size. It is therefore attractive for large systems.

The Newton method is very reliable in system solving, given good starting approximations. Heavily loaded systems with phase shifts up to 90° can be solved. The method is not troubled by ill-conditioned systems and the location of slack bus is not critical.

Due to the quadratic convergence of bus voltages, high accuracy (near exact solution) is obtained in only a few iterations. This is important for the use of load flow in short-circuit and stability studies. The method is readily extended to include tap-changing transformers, variable constraints on bus voltages, and reactive and optimal power scheduling. Network modifications are easily made.

The success of the Newton method is critical on the formulation of the problem-defining equations. Power mismatch representation is better than the current mismatch versions. To help negotiate nonlinearities in the defining functions, limits can be imposed on the permissible size of voltage corrections at each iteration. These should not be too small, however, as they may slow down the convergence for well-behaved systems.

The coefficients of the Jacobian matrix are not constant, they are functions of the voltage variables V and θ, and hence vary each iteration. However, after a few iterations, as V and θ tend to their final values the coefficients will tend to constant values.

One modification to the Newton algorithm is to calculate the Jacobian for the first two or three iterations only and then use the final one for all the following iterations. Alternatively the Jacobian can be updated every two or more iterations. Neither of these modifications greatly affects the convergence of the algorithm, but much time is saved (but not storage).

4.9. DECOUPLED NEWTON LOAD FLOW

An inherent characteristic of any practical electric power transmission system operating in the steady-state condition is the strong interdependence between active powers and bus voltage angles, and between reactive powers and voltage magnitudes. Correspondingly, the coupling between these 'P-θ' and 'Q-V' components of the problem is relatively weak. Many algorithms have been proposed which adopt this decoupling principle.[9],[10],[11]

The voltage vectors method uses a series approximation for the sine terms which appear in the system defining equations, to calculate the Jacobian elements and arrive at two decoupled equations

$$[\mathscr{P}] = [T][\theta] \tag{4.9.1}$$

$$[\mathscr{Q}] = [U][V - V_0] \tag{4.9.2}$$

where for the reference node $\theta_0 = 0$ and $V_k = V_0$. The values of $\mathscr{P}_k$ and $\mathscr{Q}_k$ represent real and reactive power quantities respectively and $[T]$ and $[U]$ are given by

$$T_{km} = -\frac{V_k V_m}{Z_{km}^2 / X_{km}} \tag{4.9.3}$$

$$T_{kk} = -\sum_{m \omega k} T_{km} \tag{4.9.4}$$

$$U_{km} = -\frac{1}{Z_{km}^2/X_{km}} \qquad (4.9.5)$$

$$U_{kk} = -\sum_{m\omega k} U_{km} \qquad (4.9.6)$$

where Z_{km} and X_{km} are the branch impedance and reactance respectively. $[U]$ is constant valued and needs be triangulated once only for a solution. $[T]$ is recalculated and triangulated each iteration.

The two equations (4.9.1) and (4.9.2) are solved alternately until a solution is obtained. These equations can be solved using Newton's method, by expressing the Jacobian equations as

$$
\begin{array}{|c|}
\hline
\Delta P \\
\hline
\Delta Q/V \\
\hline
\end{array}
=
\begin{array}{|c|c|}
\hline
T & \\
\hline
 & U \\
\hline
\end{array}
\cdot
\begin{array}{|c|}
\hline
\Delta\theta \\
\hline
\Delta V \\
\hline
\end{array}
\qquad (4.9.7)
$$

or

$$[\Delta P] = [T][\Delta\theta] \qquad (4.9.8)$$

$$[\Delta Q/V] = [U][\Delta V] \qquad (4.9.9)$$

where

$$[\Delta P] = [\Delta \mathscr{P}]$$

$$[\Delta Q/V] = [\Delta \mathscr{Q}]$$

and T and U are therefore defined in equations (4.9.3) to (4.9.6).

The most successful decoupled load flow is that based on the Jacobian matrix equation for the formal Newton method, i.e.

$$
\begin{array}{|c|}
\hline
\Delta P \\
\hline
\Delta Q \\
\hline
\end{array}
=
\begin{array}{|c|c|}
\hline
H & N \\
\hline
J & L \\
\hline
\end{array}
\cdot
\begin{array}{|c|}
\hline
\Delta\theta \\
\hline
\Delta V \\
\hline
\end{array}
\qquad (4.9.10)
$$

If the submatrices N and J are neglected, since they represented the weak coupling between 'P-θ' and 'Q-V', the following decoupled equations result

$$[\Delta P] = [H][\Delta\theta] \qquad (4.9.11)$$

$$[\Delta Q] = [L][\Delta V] \qquad (4.9.12)$$

It has been found that the latter equation is relatively unstable at some distance from the exact solution due to the nonlinear defining functions. An improvement in convergence is obtained by replacing this with the polar current-mismatch formulation[7]

$$[\Delta I] = [D][\Delta V] \qquad (4.9.13)$$

Alternatively the right-hand side of both equations (4.9.11) and (4.9.12) is divided by voltage magnitude V

$$[\Delta P/V] = [A][\Delta\theta] \tag{4.9.14}$$

$$[\Delta Q/V] = [C][\Delta V] \tag{4.9.15}$$

The equations are solved successively using the most up-to-date values of V and θ available. $[A]$ and $[C]$ are sparse, nonsymmetric in value and are both functions of V and θ.

They must be calculated and triangulated each iteration.

Further approximations that can be made are to assume that $E_k = 1.0$ p.u., for all buses, and $G_{km} \ll B_{km}$ in calculating the Jacobian elements. The off-diagonal terms then become symmetric about the leading diagonal.

The decoupled Newton method compares very favourably with the formal Newton method. While reliability is just as high for ill-conditioned problems, the decoupled method is simple and computationally efficient. Storage of the Jacobian and matrix triangulation is saved by a factor of 4, or an overall saving of 30–40 percent on the formal Newton load flow. Computation time per iteration is also less than the Newton method.

However, the convergence characteristics of the decoupled method are linear, the quadratic characteristics of the formal Newton being sacrificed. Thus, for high accuracies, more iterations are required. This is offset for practical accuracies by the fast initial convergence of the method. Typically, voltage magnitudes converge to within 0.3 percent of the final solution on the first iteration and may be used as a check for instability. Phase angles converge more slowly than voltage magnitudes but the overall solution is reached in $2 \rightarrow 5$ iterations. Adjusted solutions (the inclusion of transformer taps, phase shifters, interarea power transfers, Q and V limits) take many more iterations.

The Newton methods can be expressed as follows[12]:

$$
\begin{array}{|c|c|c|c|c|}
\hline
\Delta P/V & A_{11} & \varepsilon A_{12} & \Delta\theta \\
\hline
\Delta Q/V & \varepsilon A_{21} & A_{22} & \Delta V \\
\hline
\end{array}
\tag{4.9.16}
$$

where

$\varepsilon = 1$ for the full Newton–Raphson method

$\varepsilon = 0$ for the decoupled Newton algorithm

A Taylor series expansion of the Jacobian about $\varepsilon = 0$ results in a first-order approximation of the Newton–Raphson method whereas the decoupled method is a zero order approximation. The method has quadratic convergence properties because of the coupling, but retains storage requirements similar to that of the decoupled method.

4.10. FAST DECOUPLED LOAD FLOW

By further simplifications and assumptions, based on the physical properties of a practical system, the Jacobians of the decoupled Newton load flow can be made constant in value. This means that they need be triangulated only once per solution or for a particular network.

For ease of reference, the real and reactive power equations at a node k are reproduced here.

$$P_k = V_k \sum_{m \in k} V_m(G_{km} \cos \theta_{km} + B_{km} \sin \theta_{km}) \qquad (4.10.1)$$

$$Q_k = V_k \sum_{m \in k} V_m(G_{km} \sin \theta_{km} - B_{km} \cos \theta_{km}) \qquad (4.10.2)$$

where $\theta_{km} = \theta_k - \theta_m$.

A decoupled method which directly relates powers and voltages is derived using the series approximations for the trigonometric terms in equations (4.10.1) and (4.10.2).

$$\sin \theta = \theta - \frac{\theta^3}{6}$$

$$\cos \theta = 1 - \frac{\theta^2}{2}$$

The equations, over all buses, can be expressed in their simplified matrix form

$$[A][\theta] = [P] \qquad (4.10.3)$$

$$[C][V] = [Q] \qquad (4.10.4)$$

where P and Q are terms of real and reactive power respectively and

$$A_{kk} = V_k \sum_{m \omega k} V_m B_{km}$$

$$A_{km} = - V_k V_m B_{km}, \qquad m \neq k$$

$$C_{kk} = \sum_{m \omega k} t_{km} B_{km}$$

$$C_{km} = - B_{km}, \qquad m \neq k$$

t_{km}—tap ratio if a transformer is in the line.

A modification suggested is to replace equation (4.10.3) by

$$[\hat{A}][\hat{\theta}] = [\hat{P}]$$

where

$$\hat{A}_{km} = - B_{km}, \qquad m \neq k$$

$$\hat{A}_{kk} = \sum_{m \omega k} B_{km}$$

$$\hat{\theta}_k = \theta_k \cdot V_k$$

$$\hat{P}_k = P_k / V_k$$

Hence $[\hat{A}]$ becomes constant valued.

A similar direct method is obtained from the decoupled voltage vectors method (equations (4.9.1) and (4.9.2)). If V_m, V_k are put as 1.0 p.u. for the calculation of matrix $[T]$, then $[T]$ becomes constant and need be triangulated once only. This same simplification can be used in the decoupled voltage vectors and Newton's method of equations (4.9.8) and (4.9.9).

Fast decoupled load flow algorithms[8] are also derived from the Jacobian matrix equations of Newton's method (equations (4.9.10)), and the decoupled version (equations (4.9.11) and (4.9.12)).

If we make the assumptions

(i) $E_k, E_m = 1.0$ p.u.
(ii) $G_{km} \ll B_{km}$, and hence can be ignored (for most transmission line reactance/resistance ratios, $X/R \gg 1$)
(iii) $\cos(\theta_k - \theta_m) \doteq 1.0$
$\sin(\theta_k - \theta_m) \doteq 0.0$

since angle differences across transmission lines are small under normal loading conditions.

This leads to the decoupled equations

$$[\Delta P] = [\bar{B}][\Delta\theta] \quad \text{of order } (N-1) \tag{4.10.5}$$

$$[\Delta Q] = [\bar{B}][\Delta V] \quad \text{of order } (N-M) \tag{4.10.6}$$

where the elements of $[\bar{B}]$ are

$$\bar{B}_{km} = -B_{km} \quad \text{for } m \neq k$$

$$\bar{B}_{kk} = \sum_{m \omega k} B_{km}$$

and B_{km} are the imaginary parts of the admittance matrix. To simplify still further, line resistances may be neglected in the calculation of elements of $[\bar{B}]$.

An improvement over equations (4.10.5) and (4.10.6) is based on the decoupled equations (4.9.14) and (4.9.15) which have less nonlinear defining functions. Applying the same assumptions listed previously, we obtain the equations

$$[\Delta P/V] = [B^*][\Delta\theta] \tag{4.10.7}$$

$$[\Delta Q/V] = [B^*][\Delta V] \tag{4.10.8}$$

A number of refinements make this method very successful:

(a) Omit from the Jacobian in equation (4.10.7) the representation of those network elements that predominantly affect MVAR or reactive power flow, e.g. shunt reactances and off-nominal in-phase transformer taps. Neglect also the series resistances of lines.
(b) Omit from the Jacobian of equation (4.10.8) the angle shifting effects of phase shifters.

The resulting fast decoupled load flow equations are then,

$$[\Delta P/V] = [B'][\Delta\theta] \qquad (4.10.9)$$

$$[\Delta Q/V] = [B''][\Delta V] \qquad (4.10.10)$$

where

$$B'_{km} = -\frac{1}{X_{km}}, \quad \text{for } m \neq k$$

$$B'_{kk} = \sum_{m\omega k} \frac{1}{X_{km}}$$

$$B''_{km} = -B_{km}, \quad \text{for } m \neq k$$

$$B''_{kk} = \sum_{m\omega k} B_{km}$$

The equations are solved alternatively using the most recent values of V and θ available as shown in Fig. 4.8.[8]

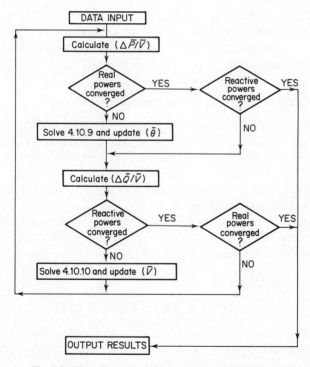

Fig. 4.8. Flow diagram of the fast-decoupled load flow

The matrices B' and B'' are real and are of order $(N-1)$ and $(N-M)$ respectively, where N is the number of busbars and M is the number of PV

busbars. B'' is symmetric in value and so is B' if phase shifters are ignored; it is found that the performance of the algorithm is not adversely affected. The elements of the matrices are constant and need to be evaluated and triangulated only once for a network.

Convergence is geometric, two to five iterations are required for practical accuracies, and more reliable than the formal Newton's method. This is because the elements of B' and B'' are fixed approximations to the tangents of the defining functions $\Delta P/V$ and $\Delta Q/V$, and are not susceptible to any 'humps' in the defining functions.

If $\Delta P/V$ and $\Delta Q/V$ are computed efficiently, then the speed for iterations of the fast decoupled method is about five times that of the formal Newton–Raphson or about two-thirds that of the Gauss–Seidel method. Storage requirements are about 60 percent of the formal Newton, but slightly more than the decoupled Newton method.

Changes in system configurations are easily effected, and while adjusted solutions take many more iterations these are short in time and the overall solution time is still low.

The fast decoupled Newton load flow can be used in optimization studies for a network and is particularly useful for accurate information of both real and reactive power for multiple load-flow studies, as in contingency evaluation for system security assessment.

4.11. CONVERGENCE CRITERIA AND TESTS[13]

The problem arises in the load-flow solution of deciding when the process has converged with sufficient accuracy. In the general field of numerical analysis, the accuracy of solution of any set of equations $F(X) = 0$ is tested by computing the 'residual' vector $F(X^p)$. The elements of this vector should all be suitably small for adequate accuracy, but how small is to a large extent a matter of experience of the requirements of the particular problem.

The normal criterion for convergence in load flow is that the busbar power-mismatches should be small, i.e. ΔQ_i and/or ΔP_i, depending upon the type of busbar i, and can take different forms, e.g.

$$|\Delta P_i| \le c_1, \quad \text{for all } PQ \text{ and } PV \text{ busbars} \tag{4.11.1}$$

$$|\Delta Q_i| \le c_2, \quad \text{for all } PQ \text{ busbars}$$

where c_1 and c_2 are small empirical constants, and $c_1 = c_2$ usually. The value of c used in practice varies from system to system and from problem to problem. In a large system, $c = 1$ MW/MVAR typically gives reasonable accuracy for most purposes. Higher accuracy, say $c = 0.1$ MW/MVAR may be needed for special studies, such as load flows preceding transient stability calculations. In smaller systems, or systems at light load, the value of c may be reduced. For approximate load flows, c may be increased, but with some danger of obtaining a meaningless solution if it becomes too large. Faced with this uncertainty, there is thus a tendency to use smaller values of c than are strictly necessary. The criterion (equation (4.11.1)) is probably the most common in use. A popular variant on it is

$$\sum_i \Delta P_i^2 + \sum_k \Delta Q_k^2 \leq c_3 \qquad (4.11.2)$$

and other similar expressions are also being used.

In the Newton–Raphson algorithms the calculation and testing of the mismatches at each iteration, are part of the algorithm.

The set of equations defining the load-flow problem has multiple solutions, only one of which corresponds to the physical mode of operation of the system. It is extremely rare for there to be more than one solution in the neighbourhood of the initial estimates for the busbar voltages $((1 + j0)$ p.u., in the absence of anything better), and apart from the possibility of data errors, a sensible-looking mathematically-converged solution is normally accepted as being the correct one. However, infrequent cases of very ill-conditioned networks and systems operating close to their stability limits arise where two or more mathematically-converged solutions of feasible appearance can be obtained by different choices of starting voltages, or by different load-flow algorithms.

A load flow problem whose data corresponds to a physically unstable system operating condition (often due to data errors, or in the investigation of unusual operating modes, or in system planning studies) usually diverges. However, the more powerful solution methods, and in particular Newton–Raphson, will sometimes produce a converged solution, and it is not always easy in such cases to recognize that the solution is a physically-unstable operating condition. Certain simple checks, e.g. on the transmission angle and the voltage drop across each line, can be included in the program to automatically monitor the solution.

Finally, a practical load-flow program should include some automatic test to discontinue the solution if it is diverging, to avoid unnecessary waste of computation, and to avoid overflow in the computer. A suitable test is to check at each iteration whether any voltage magnitude is outside the arbitrary range 0.5–1.5 p.u., since it is highly unlikely in any practical power system that a meaningful voltage solution lies outside this range.

4.12. NUMERICAL EXAMPLE

A reduced version of the New Zealand system described in Appendix II is used here to test the algorithm developed in this chapter.

The test network, illustrated in Fig. 4.9 involves the main generating and loading points of the South Island, with the h.v.d.c. convertor represented as a load, i.e. by specified P and Q.

The following computer print out illustrates the numerical input and output information for the specified conditions.

CASE STUDY NUMBER   2

THE SLACK BUS IS   6

MAXIMUM NUMBER OF ITERATIONS   10

POWER TOLERANCE   .00100

PRINTOUT INDICATOR   000000000

NUMBER OF BUSES   17

NUMBER OF LINES   20

NO OF TRANSFORMERS   6

B U S - D A T A

BUS	NAME	TYPE	VOLTS	LOAD MW	MVAR	GENERATION MW	MVAR	MINIMUM MVAR	MAXIMUM MVAR	SHUNT SUSCEPTANCE
1	INV220	0	1.0000	200.000	51.000	0.000	0.000	0.000	0.000	0.000
2	ROX220	0	1.0000	150.000	60.000	0.000	0.000	0.000	0.000	0.000
3	MAN220	0	1.0450	0.000	185.000	0.000	0.000	0.000	0.000	0.000
4	MAN014	1	1.0900	420.000	200.000	690.000	0.000	0.000	0.000	0.000
5	TIW020	0	1.0500	0.000	0.000	0.000	0.000	0.000	0.000	0.000
6	ROX011	1	1.0600	0.000	0.000	0.000	0.000	0.000	0.000	0.000
7	BEN220	0	1.0000	0.000	0.000	0.000	0.000	0.000	0.000	0.000
8	BEN016	1	1.0450	0.000	0.000	200.000	0.000	0.000	0.000	0.000
9	AVI220	0	1.0500	0.000	0.000	350.000	0.000	0.000	0.000	0.000
10	AVI011	1	1.0600	0.000	0.000	0.000	0.000	0.000	0.000	0.000
11	OHAU	1	1.0500	300.000	300.000	0.000	0.000	0.000	0.000	0.000
12	LIV220	0	1.0000	60.000	0.000	150.000	0.000	0.000	0.000	0.000
13	ISL220	1	1.0500	0.000	0.000	0.000	0.000	0.000	0.000	0.000
14	BRM220	0	1.0000	0.000	0.000	0.000	0.000	0.000	0.000	0.000
15	TEK011	0	1.0000	0.000	0.000	0.000	0.000	0.000	0.000	0.000
16	TEK220	0	1.0000	0.000	0.000	0.000	0.000	0.000	0.000	0.000
17	TWI220	0	1.0000	0.000	0.000	0.000	0.000	0.000	0.000	0.000

TRANSFORMER - DATA

BUS	NAME	BUS	NAME	RESISTANCE	REACTANCE	TAP	CODE
3	MAN220	4	MAN014	0.00060	0.01600	1.000	0
17	ROX220	6	ROX011	0.00200	0.03200	1.000	0
9	AVI220	10	AVI011	0.00400	0.04500	1.000	0
7	BEN220	8	BEN016	0.00150	0.03200	1.000	0
16	TEK220	15	TEK011	0.00300	0.05600	1.000	0

L I N E - D A T A

BUS	NAME	BUS	NAME	RESISTANCE	REACTANCE	SUSCEPTANCE
1	INV220	2	MAN220	.01300	.09000	.25000
1	INV220	2	MAN220	.01300	.09000	.25000
3	MAN220	3	MAN330	.00100	.10000	.29000
3	MAN220	3	MAN330	.00100	.10000	.29000
1	INV220	1	INV220	.00200	.04000	.04000
1	INV220	1	ROX220	.00160	.14000	.17000
2	ROX220	1	ROX220	.00160	.14000	.24000
2	ROX220	1	TWI220	.00300	.16000	.24000
7	BEN220	1	AVI220	.00400	.16000	.18000
16	LIV220	9	BEN220	.00700	.12000	.05000
9	AVI220	7	LIV220	.00400	.05000	.02000
9	AVI220	7	ISL220	.00400	.05000	.35000
12	LIV220	13	TEK220	.00300	.04000	.02000
16	TEK220	13	ISL220	.00200	.14000	.35000
17	TWI220	13	BRM220	.00200	.14000	.02000
14	BRM220	13	ISL220	.00200	.14000	.45000
17	TWI220	13	ISL220	.00200	.14000	.45000

SOLUTION CONVERGED IN 5 P-D AND 5 G-V ITERATIONS

MW LOAD MVAR		GENERATION MW MVAR		LOSSES MW MVAR		MISMATCH MW MVAR		SHUNTS MVAR
2020.000	916.000	2113.710	1420.669	93.915	504.648	-0.205	-0.019	0.000

BUS DATA

BUS	NAME	VOLTS	ANGLE	GENERATION		LOAD		SHUNT
				MW	MVAR	MW	MVAR	MVAR
1	INV220	0.936	-12.257	0.00	0.00	200.00	51.00	0.00
2	ROX220	0.982	-16.024	0.00	0.00	150.00	60.00	0.00
3	MAN220	1.002	-2.837	0.00	0.00	0.00	0.00	0.00
4	MAN014	1.045	3.119	690.00	288.73	0.00	0.00	0.00
5	TIW220	0.931	-12.527	0.00	0.00	420.00	185.00	0.00
6	ROX011	1.050	0.000	723.71	242.37	0.00	0.00	0.00
7	BEN220	0.993	-36.853	0.00	0.00	500.00	200.00	0.00
8	BEN016	1.060	-36.999	0.00	223.40	0.00	0.00	0.00
9	AVI220	0.996	-34.278	0.00	0.00	0.00	0.00	0.00
10	AVI011	1.045	-29.414	200.00	115.46	0.00	0.00	0.00
11	OHAU	1.050	-25.433	350.00	113.38	0.00	0.00	0.00
12	LIV220	0.966	-34.266	0.00	0.00	150.00	60.00	0.00
13	ISL220	1.000	-45.170	0.00	437.32	500.00	300.00	0.00
14	BRM220	0.994	-44.734	0.00	0.00	100.00	60.00	0.00
15	TEK011	1.008	-26.722	150.00	0.00	0.00	0.00	0.00
16	TEK220	1.007	-31.466	0.00	0.00	0.00	0.00	0.00
17	TWI220	1.007	-31.265	0.00	0.00	0.00	0.00	0.00

THE MAXIMUM MISMATCH IS   0.0881   ON BUS   3
THE SLACK BUS GENERATION IS 723.709   242.372

LINE AND TRANSFORMER DATA

BUS	NAME	MW	MVAR
3	MAN220	-174.878	-40.447
3	MAN220	-174.878	-40.447
5	TIW220	49.339	39.647
5	TIW220	49.339	39.647
2	ROX220	51.092	-49.397
MISMATCH		-0.014	-0.004
1	INV220	-50.593	39.238
17	TWI220	184.156	-25.506
17	TWI220	184.156	-25.506
12	LIV220	245.348	-17.183
6	ROX011	-713.143	-31.034
MISMATCH		0.076	-0.009
1	INV220	179.544	49.242
1	INV220	179.544	49.242
5	TIW220	163.865	54.142
5	TIW220	163.865	54.142
4	MAN014	-686.907	-206.766
MISMATCH		0.088	-0.003
3	MAN220	689.980	288.734
MISMATCH		0.020	0.000
3	MAN220	-160.722	-49.833
3	MAN220	-160.722	-49.833
1	INV220	-49.244	-42.659
1	INV220	-49.244	-42.659
MISMATCH		-0.067	-0.016
2	ROX220	723.710	242.372
MISMATCH		0.000	0.000
17	TWI220	-323.190	6.631
9	AVI220	-88.638	1.281
9	AVI220	-88.638	1.281
8	BEN016	0.527	-209.188
MISMATCH		-0.061	-0.005
7	BEN220	0.006	223.402
MISMATCH		-0.006	0.000
12	LIV220	21.370	92.023
7	BEN220	88.957	0.732
7	BEN220	88.957	0.732
10	AVI011	-199.261	-93.486
MISMATCH		-0.023	-0.000
9	AVI220	199.993	115.461
MISMATCH		0.007	0.000
17	TWI220	350.004	113.378
MISMATCH		-0.004	0.000
2	ROX220	-226.600	75.096
9	AVI220	-20.708	-93.997
13	ISL220	97.390	-41.105
MISMATCH		-0.082	0.006
12	LIV220	-94.143	26.752
16	TEK220	-176.860	26.134
14	BRM220	-61.680	67.194
17	TWI220	-167.232	17.242
MISMATCH		-0.085	0.000
17	TWI220	-161.844	11.297
13	ISL220	61.853	-71.300
MISMATCH		-0.009	0.002
16	TEK220	149.983	0.000
MISMATCH		0.017	-0.000
17	TWI220	-34.170	5.862
13	ISL220	183.497	-18.253
15	TEK011	-149.320	12.390
MISMATCH		-0.007	0.001
2	ROX220	-178.498	51.270
2	ROX220	-178.498	51.270
7	BEN220	327.435	18.207
16	TEK220	34.194	-7.772
14	BRM220	167.371	-17.686
13	ISL220	173.141	-21.209
11	OHAU	-345.093	-74.090
MISMATCH		-0.052	0.010

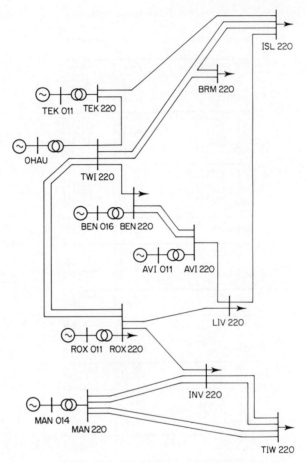

Fig. 4.9. Reduced primary a.c. system for the South Island of
New Zealand

4.13. REFERENCES

1. B. Stott, 1974. 'Review of load-flow calculation methods', *Proc. IEEE*, **62** (7), 916–929.
2. J. B. Ward and H. W. Hale, 1956. 'Digital computer solution of power-flow problems', *Trans. AIEE.*, **PAS-75**, 398–404.
3. H. E. Brown, G. K. Carter, H. H. Happ and C. E. Person, 1963. 'Power-flow solution by impedance matrix iterative method', *IEEE Trans.* **PAS-82**, 1–10.
4. J. E. Van Ness and J. H. Griffin, 1961. 'Elimination methods for load-flow studies', *Trans. AIEE.*, **PAS-80**, 299–304.
5. W. F. Tinney, C. E. Hart, 'Power flow solution by Newton's method', *IEEE Trans.*, **PAS-86** (11), 1449–1460.
6. E. C. Ogbuobiri, W. F. Tinney and J. W. Walker, 1970. 'Sparsity-directed decomposition for Gaussian elimination on matrices', *IEEE*, Trans., **PAS-89** (1), 141–150.
7. B. Stott and E. Hobson, 1971. 'Solution of large power-system networks by ordered elimination: a comparison of ordering schemes', *Proc. IEE*, **118** (1), 125–134.

8. W. F. Tinney and J. W. Walker, 1967. 'Direct solutions of sparse network equations by optimally ordered triangular factorization', *Proc. IEEE*, **55** (11), 1801–1809.
9. S. T. Despotovic, 1974. 'A new decoupled load-flow method', *IEEE Trans.*, **PAS-93** (3), 884–891.
10. B. Stott, 1972. 'Decoupled Newton load flow', *IEEE Trans.*, **PAS-91**, 1955–1959.
11. B. Stott and O. Alsac, 1974. 'Fast decoupled load flow', *IEEE Trans.*, **PAS-93** (3), 859–869.
12. J. Medanic and B. Avramovic, 1975. 'Solution of load-flow problems in power systems by ε-coupling method', *Proc IEE*, **122** (8), 801–805.
13. B. Stott, 1972. 'Power-system load flow' (M.Sc. lecture notes). University of Manchester Institute of Science and Technology.

5

Three-Phase Load Flow

5.1. INTRODUCTION

For most purposes in the steady-state analysis of power systems, the system unbalance can be ignored and the single-phase analysis described in Chapter 4 is adequate. However, in practice it is uneconomical to balance the load completely or to achieve perfectly balanced transmission system impedances, as a result of untransposed high-voltage lines and lines sharing the same right of way for considerable lengths.

Among the effects of power-system unbalance are: negative sequence currents causing machine rotor overheating, zero sequence currents causing relay maloperations and increased losses due to parallel untransposed lines.

The use of long-distance transmission motivated the development of analytical techniques for the assessment of power-system unbalance. Early techniques[1],[2] were restricted to the case of isolated unbalanced lines operating from known terminal conditions. However, a realistic assessment of the unbalanced operation of an interconnected system, including the influence of any significant load unbalance, requires the use of three-phase load-flow algorithms.[3],[4],[5] The object of the three-phase load flow is to find the state of the three-phase power system under the specified conditions of load, generation and system configuration.

The basic three-phase models of system plant and the rules for their combination into overall network admittance matrices, discussed in Chapter 2, are used as the framework for the three-phase load flow described in this chapter.

The storage and computational requirements of a three-phase load-flow program are much greater than those of the corresponding single-phase case. The need for efficient algorithms is therefore significant even though, in contrast to single-phase analysis, the three-phase load flow is likely to remain a planning, rather than an operational exercise.

The basic characteristics of the Fast Decoupled Newton–Raphson algorithms described in Chapter 4, have recently been shown[6] to apply equally to the three-phase load-flow problem. Consequently, this algorithm is now used as a basis for the development of an efficient three-phase load-flow program. When the program is used for post-operational studies of important unbalanced situations on the power system, additional practical features such as automatic transformer tapping and generator VAR limiting are necessary.

5.2. NOTATION

A clear and unambiguous identification of the three-phase vector and matrix elements requires a suitable symbolic notation using superscripts and subscripts.

The a.c. system is considered to have a total of n busbars where:

$n = nb + ng$
nb is the number of actual system busbars
ng is the number of synchronous machines

Subscripts i, j, etc. refer to system busbars as shown in the following examples:

$i = 1, nb$ identifies all actual system busbars, i.e. all load busbars plus all generator terminal busbars.
$i = nb + 1, nb + ng - 1$ identifies all generator internal busbars with the exception of the slack machine.
$i = nb + ng$ identifies the internal busbar at the slack machine.

The following subscripts are also used for clarity:

reg—refers to a voltage regulator
int—refers to an internal busbar at a generator
gen—refers to a generator

Superscripts p, m identify the three phases at a particular busbar.

5.3. FORMULATION OF THE THREE-PHASE LOAD-FLOW PROBLEM

Synchronous machine modelling

Synchronous machines are designed for maximum symmetry of the phase windings and are therefore adequately modelled by their sequence impedances. Such impedances contain all the information that is required to analyse the steady-state unbalanced behaviour of the synchronous machine.

The representation of the generator in phase components may be derived from the sequence impedance matrix $(Z_g)_{012}$ as follows:

$$[Z_g]_{abc} = [T_s][Z_g]_{012}[T_s]^{-1} \tag{5.3.1}$$

$$= [T_s][Z_g]_{012}[T_s]^* \tag{5.3.2}$$

where

$$[T_s] = . \begin{bmatrix} 1 & 1 & 1 \\ 1 & a^2 & a \\ 1 & a & a^2 \end{bmatrix} \tag{5.3.3}$$

and 'a' is the complex operator $e^{j2\pi/3}$. The phase component impedance matrix is thus

$$[Z_g]_{abc} = \begin{array}{|c|c|c|} \hline Z_0 + Z_1 + Z_2 & Z_0 + aZ_1 + a^2Z_2 & Z_0 + a^2Z_1 + aZ_2 \\ \hline Z_0 + a^2Z_1 + aZ_2 & Z_0 + Z_1 + Z_2 & Z_0 + aZ_1 + a^2Z_1 \\ \hline Z_0 + aZ_1 + a^2Z_2 & Z_0 + a^2Z_1 + aZ_2 & Z_0 + Z_1 + Z_2 \\ \hline \end{array} \qquad (5.3.4)$$

The phase component model of the generator is illustrated in Fig. 5.1(a) The machine excitation acts symmetrically on the three phases and the voltages at the internal or excitation busbar form a balanced three-phase set, i.e.

$$E_k^a = E_k^b = E_k^c \qquad (5.3.5)$$

and

$$\theta_k^a = \theta_k^b + \frac{2\pi}{3} = \theta_k^c - \frac{2\pi}{3} \qquad (5.3.6)$$

For three-phase load flow the voltage regulator must be accurately modelled as it influences the machine operation under unbalanced conditions. The voltage regulator monitors the terminal voltages of the machine and controls the excitation voltage according to some predetermined function of the terminal voltages. Often the positive sequence is extracted from the three-phase voltage measurement using a sequence filter.

Before proceeding further it is instructive to consider the generator modelling from a symmetrical component frame of reference. The sequence network model of the generator is illustrated in Fig. 5.1(b). As the machine excitation acts symmetrically on the three-phases, positive sequence voltages only are present at the internal busbar.

The influence of the generator upon the unbalanced system is known if the voltages at the terminal busbar are known. In terms of sequence voltages, the positive sequence voltage may be obtained from the excitation and the positive sequence voltage drop caused by the flow of positive sequence currents through the positive sequence reactance. The negative and zero sequence voltages are derived from the flow of their respective currents through their respective impedances. It is important to note that the negative and zero sequence voltages are not influenced by the excitation or positive sequence impedance.

There are infinite combinations of machine excitation and machine positive sequence reactance which will satisfy the conditions at the machine terminals and give the correct positive sequence voltage. Whenever the machine excitation must be known (as in fault studies) the actual positive sequence impedance must be used. For load flow however, the excitation is not of any particular interest and the positive sequence impedance may be arbitrarily assigned to any value.[3] The positive sequence impedance is usually set to zero for these studies.

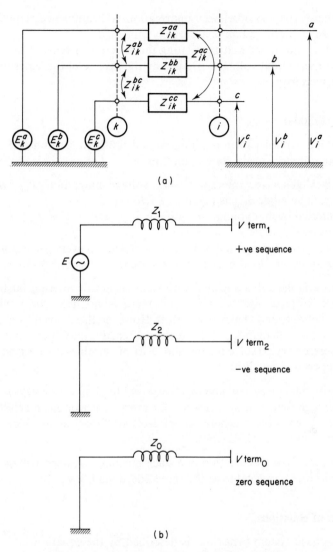

Fig. 5.1. Synchronous machine models. (a) Phase component representation. (b) Symmetrical component representation

Thus the practice with regard to three-phase load flow in phase coordinates, is to set the positive sequence reactance to a small value in order to reduce the excitation voltage to the same order as the usual system voltages with a corresponding reduction in the angle between the internal busbar and the terminal busbar. Both these features are important when a fast decoupled algorithm is used.

Therefore, in forming the phase component generator model using equation (5.3.4), an arbitrary value may be used for Z_1 but the actual values are used for Z_0

and Z_2. There is no loss of relevant information as the influence of the generator upon the unbalanced system is accurately modelled.

The nodal admittance matrix, relating the injected currents at the generator busbars to their nodal voltages, is given by the inverse of the series impedance matrix derived from equation (5.3.4).

Specified variables

The following variables form a minimum and sufficient set to define the three-phase system under steady-state operation:

— The slack generator internal busbar voltage magnitude $V_{\text{int}j}$ where $j = nb + ng$. (The angle $\theta_{\text{int}j}$ is taken as a reference.)
— The internal busbar voltage magnitude $V_{\text{int}j}$ and angles $\theta_{\text{int}j}$ at all other generators, i.e. $j = nb + 1, nb + ng - 1$.
— The three voltage magnitudes (V_i^p) and angles (θ_i^p) at every generator terminal busbar and every load busbar in the system, i.e. $i = 1, nb$ and $p = 1, 3$.

Only two variables are associated with each generator internal busbar as the three-phase voltages are balanced and there is no need for retaining the redundant voltages and angles as variables. However, these variables are retained to facilitate the calculation of the real and reactive power mismatches. The equations necessary to solve for the above set of variables are derived from the specified operating conditions, i.e.

— The individual phase real and reactive power loading at every system busbar.
— The voltage regulator specification for every synchronous machine.
— The total real power generation of each synchronous machine, with the exception of the slack machine.

The usual load-flow specification of a slack machine, i.e. fixed voltage in phase and magnitude, is applicable to the three-phase load flow.

Derivation of equations

The three-phase system behaviour is described by the equation

$$[I] - [Y][V] = 0 \tag{5.3.7}$$

where the system admittance matrix $[Y]$, as developed in Chapter 2, represents each phase independently and models all inductive and capacitive mutual couplings between phases and between circuits. The mathematical statement of the specified conditions is derived in terms of the system admittance matrix

$$[Y] = [G] + j[B]$$

as follows:

(i) For each of the three phases (p) at every load and generator terminal busbar (i),
$$\Delta P_i^p = (P_i^p)^{sp} - P_i^p$$

$$= (P_i^p)^{sp} - V_i^p \sum_{k=1}^{n} \sum_{m=1}^{3} V_k^m [G_{ik}^{pm} \cos \theta_{ik}^{pm} + B_{ik}^{pm} \sin \theta_{ik}^{pm}] \tag{5.3.8}$$

and

$$\Delta Q_i^p = (Q_i^p)^{sp} - Q_i^p$$

$$= (Q_i^p)^{sp} - V_i^p \sum_{k=1}^{n} \sum_{m=1}^{3} V_k^m [G_{ik}^{pm} \sin \theta_{ik}^{pm} - B_{ik}^{pm} \cos \theta_{ik}^{pm}] \tag{5.3.9}$$

(ii) For every generator j,

$$(\Delta V_{\text{reg}})_j = f(V_k^1, V_k^2, V_k^3) \tag{5.3.10}$$

where k is the bus number of the jth generator's terminal busbar.

(iii) For every generator j, with the exception of the slack machine, i.e. $j \neq nb + ng$.

$$(\Delta P_{\text{gen}})_j = (P_{\text{gen}}^{sp})_j - (P_{\text{gen}})_j$$

$$= (P_{\text{gen}}^{sp})_j - \sum_{p=1}^{3} V_{\text{int } j} \sum_{k=1}^{n} \sum_{m=1}^{3} V_k^m [G_{jk}^{pm} \cos \theta_{jk}^{pm} + B_{jk}^{pm} \sin \theta_{jk}^{pm}]$$

$$\tag{5.3.11}$$

where, although the summation for k is over all system busbars, the mutual terms G_{jk} and B_{jk} are nonzero only when k is the terminal busbar of the jth generator.

It should be noted that the real power specified for the generator is the total real power at the internal or excitation busbar whereas in actual practice the specified quantity is the power leaving the terminal busbar. This in effect means that the generator's real power loss is ignored.

The generator losses have no significant influence on the system operation and may be calculated from the sequence impedances at the end of the load-flow solution, when all generator sequence currents have been found. Any other method would require the real power mismatch to be written at busbars remote from the variable in question, that is, the angle at the internal busbar. In addition, inspection of equations (5.3.8) and (5.3.11) will show that the equations are identical except for the summation over the three phases at the generator internal busbar.

That is, the sum of the powers leaving the generator may be calculated in exactly the same way and by the same subroutines as the power mismatches at other system busbars. This is possible because the generator internal busbar is not connected to any other element in the system. Inspection of the Jacobian submatrices derived later will show that this feature is retained throughout the study. In terms of programming the generators present no additional complexity.

Equations (5.3.8) to (5.3.11) form the mathematical formulation of the three-phase load flow as a set of independent algebraic equations in terms of the system variables.

The solution to the load-flow problem is the set of variables which, upon substitution, make the left-hand side mismatches in equations (5.3.8) to (5.3.11) equal to zero.

5.4. FAST DECOUPLED THREE-PHASE ALGORITHM

The standard Newton–Raphson algorithm may be used to solve equations (5.3.8) to (5.3.11). This involves an iterative solution of the matrix equation

$$
\begin{bmatrix}
\Delta P \\
\Delta P_{\text{gen}} \\
\Delta Q \\
\Delta V_{\text{reg}}
\end{bmatrix}
=
\begin{bmatrix}
A & E & I & M \\
B & F & J & N \\
C & G & K & P \\
D & H & L & R
\end{bmatrix}
\cdot
\begin{bmatrix}
\Delta \theta \\
\Delta \theta_{\text{int}} \\
\Delta V/V \\
\Delta V_{\text{int}}/V_{\text{int}}
\end{bmatrix}
\tag{5.4.1}
$$

for the right-hand side vector of variable updates. The right-hand side matrix in equation (5.4.1) is the Jacobian matrix of first order partial derivatives.

Following decoupled single-phase load-flow practice, the effects of $\Delta \theta$ on reactive power flows and ΔV on real power flows are ignored. Equation (5.4.1) may therefore be simplified by assigning

$$[I] = [M] = [J] = [N] = 0$$

and

$$[C] = [G] = 0$$

In addition, the voltage regulator specification is assumed to be in terms of the terminal voltage magnitudes only and therefore,

$$[D] = [H] = 0$$

Equation (5.4.1) may then be written in decoupled form as:

$$
\begin{bmatrix}
\Delta P_i^p \\
\Delta P_{\text{gen} j}
\end{bmatrix}
=
\begin{bmatrix}
A & E \\
B & F
\end{bmatrix}
\begin{bmatrix}
\Delta \theta_k^m \\
\Delta \theta_{\text{int} l}
\end{bmatrix}
\tag{5.4.2}
$$

for $i, k = i, nb$ and $j, l = 1, ng - 1$ (i.e. excluding the slack generator)

$$
\begin{bmatrix}
\Delta Q_i^p \\
\Delta V_{\text{reg} j}
\end{bmatrix}
=
\begin{bmatrix}
K & P \\
L & R
\end{bmatrix}
\begin{bmatrix}
\Delta V_k^m / V_k^m \\
\Delta V_{\text{int} l} / V_{\text{int} l}
\end{bmatrix}
\tag{5.4.3}
$$

for $i, k = 1, nb$ and $j, l = 1, ng$ (i.e. including the slack generator).

To enable further development of the algorithm it is necessary to consider the Jacobian submatrices in more detail.

In deriving these Jacobians from equations (5.3.8) to (5.3.11) it must be remembered that,

$$V_l^1 = V_l^2 = V_l^3 = V_{\text{int} l}$$

$$\theta_l^1 = \theta_l^2 - \frac{2\pi}{3} = \theta_l^3 + \frac{2\pi}{3} = \theta_{\text{int} l}$$

when l refers to a generator internal busbar.

The coefficients of matrix equation (5.4.2) are:

$$[A_{ik}^{pm}] = [\partial \Delta P_i^p / \partial \theta_k^m]$$

or

$$A_{ik}^{pm} = V_i^p V_k^m [G_{ik}^{pm} \sin \theta_{ik}^{pm} - B_{ik}^{pm} \cos \theta_{ik}^{pm}]$$

except for

$$A_{kk}^{mm} = - B_{kk}^{mm}(V_k^m)^2 - Q_k^m$$

$$[B_{jk}^m] = [\partial \Delta P_{\text{gen}\,j} / \partial \theta_k^m]$$

$$= \sum_{p=1}^{3} V_{\text{int}\,j} V_k^m [G_{jk}^{pm} \sin \theta_{jk}^{pm} - B_{jk}^{pm} \cos \theta_{jk}^{pm}]$$

$$[E_{il}^p] = [\partial P_i^p / \partial \theta_{\text{int}\,l}]$$

$$= \sum_{m=1}^{3} V_{\text{int}\,l} V_i^p [G_{il}^{pm} \sin \theta_{il}^{pm} - B_{il}^{pm} \cos \theta_{il}^{pm}]$$

$$[F_{jl}] = [\partial P_{\text{gen}\,j} / \partial \theta_{\text{int}\,l}]$$

where $[F_{jl}] = 0$ for all $j \neq l$ because the jth generator has no connection with the lth generator's internal busbar, and

$$[F_{ll}] = \sum_{p=1}^{3} (- B_{ll}^{pp}(V_{\text{int}\,l})^2 - Q_l^p)$$

$$+ \sum_{\substack{m=1 \\ m \neq p}}^{3} \sum_{p=1}^{3} (V_{\text{int}\,l})^2 [G_{ll}^{pm} \sin \theta_{ll}^{pm} - B_{ll}^{pm} \cos \theta_{ll}^{pm}]$$

The coefficients of matrix equation (5.4.3) are

$$- [K_{ik}^{pm}] = V_k^m [\partial \Delta Q_i^p / \partial V_k^m]$$

where

$$K_{ik}^{pm} = V_k^m V_i^p [G_{ik}^{pm} \sin \theta_{ik}^{pm} - B_{ik}^{pm} \cos \theta_{ik}^{pm}]$$

except

$$K_{kk}^{mm} = - B_{kk}^{mm}(V_k^m)^2 + Q_k^m$$

$$- [L_{jk}^m] = V_k^m [\partial \Delta V_{\text{reg}\,j} / \partial V_k^m]$$

let $[L_{jk}^m] = V_k^m [L_{jk}^m]'$ where k is the terminal busbar of the jth generator and $L_{jk}^m = 0$ otherwise.

$$- [P_{il}^p] = V_{\text{int}\,l} [\partial \Delta Q_i^p / V_{\text{int}\,l}]$$

$$= V_{\text{int}\,l} \sum_{m=1}^{3} V_i^p [G_{il}^{pm} \sin \theta_{il}^{pm} - B_{il}^{pm} \cos \theta_{il}^{pm}]$$

$$- [R_{jl}] = [\partial \Delta V_{\text{reg}\,j} / \partial V_{\text{int}\,l}]$$

$$= 0 \text{ for all } j, l \text{ as the voltage regulator}$$
specification does not explicitly
include the variables V_{int}.

Although the above expressions appear complex, their meaning and derivation are similar to those of the usual single phase Jacobian elements.

Jacobian approximations

Approximations similar to those applied to the single-phase load flow are applicable to the Jacobian elements as follows:

(i) at all nodes (i.e. all phases of all busbars)

$$Q_k^m << B_{kk}^{mm}(V_k^m)^2$$

(ii) between connected nodes of the same phase,

$$\cos\theta_{ik}^{mm} \approx 1 \quad \text{i.e. } \theta_{ik}^{mm} \quad \text{is small}$$

and

$$G_{ik}^{mm}\sin\theta_{ik}^{mm} << B_{ik}^{mm}$$

(iii) Moreover the phase-angle unbalance at any busbar will be small and hence an additional approximation applies to the three-phase system, i.e.

$$\theta_{kk}^{pm} \approx \pm 120° \quad \text{for } p \neq m$$

(iv) Finally, as a result of (ii) and (iii) the angle between different phases of connected busbars will be approximately 120°, i.e.

$$\theta_{ik}^{pm} \approx \pm 120° \quad \text{for } p \neq m$$

or

$$\cos\theta_{ik}^{pm} \approx -0.5$$

and

$$\sin\theta_{ik}^{pm} \approx \pm 0.866$$

These values are modified for the $\pm 30°$ phase-shift inherent in the star–delta connection of three-phase transformers.

The final approximation (iv), necessary if the Jacobians are to be kept constant, is the least valid, as the cosine and sine values change rapidly with small angle variations around 120 degrees. This accounts for the slower convergence of the phase unbalance at busbars as compared with that of the voltage magnitudes and angles.

It should be emphasised that these approximations apply to the Jacobian elements only, i.e. they do not prejudice the accuracy of the solution nor do they restrict the type of problem which may be attempted.

Applying approximations (i) to (iv) to the Jacobians and substituting into equations (5.4.2) and (5.4.3) yields,

$$
\begin{bmatrix} \Delta P_i^p \\ \Delta P_{\text{gen}\,j} \end{bmatrix} = \begin{bmatrix} \begin{bmatrix} V_i^p M_{ik}^{pm} V_k^m \end{bmatrix} & \begin{bmatrix} \displaystyle\sum_{m=1}^{3} V_i^p M_{il}^{pm} V_{\text{int}l} \end{bmatrix} \\ \begin{bmatrix} \displaystyle\sum_{p=1}^{3} V_{\text{int}\,j} M_{jk}^{pm} V_k^m \end{bmatrix} & \begin{bmatrix} \displaystyle\sum_{m=1}^{3} \displaystyle\sum_{p=1}^{3} V_{\text{int}\,j} M_{jl}^{pm} V_{\text{int}l} \end{bmatrix} \end{bmatrix} \begin{bmatrix} \Delta\theta_k^m \\ \Delta\theta_{\text{int}l} \end{bmatrix}
$$

(5.4.4)

and

$$
\begin{bmatrix} \Delta Q_i^p \\ \Delta V_{\text{reg}\,j} \end{bmatrix} = \begin{bmatrix} \begin{bmatrix} V_i^p M_{ik}^{pm} V_k^m \end{bmatrix} & \begin{bmatrix} \displaystyle\sum_{m=1}^{3} V_i^p M_{il}^{pm} V_{\text{int}l} \end{bmatrix} \\ V_k^m [L'] & [0] \end{bmatrix} \begin{bmatrix} \Delta V_k^m / V_k^m \\ \Delta V_{\text{int}l} / V_{\text{int}l} \end{bmatrix}
$$

(5.4.5)

where

$$
M_{ik}^{pm} = G_{ik}^{pm} \sin \theta_{ik}^{pm} - B_{ik}^{pm} \cos \theta_{ik}^{pm}
$$

with

$$
\theta_{kk}^{mm} = 0
$$

$$
\theta_{ik}^{mm} = 0
$$

$$
\theta_{ik}^{pm} = \pm 120° \quad \text{for } p \neq m
$$

All terms in the matrix $[M]$ are constant, being derived solely from the system admittance matrix. Matrix $[M]$ is the same as matrix $[-B]$ except for the off-diagonal terms which connect nodes of different phases. These are modified by allowing for the nominal 120° angle and also including the $G_{ik}^{pm} \sin \theta_{ik}^{pm}$ terms.

The similarity in structure of all Jacobian submatrices reduces the programming complexity normally found in three-phase load flows. This uniformity has been achieved primarily by the method used to implement the three-phase generator constraints.

The above derivation closely parallels the single-phase fast decoupled algorithm, but the added complexity of the notation obscures this feature. At the present stage the Jacobian elements in equations (5.4.4) and (5.4.5) are identical except for those terms which involve the additional features of the generator modelling.

These functions are more linear in terms of the voltage magnitude $[\bar{V}]$ than are the functions $[\Delta \bar{P}]$ and $[\Delta \bar{Q}]$. In the Newton–Raphson and related constant Jacobian methods the reliability and speed of convergence[7] improve with the linearity of the defining functions. With this aim, equations (5.4.4) and (5.4.5) are modified as follows:

— The left-hand side defining functions are redefined as $[\Delta P_i^p / V_i^p]$, $[\Delta P_{\text{gen}\,j} / V_{\text{int}\,j}]$ and $[\Delta Q_i^p / V_i^p]$
— In equation (5.4.4), the remaining right-hand side V terms are set to 1 p.u.
— In equation (5.4.5), the remaining right-hand side V terms are cancelled by the corresponding terms in the right-hand side vector.

These modifications yield the following expressions:

$$
\begin{bmatrix} \Delta P_i^p / V_i^p \\[2ex] \Delta P_{\text{gen}\,j} / V_{\text{int}\,j} \end{bmatrix} = \begin{bmatrix} M_{ik}^{pm} & \displaystyle\sum_{m=1}^{3} M_{il}^{pm} \\[3ex] \displaystyle\sum_{p=1}^{3} M_{jk}^{pm} & \displaystyle\sum_{p=1}^{3}\sum_{m=1}^{3} M_{jl}^{pm} \end{bmatrix} \begin{bmatrix} \Delta\theta_k^p \\[2ex] \Delta\theta_{\text{int}\,l} \end{bmatrix}
\qquad (5.4.6)
$$

$$
[B']
$$

$$
\begin{bmatrix} \Delta Q_i^p / V_i^p \\[2ex] \Delta V_{\text{reg}\,j} \end{bmatrix} = \begin{bmatrix} M_{ik}^{pm} & \displaystyle\sum_{m=1}^{3} M_{il}^{pm} \\[3ex] [L_{jk}^m]' & 0 \end{bmatrix} \begin{bmatrix} \Delta V_k^m \\[2ex] \Delta V_{\text{int}\,l} \end{bmatrix}
\qquad (5.4.7)
$$

$$
[B'']
$$

Recalling that $[L_{jk}^m]' = [\partial\Delta V_{\text{reg}\,j}/\partial V_k^m]$, as V_{reg} is normally a simple linear function of the terminal voltages, $[L']$ will be a constant matrix.

Therefore, the Jacobian matrices $[B']$ and $[B'']$ in equations (5.4.6) and (5.4.7) have been approximated to constants.

Zero diagonal elements in equation (5.4.7) may result from the ordering of the equations and variables. This feature causes no problems if these diagonals are not used as pivots until the rest of the matrix has been factorized (by which time, fill-in terms will have appeared on the diagonal). This causes a minor loss of efficiency as it inhibits optimal ordering for the complete matrix. Although this could be avoided by reordering the equations, the extra program complexity is not justified.

Based on the reasoning of Stott and Alsac,[8] which proved successful in the single-phase load flow the $[B']$ matrix in equation (5.4.6) is further modified by omitting the representation of those elements that predominantly affect MVAR flows.

The capacitance matrix and its physical significance is illustrated in Fig. 5.2, for a single three-phase line. With n capacitively-coupled parallel lines the matrix will be $3n \times 3n$.

In single-phase load flows the shunt capacitance is the positive sequence capacitance which is determined from both the phase-to-phase and the phase-to-earth capacitances of the line. It therefore appears that the entire shunt capacitance matrix predominantly affects MVAR flows only. Thus, following single-phase fast decoupling practice the representation of the entire shunt capacitance matrix is omitted in the formulation of $[B']$. This increases dramatically the rate of real power convergence.

With capacitively coupled three-phase lines the interline capacitance influences the positive sequence shunt capacitance. However, as the values of interline capacitances are small in comparison with the self capacitance of the phases, their inclusion makes no noticeable difference. The effective tap of $\sqrt{3}$ introduced by the star–delta transformer connection is interpreted as a nominal tap and is therefore included when forming the $[B']$ matrix.

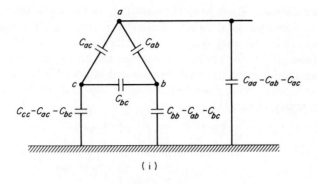

(i)

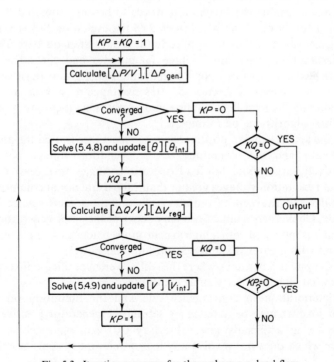

(ii)

Fig. 5.2. Shunt capacitance matrices

Fig. 5.3. Iteration sequence for three-phase a.c. load flow

A further difficulty arises from the modelling of the star-g/delta transformer connection. The equivalent circuit, illustrated in Section 2.6 shows that large shunt admittances are effectively introduced into the system. When these are excluded from $[B']$, as for a normal shunt element, divergence results. The entire transformer model, must therefore be included in both $[B']$ and $[B'']$.

With the modifications described above the two final algorithmic equations may be concisely written, i.e.

$$\begin{bmatrix} \Delta P/V \\ \Delta P_{\text{gen}}/V_{\text{int}} \end{bmatrix} = \begin{bmatrix} B'_m \end{bmatrix} \begin{bmatrix} \Delta \theta \\ \Delta \theta_{\text{int}} \end{bmatrix} \tag{5.4.8}$$

$$\begin{bmatrix} \Delta Q/V \\ \Delta V_{\text{reg}} \end{bmatrix} = \begin{bmatrix} B''_m \end{bmatrix} \begin{bmatrix} \Delta V \\ \Delta V_{\text{int}} \end{bmatrix} \tag{5.4.9}$$

The constant Jacobians $[B'_m]$ and $[B''_m]$ correspond to fixed approximated tangent slopes to the multidimensional surfaces defined by the left hand-side defining functions.

Equations (5.4.8) and (5.4.9) are then solved according to the iteration sequence illustrated in Fig. 5.3.

Generator models and the fast decoupled algorithm

The derivation of the fast decoupled algorithm involves the use of several assumptions to enable the Jacobian matrices to be approximated to constant. The same assumptions have been applied to the excitation busbars associated with the generator model as are applied to the usual system busbars. The validity of the assumptions regarding voltage magnitudes and the angles between connected busbars depends upon the machine loading and positive sequence reactance. As discussed in Section 5.3 this reactance may be set to any value without altering the load-flow solution and a value may therefore be selected to give the best algorithmic performance.

When the actual value of positive sequence reactance is used the angle across the generator and the magnitude of the excitation voltage both become comparatively large under full-load operation. Angles in excess of forty-five degrees and excitation voltages greater than 2.0 p.u. are not uncommon. Despite this considerable divergence from assumed conditions, convergence is surprisingly good. Convergence difficulties may occur at the slack generator and then only when it is modelled with a high synchronous reactance (1.5 p.u. on machine rating) and with greater than 70 percent full load power.

All other system generators, where the real power is specified, converge reliably but somewhat slowly under similar conditions.

This deterioration in convergence rate and the limitation on the slack generator loading may be avoided by setting the generator positive sequence reactance to an artificially low value (say 0.01 p.u. on machine rating), a procedure which does not involve any loss of relevant system information.

5.5. STRUCTURE OF THE COMPUTER PROGRAM

The main components of the computer program are illustrated in Fig. 5.4. The approximate number of FORTRAN statements for each block is indicated in parenthesis. The main features of each block are described in the following sections.

Data input

The input data routine implements the system modelling techniques described in Chapter 2 and forms the system admittance model from the raw data for each system component. Examples of the raw data are given in Section 5.7. with reference to a particular test system.

The structure and content of the constant Jacobians B' and B'' are based upon the system admittance matrix and are thus formed simultaneously with this matrix.

Both the system admittance matrix and the Jacobian matrices are stored and processed using sparsity techniques which are structured in 3×3 matrix blocks to take full advantage of the inherent block structure of the three-phase system matrices.

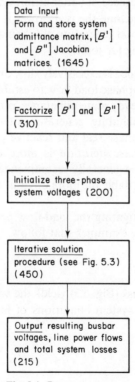

Fig. 5.4. Program structure

Factorization of constant Jacobians

The heart of the load-flow program is the repeat solutions of equations (5.4.8) and (5.4.9) as illustrated in Fig. 5.3. These equations are solved using sparsity techniques and near optimal ordering as discussed in Chapter 4 (Section 4.7) or like those embodied in Zollenkopf's bifactorization.[9] The constant Jacobians are factorized before the iteration sequence is initiated. The solution of each equation within the iterative procedure is relatively fast, consisting only of the forward and back substitution processes.

Starting values

Starting values are assigned as follows:

— The nonvoltage controlled busbars are assigned 1 p.u. on all phases.
— At generator terminal busbars all voltages are assigned values according to the voltage regulator specifications.
— All system busbar angles are assigned 0, $-120°$, $+120°$ for the three phases respectively.
— The generator internal voltages and angles are calculated from the specified real power and, in the absence of better estimates, by assuming zero reactive power. For the slack machine the real power is estimated as the difference between total load and total generation plus a small percentage (say 8 percent) of the total load to allow for losses.

For cases where convergence is excessively slow or difficult it is recommended to use the results of a single-phase load flow to establish the starting values. The values will, under normal steady-state unbalance, provide excellent estimates for all voltages and angles including generator internal conditions which are calculated from the single-phase real and reactive power conditions.

Moreover, as a three-phase iteration is more costly than a single-phase iteration, this practice can be generally recommended to provide more efficient overall convergence and to enable the more obvious data errors to be detected at an early stage.

For the purpose of investigating the load-flow performance, flat voltage and angle values are used in the examples that follow.

Iterative solution

The iterative solution process (Fig. 5.3) yields the values of the system voltages which satisfy the specified system conditions of load, generation and system configuration.

Output results

The three-phase busbar voltages, the line power flows and the total system losses are calculated and printed out. An example is given in Table 5.6. In addition the sequence components of busbar voltages are also calculated as these provide a more direct measure of the unbalance present in the system under study.

5.6. PERFORMANCE OF THE ALGORITHM

This section attempts to identify those features which influence the convergence with particular reference to several small to medium-sized test systems.

The performance of the 'three-phase' algorithm is examined under both balanced and unbalanced conditions, and comparisons are made with the performance of the single-phase fast-decoupled algorithm.

Performance under balanced conditions

A symmetrical three-phase system, operating with balanced loading, is accurately modelled by the positive sequence system and either a three-phase or a single-phase load flow may be used to analyse the system. Under these conditions it is possible to compare the performance of the three-phase and single-phase fast-decoupled algorithms.

The three-phase system transmission lines are represented by balanced full 3 × 3 matrices. Transformers are modelled with balanced parameters on all phases and generators are modelled by their phase parameter matrices as derived from their sequence impedances.

Typical numbers of iterations to convergence for both the single-phase and three-phase algorithms, given in Table 5.1, indicate that the algorithms behave identically. Features such as the transformer connection and the negative and zero sequence generator impedances have no effect on the convergence rate of the three-phase system under balanced conditions. This is not unexpected as, under balanced conditions, only the positive sequence network has any power flow and there is no coupling between sequence networks. The negative and zero sequence information inherent in the three-phase system model of the balanced systems, has no influence on system operation and this is reflected into the performance of the algorithm.

Table 5.1. Convergence results

Case	Number of busbars	Single-phase load flow	Balanced three phase load flow		Typical three-phase unbalance
			$\lambda\lambda$	$\lambda\Delta$	
1	5	4, 3	4, 3	4, 3	6, 6
2	6	3, 3	3, 3	3, 3	8, 8
3	14	3, 3	3, 3	3, 3	6, 5
4	17	3, 3	3, 3	3, 3	8, 7
5	30	3, 3	3, 3	3, 3	6, 6

Convergence tolerance is 0.1 MW/MVAR. The numerical results, (i, j), should be interpreted as follows:

i—refers to the number of real power-angle update iterations.

j—refers to the number of reactive power-voltage update iterations.

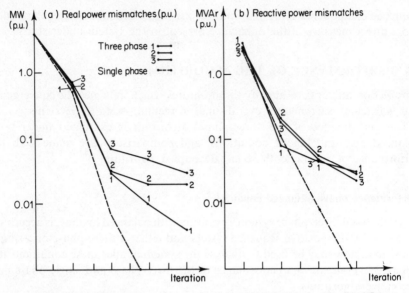

Fig. 5.5. Power convergence patterns for three-phase and single-phase load flow

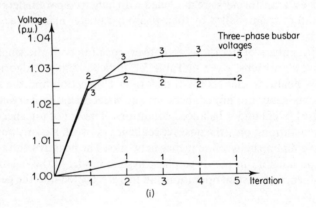

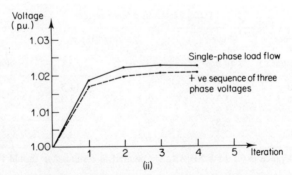

Fig. 5.6. Voltage convergence patterns for three-phase and single-phase load flows. (i) Three-phase voltages. (ii) Single-phase and three-phase positive sequence voltages

Performance with unbalanced systems

The number of iterations to convergence for the same test systems, under realistic steady state unbalanced operation, are also given in Table 5.1. The convergence rate deteriorates as compared with the balanced case, requiring on average twice as many iterations.

The graphs of Fig. 5.5 show that initial convergence of the three-phase mismatches is very close to that of the single-phase load flow. However, as the solution is approached the three-phase convergence becomes slower. It appears that, although the voltage and angle unbalance are introduced from the first iteration, they have only a secondary effect on the convergence until the positive sequence power flows are approaching convergence.

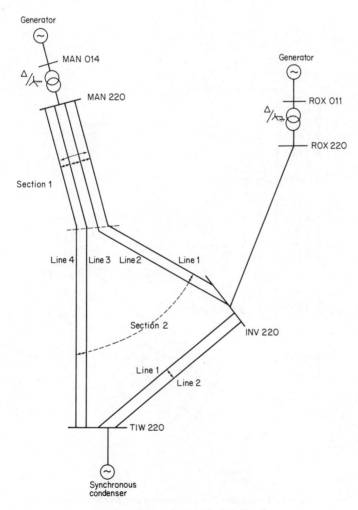

Fig. 5.7. Test system single-line diagram

This feature is further illustrated in Fig. 5.6(i) where the convergence pattern of the three-phase voltages is shown. The convergence pattern of the positive sequence component of the unbalanced voltages is shown in Fig. 5.6(ii) together with the convergence pattern of the voltage at the same busbar for the corresponding single-phase load flow. The latter figure illustrates that the positive sequence voltage of the three-phase unbalanced load flow has an almost identical convergence pattern to the corresponding single-phase fast-decoupled load flow.

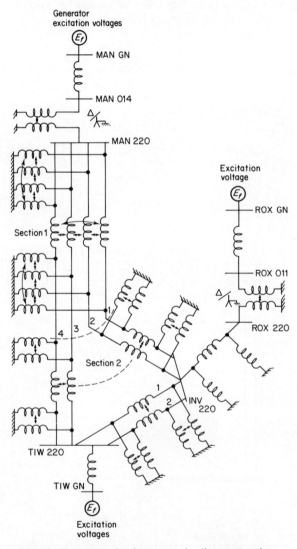

Fig. 5.8. Test system 3 × 3 compound coil representation

The final convergence of the system unbalance is somewhat slow but is reliable.

The following features are peculiar to a three-phase load flow and their influence on convergence is of interest:

— Asymmetry of the system parameters.
— Unbalance of the system loading.
— Influence of the transformer connection.
— Mutual coupling between parallel transmission lines.

These features have been examined with reference to a small six-bus test system.

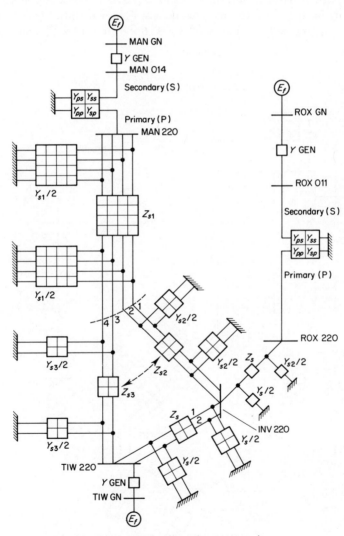

Fig. 5.9. Test system 3 × 3 matrix representation

5.7. TEST SYSTEM AND RESULTS

A single-line diagram of the test system under consideration is illustrated in Fig. 5.7. Some features of interest are:

— An example of a line sectionalization is included. One section contains four mutually coupled three-phase power lines. The other section contains two sets of two mutually coupled three-phase lines.
— All parallel lines are represented in their unbalanced mutually coupled state.
— Both transformers are star–delta connected with the star neutrals solidly earthed. Tap ratios are present on both primary and secondary sides.

The system is redrawn in Fig. 5.8 using 3×3 compound coil notation and substituting for the generator and line models. Following this, Fig. 5.9 illustrates the system graphically in terms of 3×3, 6×6 and 12×12 matrix blocks,

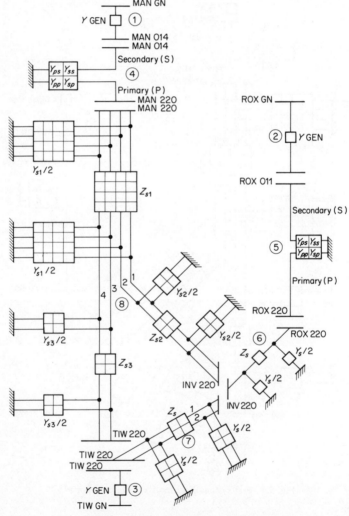

Fig. 5.10. Test system exploded into eight systems

representing the various system elements. The matrix quantities illustrated in Fig. 5.9 are given by, or derived from, the input data to the load-flow program.

For the purposes of input data organization and the formation of the system admittance matrix, the system is divided into eight natural subsystems. These are illustrated in the exploded system diagram of Fig. 5.10.

Once the matrices defined in Figs. 5.9 and 5.10 are known, the admittance matrix for each subsystem can be formed following the procedures outlined in Chapter 2. The subsystems are then combined to form the overall system admittance matrix.

The input data, which enables all the matrices in Fig. 5.10 to be formed, is listed for each subsystem in the following sections. The data is all in p.u. to a base of 33.3 MVA.

Generator data—subsystems 1 and 2

Subsystems 1 and 2 represent two synchronous generators. The input data to the computer program consists of the three-sequence impedances, the voltage regulator specification and the total real power generation at all generators except one which is the slack machine.

Table of generator data

Generator No. Name	Zero Impedance R0	X0	Pos. Impedance R1	X1	Neg. Impedance R2	X2	P p.u.	Voltage regulator $V_{\text{phase}\,\alpha}$
1 MAN014	0.0	0.080	0.0	0.010	0.0	0.021	15.000	1.045
2 ROX011	0.0	0.150	0.0	0.010	0.0	0.091	SLACK	1.050

The effect of subsystem 3 (the synchronous condenser) is not included in the numerical example.

Transformer data—subsystems 4 and 5

The input data for a transformer subsystem consists of:

(a) Leakage impedance in p.u. $(r + jx)$.
(b) Transformer type (specified in case descriptions).
(c) Primary and secondary tap ratios.

The data for the two transformers is summarized in the following table.

Busbar names		Leakage reactance	Tap ratio prim.
primary	secondary		
MAN220	MAN014	$0.0006 + j0.0164$	0.045
ROX220	ROX011	$0.0020 + j0.038$	0.022

The subsystem matrices are formed as described in Section 2.6.

Line data—subsystems 6, 7 and 8

The series impedance and shunt admittance matrices must be read into the computer program.

Subsystem 6

This subsystem consists of a single-balanced line between the two terminal busbars. The phases are taken as uncoupled and the matrices are given below.
Terminal Busbars INV 220–ROX 220

	a		
a	$0.006 + j0.045$	$0.002 + j0.015$	$0.001 + j0.017$
$[Z_S] = $ b	$0.002 + j0.015$	$0.006 + j0.050$	$0.002 + j0.017$
c	$0.001 + j0.017$	$0.002 + j0.017$	$0.007 + j0.047$

	a		
a	$0.0 + j0.35$	$0.0 - j0.6$	$0.0 - j0.04$
$[Y_S] = $ b	$0.0 - j0.06$	$0.0 + j0.352$	$0.0 - j0.06$
c	$0.0 - j0.04$	$0.0 - j0.06$	$0.0 + j0.34$

Both these matrices are in p.u. for the total line length.

Subsystem 7

This subsystem consists of a pair of parallel, mutually coupled three-phase lines. These lines are represented in their natural coupled unbalanced state.
Terminal busbars
Line 1   INV220 – TIW220
Line 2   INV220 – TIW220
The series impedance matrix for the total length (Z_s) is:

		Line 1			Line 2		
		a	b	c	a	b	c
Line 1	a	$0.0023 + j0.0147$					
	b	$0.0012 + j0.008$	$0.0021 + j0.015$				
	c	$0.0011 + j0.007$	$0.001 + j0.008$	$0.0024 + j0.0148$			
Line 2	a	$0.0009 + j0.0062$	$0.0008 + j0.0061$	$0.0008 + j0.0058$	$0.0023 + j0.0147$		
	b	$0.0008 + j0.0061$	$0.0007 + j0.0059$	$0.0007 + j0.0056$	$0.0014 + j0.009$	$0.0026 + j0.015$	
	c	$0.0008 + j0.0058$	$0.0007 + j0.0056$	$0.0006 + j0.0054$	$0.0012 + j0.009$	$0.001 + j0.009$	$0.0021 + j0.013$

The shunt admittance matrix for the total line length is:

		Line 1			Line 2		
		a	b	c	a	b	c
Line 1	a	$+j0.045$					
	b	$-j0.008$	$+j0.040$				
	c	$-j0.009$	$-j0.011$	$+j0.035$			
Line 2	a	$-j0.007$	$-j0.003$	$-j0.003$	$+j0.044$		
	b	$-j0.003$	$-j0.005$	$-j0.002$	$-j0.01$	$+j0.040$	
	c	$-j0.002$	$-j0.002$	$-j0.004$	$-j0.01$	$-j0.011$	$+j0.036$

The lower diagonal half only is shown as all line matrices are symmetrical.

Subsystem 8

Sectionalized mutually coupled lines: Section 1 consists of four mutually coupled three-phase lines and has 12×12 characteristic matrices, $[Z_{S1}]$ and $[Y_{S1}]$, as indicated in the system diagrams. These are given in Figs. 5.11 and 5.12 in per unit length of line and Section 1 is taken as having a length of 0.75 units.

Section 2 consists of two sets of two mutually coupled three-phase lines. To ensure consistent dimensionality with Section 1, the second section is considered as being composed of four mutually coupled three-phase lines, the elements representing the coupling between the two separate double-circuit lines being set to zero. The characteristic matrices for Section 2 become,

$$
\begin{array}{c}
[Z_s] \\
12 \times 12 \\
\text{Section 2}
\end{array}
=
\begin{array}{c}
\begin{array}{cccc} 1 & 2 & 3 & 4 \end{array} \\
\begin{array}{c} 1 \\ 2 \\ 3 \\ 4 \end{array}
\begin{array}{|c|c|}
\hline
[Z_{s2}] & [0] \\
\hline
[0] & [Z_{s3}] \\
\hline
\end{array}
\end{array}
$$

$$
\begin{array}{c}
[Y_s] \\
12 \times 12 \\
\text{Section 2}
\end{array}
=
\begin{array}{c}
\begin{array}{cccc} 1 & 2 & 3 & 4 \end{array} \\
\begin{array}{c} 1 \\ 2 \\ 3 \\ 4 \end{array}
\begin{array}{|c|c|}
\hline
[Y_{s2}] & [0] \\
\hline
[0] & [Y_{s3}] \\
\hline
\end{array}
\end{array}
$$

where $[0]$ is a matrix of zeros. The submatrices are labelled as those in Fig. 5.8.

These 12×12 matrices are given in Fig. 5.13 and Fig. 5.14. in per unit length of line. Section 2 is taken as having a length of 0.25 units.

Once the overall admittance matrix for the combined sections has been found it must be stored in full. This is to enable calculation of the power flows in the four individual lines. The matrix is modified, as described in Section 2.5. for the terminal connections. This modified matrix is the subsystem admittance matrix to be combined into the overall system admittance matrix.

	Line 1 a	Line 1 b	Line 1 c	Line 2 a	Line 2 b	Line 2 c	Line 3 a	Line 3 b	Line 3 c	Line 4 a	Line 4 b	Line 4 c
Line 1 a	0.0156 +j0.1088											
Line 1 b	0.008 +j0.032	0.015 +j0.1080										
Line 1 c	0.007 +j0.022	0.008 +j0.032	0.0160 +j0.1095									
Line 2 a	0.003 +j0.025	0.004 +j0.025	0.002 +j0.025	0.0156 +j0.1088								
Line 2 b	0.0025 +j0.022	0.0042 +j0.028	0.004 +j0.028	0.008 +j0.032	0.0150 +j0.1080							
Line 2 c	0.002 +j0.025	0.004 +j0.028	0.0042 +j0.028	0.007 +j0.032	0.008 +j0.032	0.0160 +j0.1095						
Line 3 a	0.0015 +j0.012	0.0012 +j0.01	0.001 +j0.011	0.001 +j0.01	0.0008 +j0.007	0.0009 +j0.007	0.0133 +j0.0904					
Line 3 b	0.0012 +j0.01	0.0015 +j0.012	0.0012 +j0.01	0.0008 +j0.007	0.001 +j0.01	0.0008 +j0.007	0.006 +j0.04	0.140 +j0.08				
Line 3 c	0.001 +j0.011	0.0012 +j0.015	0.0015 +j0.012	0.0009 +j0.007	0.0008 +j0.007	0.001 +j0.01	0.005 +j0.02	0.006 +j0.04	0.0130 +j0.085			
Line 4 a	0.0009 +j0.009	0.0008 +j0.01	0.0008 +j0.009	0.0008 +j0.006	0.0006 +j0.004	0.0005 +j0.003	0.003 +j0.025	0.002 +j0.02	0.002 +j0.01	0.0133 +j0.0904		
Line 4 b	0.0008 +j0.01	0.0006 +j0.008	0.0006 +j0.008	0.0006 +j0.004	0.0008 +j0.006	0.0006 +j0.004	0.002 +j0.02	0.003 +j0.025	0.002 +j0.02	0.006 +j0.04	0.0140 +j0.08	
Line 4 c	0.0008 +j0.009	0.0006 +j0.008	0.0006 +j0.008	0.0005 +j0.003	0.0006 +j0.004	0.0008 +j0.006	0.002 +j0.01	0.002 +j0.02	0.003 +j0.025	0.005 +j0.03	0.006 +j0.04	0.0130 +j0.085

Note: Lower diagonal only given as matrix is symmetrical.

Fig. 5.11. Series impedance matrix $[X_{S1}]$ for Section 1

		Line 1			Line 2			Line 3			Line 4		
		a	b	c	a	b	c	a	b	c	a	b	c
Line 1	a	$+j0.2967$											
	b	$-j0.06$	$+j0.299$										
	c	$-j0.05$	$-j0.06$	$+j0.31$									
Line 2	a	$-j0.04$	$-j0.03$	$-j0.035$	$+j0.2967$								
	b	$-j0.045$	$-j0.035$	$-j0.032$	$-j0.06$	$+j0.299$							
	c	$-j0.04$	$-j0.032$	$-j0.028$	$-j0.05$	$-j0.06$	$+j0.3$						
Line 3	a	$-j0.02$	$-j0.22$	$-j0.018$	$-j0.018$	$-j0.012$	$-j00.009$	$+j0.2569$					
	b	$-j0.022$	$-j0.018$	$-j0.15$	$-j0.012$	$-j0.012$	$-j0.01$	$-j0.50$	$+j0.26$				
	c	$-j0.018$	$-j0.015$	$-j0.018$	$-j0.009$	$-j0.01$	$-j0.014$	$-j0.045$	$-j0.042$	$+j0.251$			
Line 4	a	$-j0.15$	$-j0.009$	$-j0.009$	$-j0.01$	$-j0.009$	$-j0.008$	$-j0.043$	$-j0.04$	$-j0.032$	$+j0.2569$		
	b	$-j0.009$	$-j0.008$	$-j0.009$	$-j0.009$	$-j0.008$	$-j0.007$	$-j0.032$	$-j0.038$	$-j0.028$	$-j0.050$	$+j0.026$	
	c	$-j0.009$	$-j0.008$	$-j0.008$	$-j0.008$	$-j0.007$	$-j0.006$	$-j0.028$	$-j0.032$	$-j0.025$	$-j0.045$	$-j0.041$	$+j0.251$

Fig. 5.12. Shunt admittance matrix $[Y_{s1}]$ for Section 1

	Line 1 a	Line 1 b	Line 1 c	Line 2 a	Line 2 b	Line 2 c	Line 3 a	Line 3 b	Line 3 c	Line 4 a	Line 4 b	Line 4 c
Line 1 a	0.0156 +j0.1088											
Line 1 b	0.008 +j0.032	0.015 +j0.1080										
Line 1 c	0.007 +j0.022	0.008 +j0.032	0.0160 +j0.1095									
Line 2 a	0.003 +j0.025	0.004 +j0.025	0.002 +j0.025	0.0156 +j0.1088								
Line 2 b	0.0025 +j0.022	0.0042 +j0.028	0.004 +j0.028	0.008 +j0.032	0.0150 +j0.1080							
Line 2 c	0.002 +j0.025	0.004 +j0.028	0.0042 +j0.028	0.007 +j0.032	0.008 +j0.032	0.0160 +j0.1095						
Line 3 a							0.0133 +j0.0904					
Line 3 b							0.006 +j0.04	0.0140 +j0.08				
Line 3 c							0.005 +j0.03	0.006 +j0.04	0.0130 +j0.085			
Line 4 a							0.003 +j0.025	0.002 +j0.02	0.002 +j0.01	0.0133 +j0.0904		
Line 4 b							0.002 +j0.02	0.003 +j0.025	0.002 +j0.02	0.006 +j0.04	0.0140 +j0.08	
Line 4 c							0.002	0.002	0.003	0.005 +j0.04	0.006	0.0130

		Line 1			Line 2			Line 3			Line 4		
		a	b	c	a	b	c	a	b	c	a	b	c
	a	+j0.2967											
Line 1	b	−j0.06	+j0.299										
	c	−j0.05	−j0.06	+j0.31									
	a	−j0.04	−j0.03	−j0.035	+j0.2967								
Line 2	b	−j0.045	−j0.035	−j0.032	−j0.06	+j0.299							
	c	−j0.04	−j0.032	−j0.028	−j0.05	−j0.06	+j0.3						
	a							+j0.2569					
Line 3	b							−j0.50	+j0.26				
	c							−j0.045	−j0.042	+j0.251			
	a							−j0.043	−j0.04	−j0.032	+j0.2569		
Line 4	b							−j0.032	−j0.038	−j0.028	−j0.050	+j0.026	
	c							−j0.028	−j0.032	−j0.025	−j0.045	−j0.041	+j0.251

Note: Lower diagonal only shown as matrix is symmetrical.

Fig. 5.14. Shunt capacitance matrix $[Y_S]$ for Section 2

Test cases and typical results

The following cases have been examined:

(i) Balanced system with balanced loading and no mutual coupling between parallel three-phase lines. Generator transformers are star-g/star-g.

(ii) As for case (i) but with balanced mutual coupling introduced for all parallel three-phase lines as indicated in Fig. 5.7.

(iii) As for case (ii) but with balanced loading.

(iv) As for case (ii) but with system unbalance introduced by line capacitance unbalance only.

(v) As for case (ii) but with system unbalance introduced by line series impedance unbalance only.

(vi) Combined system capacitance and series impedance unbalance with balanced loading. Generator transformers star-g/star-g.

(vii) As for case (vi) but with unbalanced loading.

(viii) As for case (vii) but with delta/star-g for the generator transformers.

(ix) As for case (viii) but with large unbalanced real power loading at INV220.

(x) As for case (viii) but with large unbalanced reactive power loading at INV220.

The number of iterations to convergence, given in Table 5.2, clearly indicates that system unbalance causes a deterioration in convergence. Such deterioration is largely independent of the cause of the unbalance, but it is very dependent on the severity or degree of the unbalance.

In all these cases the degree of system unbalance is significant as may be assessed from the sequence components of the busbar voltages, which are given in Table 5.3 for cases (vii), (viii), and (x). The latter case is only included to demonstrate the convergence properties of the algorithm.

It is noteworthy that the initial convergence of the algorithm is fast even in cases of extreme steady-state unbalance. The reliability of the algorithm is not

Table 5.2. Number of iterations to convergence for six bus test system

Case	Convergence tolerance (MW/MVAR)		
	10.0	1.0	0.1
i	2,1	2,2	3,3
ii	2,1	2,2	3,3
iii	2,1	6,5	10,10
iv	2,1	5,4	8,8
v	2,1	5,4	9,9
vi	2,1	5,4	9,9
vii	2,1	4,3	10,9
viii	2,1	3,3	8,7
ix	4,3	11,9	17,16
x	4,3	10,10	16,16

Table 5.3. Sequence components of busbar voltages

Busbar	+ ve sequence		− ve sequence		Zero sequence	
	V_1	θ_1	V_2	θ_2	V_0	θ_0
INV220	1.020	− 0.16	0.028	2.42	0.021	− 0.85
ROX220	1.037	− 0.13	0.028	2.37	0.025	− 1.13
MAN220	1.058	− 0.09	0.015	1.84	0.014	− 0.77
MAN014	1.039	− 0.01	0.008	1.85	0.012	− 0.76
TIW220	1.015	− 0.17	0.028	2.40	0.021	− 0.74
ROX011	1.055	− 0.03	0.019	2.39	0.019	− 1.12
MAN.GN	1.056	0.03	0.0	—	0.0	—
ROX.GN	1.066	0.0	0.0	—	0.0	—

Case (vii)

Busbar	+ ve sequence		− ve sequence		Zero sequence	
	V_1	θ_1	V_2	θ_2	V_0	θ_0
INV220	1.034	0.36	0.023	− 3.12	0.004	0.23
ROX220	1.049	0.40	0.023	3.04	0.005	− 0.80
MAN220	1.071	0.43	0.015	2.39	0.001	0.20
MAN014	1.050	− 0.01	0.006	2.93	0.0	—
TIW220	1.029	0.36	0.023	3.11	0.005	0.69
ROX011	1.064	− 0.02	0.016	− 2.70	0.0	—
MAN.GN	1.067	0.03	0.0	—	0.0	—
ROX.GN	1.074	0.0	0.0	—	0.0	—

Case (viii)

Busbar	+ ve sequence		− ve sequence		Zero sequence	
	V_1	θ_1	V_2	θ_2	V_0	θ_0
INV220	1.011	0.37	0.100	− 2.69	0.083	− 2.62
ROX220	1.043	0.40	0.086	− 2.70	0.031	− 2.36
MAN220	1.065	0.44	0.058	− 2.65	0.017	− 2.50
MAN014	1.061	− 0.01	0.032	− 2.11	0.0	—
TIW220	1.007	0.36	0.098	− 2.68	0.080	− 2.59
ROX011	1.081	− 0.02	0.060	− 2.16	0.0	—
MAN.GN	1.086	0.03	0.0	—	0.0	—
ROX.GN	1.096	0.0	0.0	—	0.0	—

Case (x)

prejudiced by significant unbalance although convergence to small tolerances becomes slow.

The influence of the three-phase transformer connection may be seen in the sequence voltages of cases (vii) and (viii). The star-g/delta connection provides no through path for zero-sequence currents and the zero-sequence machine current is zero. This is reflected in the zero-sequence voltages at the machine terminal voltages.

The sequence voltages also illustrate the position of angle reference at the slack generator internal busbar. In addition, it may be seen that at all generator internal busbars the negative and zero-sequence voltages are zero reflecting the balanced and symmetrical nature of the machine excitations.

As an example of the numerical results, the busbar loadings for case (viii) are given in Table 5.4 and the resulting busbar voltages and line power flows are presented in Tables 5.5 and 5.6.

Besides the significant unbalance other features to be noticed are:

— The approximate 30° phase shift due to the star–delta connected transformers.
— Balanced voltages at the generator-internal busbars.
— Balanced angles at the generator-internal busbars.
— An apparent gain in active power flow in any one phase. This power flows through the mutual coupling terms between phases. The overall active power shows a net loss as expected for a realistic system.

Table 5.4. Table of busbar data

No.	Busbar Name	Phase A P-Load	Phase A Q-Load	Phase B P-Load	Phase B Q-Load	Phase C P-Load	Phase C Q-Load
1	INV220	50.000	15.000	45.000	14.000	48.300	16.600
2	ROX220	48.000	20.000	47.000	12.000	51.300	28.300
3	MAN220	0.0	0.0	0.0	0.0	0.0	0.0
4	MAN014	0.0	0.0	0.0	0.0	0.0	0.0
5	TIW220	150.000	80.000	157.000	78.000	173.000	72.000
6	ROX011	0.0	0.0	0.0	0.0	0.0	0.0

Table 5.5. Busbar results

No.	Busbar Name	Phase A Volt	Phase A Ang	Phase B Volt	Phase B Ang	Phase C Volt	Phase C Ang	Generation Total	
1	INV220	1.0173	21.36	1.0509	−98.16	1.0351	139.44	0.0	0.0
2	ROX220	1.0319	23.30	1.0730	−96.18	1.0449	141.76	0.0	0.0
3	MAN220	1.0693	25.34	1.0816	−95.21	1.0641	144.34	0.0	0.0
4	MAN014	1.0450	−0.79	1.0545	−120.64	1.0522	118.84	0.0	0.0
5	TIW220	1.0137	21.08	1.0434	−98.61	1.0316	138.98	0.0	0.0
6	ROX011	1.0500	−1.79	1.0653	−120.57	1.0771	118.12	0.0	0.0
7	MAN.GN	1.0669	1.69	1.0669	−118.31	1.0669	121.69	500.000	185.804
8	ROX.GN	1.0738	0.0	1.0738	−120.00	1.0738	120.00	281.277	108.106

Table 5.6 Computed print-out power flows

Sending end Busbar		Receiving end Busbar		Sending end		Receiving end	
No.	Name	No.	Name	MW	MVAR	MW	MVAR
4	MAN014	7	MAN.GN	−163.583	−62.676	164.077	71.179
				−160.184	−47.925	159.968	55.047
				−176.232	−50.050	175.955	59.577
6	ROX011	8	ROX.GN	−95.416	−37.762	96.329	41.642
				−87.270	−34.303	87.620	35.449
				−98.590	−27.893	97.327	31.011
3	MAN220	5	TIW220	34.710	10.919	−34.135	−19.871
				33.977	8.911	−32.640	−19.255
				38.172	6.260	−37.730	−15.979
3	MAN220	5	TIW220	36.209	16.598	−36.075	−24.989
				29.544	4.985	−28.602	−16.504
				40.282	4.235	−39.851	−13.410
3	MAN220	1	INV220	41.950	7.870	−41.154	−14.703
				50.720	8.167	−49.293	−16.798
				47.746	18.296	−48.539	−24.434
3	MAN220	1	INV220	44.368	6.728	−43.347	−13.739
				52.547	9.863	−51.290	−18.097
				48.269	16.689	−48.704	−23.079
1	INV220	5	TIW220	35.058	10.315	−34.987	−11.785

Table 5.6. (*contd.*)

Sending end Busbar		Receiving end Busbar		Sending end		Receiving end	
				MW	MVAR	MW	MVAR
1	INV220	5	TIW220	43.883	22.383	−43.806	−23.593
				34.740	18.915	−34.720	−20.106
1	INV220		TIW220	44.852	22.010	−44.801	−23.424
				52.175	17.444	−51.939	−18.659
				60.745	21.385	−60.691	−22.413
1	INV220	2	ROX220	−22.706	−9.412	22.491	3.271
				−20.242	−9.462	20.467	2.548
				−23.275	−4.725	23.737	−1.362
1	INV220	2	ROX220	−22.706	−9.412	22.491	3.271
				−20.242	−9.462	20.467	2.548
				−23.275	−4.725	23.737	−1.362
3	MAN220	4	MAN014	−157.242	−42.113	163.587	62.660
				−166.786	−31.943	160.186	47.958
				−174.468	−45.462	176.229	50.033
2	ROX220	6	ROX011	−92.984	−26.544	95.416	37.757
				−87.935	−17.105	87.271	34.312
				−98.772	−25.566	98.588	27.888

Total Generation 781.27 MW   293.91 MVAR

Total Load 768.60 MW   335.90 MVAR

System Losses 11.67 MW   − 41.98 MVAR

Mismatch 0.0013 MW   − 0.0096 MVAR

5.8. REFERENCES

1. A. Holley, C. Coleman and R. B. Shipley, 1964. 'Untransposed e.h.v. line computations', *Trans. IEEE*, **PAS-83**, 291.
2. M. H. Hesse, 1966. 'Circulating currents in parallel untransposed multicircuit lines: I—Numberical evaluations', *Trans. IEEE*, **PAS-85**, 802. 'II—Methods of eliminating current unbalance', *Trans. IEEE*, **PAS-95**, 812.
3. A. H. El-Abiad and D. C. Tarisi, 1967. 'Load-flow solution of untransposed EHV networks', *PICA, Pittsburgh, Pa*, 377–384.
4. R. G. Wasley and M. A. Slash, 1974. 'Newton–Raphson Algorithm for three-phase load flow', *Proc. IEE*, **121** (7), 630.
5. K. A. Birt, J. J. Graffy, J. D. McDonald and A. H. El-Abiad, 1976, 'Three-phase load-flow program', *Trans. IEEE*, **PAS-95**, 59.
6. J. Arrillaga and B. J. Harker, 1978. 'Fast decoupled three-phase load flow', *Proc. IEE*, **125** (8), 734–740.
7. B. Stott, Load flows for a.c. and a.c.–d.c. systems.' Ph.D. thesis, University of Manchester, 1971.
8. B. Stott and O. Alsac, 'Fast decoupled load flow', *Trans. IEEE* **PAS-93**, 1978, 859.
9. K. Zollenkopf, 1970. 'Bifactorization—basic computational algorithm and programming techniques', Conference on Large Sets of Sparse Linear Equations, Oxford, 75–96.

A. C.–D. C. Load Flow

6.1. INTRODUCTION

H.V.D.C. transmission is now an acceptable alternative to a.c. and is proving an economical solution not only for very long distance but also for underground and submarine transmission as well as a means of interconnecting systems of different frequency or with problems of stability or fault level.

The growing number of schemes in existence and under consideration demands corresponding modelling facilities for planning and operational purposes.

The basic load flow has to be substantially modified to be capable of modelling the operating state of the combined a.c. and d.c. systems under the specified conditions of load, generation and d.c. system control strategies.

Having established the superiority of the fast decoupled a.c. load flow[1] the integration of h.v.d.c. transmission is now described with reference to such an algorithm. This is carried out using sequential and unified approaches.

In the sequential approach[2],[3] the a.c. and d.c. equations are solved separately and thus the integration into existing load-flow programs is carried out without significant modification or restructuring of the a.c. solution technique. For the a.c. iterations each converter is modelled simply by the equivalent real or reactive power injection at the terminal busbar. The terminal busbar voltages obtained from the a.c. iteration are then used to solve the d.c. equations and consequently, new power injections are obtained. This process continues iteratively to convergence.

Alternatively, the more sophisticated unified methods[4],[5],[6] give full recognition to the interdependence between a.c. and d.c. system equations and simultaneously solve the complete set of equations.

6.2. FORMULATION OF THE PROBLEM

The operating state of the combined power system is defined by the vector

$$[\bar{V}, \theta, \bar{x}]^T$$

where

$\bar{V}$ is a vector of the voltage magnitudes at all a.c. system busbars.

124

$\bar{\theta}$ is a vector of the angles at all a.c. system busbars (except the reference bus which is assigned $\theta = 0$).

$\bar{x}$ is a vector of d.c. variables.

Chapter 4 has described the use of $\bar{V}$ and $\bar{\theta}$ as a.c. system variables and the selection of d.c. variables $\bar{x}$ is discussed in Section 6.3.

The development of a Newton–Raphson based algorithm requires the formulation of n independent equations in terms of the n variables.

The equations which relate to the a.c. system variables are derived from the specified a.c. system operating conditions. The only modification required to the usual real and reactive power mismatches occurs for those equations which relate to the converter terminal busbars. These equations become:

$$P_{\text{term}}^{sp} - P_{\text{term}}(ac) - P_{\text{term}}(dc) = 0 \qquad (6.2.1)$$

$$Q_{\text{term}}^{sp} - Q_{\text{term}}(ac) - Q_{\text{term}}(dc) = 0 \qquad (6.2.2)$$

where

$P_{\text{term}}(ac)$ is the injected power at the terminal busbar as a function of the a.c. system variables

$P_{\text{term}}(dc)$ is the injected power at the terminal busbar as a function of the d.c. system variables

P_{term}^{sp} is the usual a.c. system load at the busbar.

and similarly for $Q_{\text{term}}(dc)$ and $Q_{\text{term}}(ac)$

The injected powers $Q_{\text{term}}(dc)$ and $P_{\text{term}}(dc)$ are functions of the convertor a.c. terminal busbar voltage and of the d.c. system variables, i.e.

$$P_{\text{term}}(dc) = f(V_{\text{term}}, \bar{x}) \qquad (6.2.3)$$

$$Q_{\text{term}}(dc) = f(V_{\text{term}}, \bar{x}) \qquad (6.2.4)$$

The equations derived from the specified a.c. system conditions may therefore be summarized as:

$$\begin{bmatrix} \Delta \bar{P}(\bar{V}, \bar{\theta}) \\ \Delta \bar{P}_{\text{term}}(\bar{V}, \bar{\theta}, \bar{x}) \\ \Delta \bar{Q}(\bar{V}, \bar{\theta}) \\ \Delta \bar{Q}_{\text{term}}(\bar{V}, \bar{\theta}, \bar{x}) \end{bmatrix} = 0 \qquad (6.2.5)$$

where the mismatches at the converter terminal busbars are indicated separately.

A further set of independent equations are derived from the d.c. system conditions. These are designated,

$$\bar{R}(V_{\text{term}}, \bar{x})_k = 0 \qquad (6.2.6)$$

for $k = 1$, number of converters present.

The d.c. system equations (6.2.3), (6.2.4) and (6.2.6) are made independent of the a.c. system angles $\bar{\theta}$ by selecting a separate angle reference for the d.c. system

126

variables as defined in Fig. 6.2. This improves the algorithmic performance by effectively decoupling the angle dependence of a.c. and d.c. systems.

The general a.c.–d.c. load flow problem may therefore be summarized as the solution of:

$$
\begin{bmatrix}
\Delta \bar{P}(\bar{V}, \theta) \\
\Delta \bar{P}_{\text{term}}(\bar{V}, \theta, \bar{x}) \\
\Delta \bar{Q}(\bar{V}, \theta) \\
\Delta \bar{Q}_{\text{term}}(\bar{V}, \theta, \bar{x}) \\
\bar{R}(V_{\text{term}}, \bar{x})
\end{bmatrix} = 0
\tag{6.2.7}
$$

where the subscript 'term' refers to the converter a.c. terminal busbar.

6.3. D.C. SYSTEM MODEL

The selection of variables $\bar{x}$ and formulation of the equations require several basic assumptions which are generally accepted[1] in the analysis of steady state d.c. converter operation. These are:

(i) The three a.c. voltages at the terminal busbar are balanced and sinusoidal.
(ii) The converter operation is perfectly balanced.
(iii) The direct current and voltage are smooth.
(iv) The converter transformer is lossless and the magnetizing admittance is ignored.

Converter variables

Under balanced conditions similar converter bridges attached to the same a.c. terminal busbar will operate identically regardless of the transformer connection. They may therefore be replaced by an equivalent single bridge for the purpose of single-phase load flow analysis. With reference to Fig. 6.1 the set of variables illustrated, representing fundamental frequency or d.c. quantities permits a full description of the converter system operation.

An equivalent circuit for the converter is shown in Fig. 6.2 which includes the modification explained in Section 6.2 as regards the position of angle reference.

The variables, defined with reference to Fig. 6.2 are:

$V_{\text{term}}\underline{/\phi}$ converter terminal busbar nodal voltage (phase angle referred to converter reference)

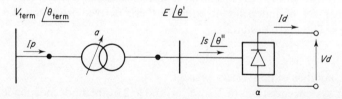

Fig. 6.1. Basic d.c. convertor (angles refer to a.c. system reference)

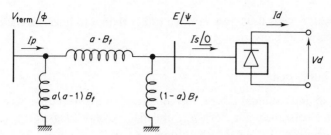

Fig. 6.2. Single-phase equivalent circuit for basic converter. (Angles referred to d.c. reference)

E/ψ fundamental frequency component of the voltage waveform at the converter transformer secondary

I_p, I_s fundamental frequency component of the current waveshape on the primary and secondary of the converter transformer respectively

α firing delay angle

a transformer off-nominal tap ratio

V_d average d.c. voltage

I_d converter direct current

These ten variables, nine associated with the converter, plus the a.c. terminal voltage magnitude V_{term}, form a possible choice of $\bar{x}$ for the formulation of equations (6.2.3), (6.2.4) and (6.2.6).

The minimum number of variables required to define the operation of the system is the number of independent variables. Any other system variable or parameter (e.g. P_{dc} and Q_{dc}) may be written in terms of these variables.

Two independent variables are sufficient to model a d.c. converter, operating under balanced conditions, from a known terminal voltage source. However, the control requirements of h.v.d.c. converters are such that a range of variables, or functions of them (e.g. constant power), are the specified conditions. If the minimum number of variables are used, then the control specifications must be translated into equations in terms of these two variables. These equations will often contain complex non-linearities, and present difficulties in their derivation and program implementation. In addition, the expressions used for P_{dc} and Q_{dc} in equations (6.2.1.) and (6.2.2.) may be rather complex and this will make the programming of a unified solution more difficult.

For these reasons, a nonminimal set of variables is recommended, i.e. all variables which are influenced by control action are retained in the model. This is in contrast to a.c. load flows where, due to the restricted nature of control specifications, the minimum set is normally used.

The following set of variables permits simple relationships for all the normal control strategies.

$$[\bar{x}] = [V_d, I_d, a, \cos \alpha, \phi]^T$$

Variable ϕ is included to ensure a simple expression for Q_{dc}. While this is important in the formulation of the unified solution, variable ϕ may be omitted with the sequential solution as it is not involved in the formulation of any control specification; $\cos \alpha$ is used as a variable rather than α to linearize the equations and thus improve convergence.

D.C. per unit system

To avoid per unit to actual value translations and to enable the use of comparable convergence tolerances for both a.c. and d.c. system mismatches, a per unit system is also used for the d.c. quantities.

Computational simplicity is achieved by using common power and voltage base parameters on both sides of the converter, i.e. the a.c. and d.c. sides. Consequently, in order to preserve consistency of power in per unit, the direct current base, obtained from $(MVA_B)/V_B$, has to be $\sqrt{3}$ times larger than the a.c. current base.

This has the effect of changing the coefficients involved in the a.c.–d.c. current relationships. For a perfectly smooth direct current and neglecting the commutation overlap, the r.m.s. fundamental components of the phase current is related to I_d by the approximation,

$$I_s = \frac{\sqrt{6}}{\pi} \cdot I_d \qquad (6.3.1)$$

Translating equation (6.3.1) to per unit yields:

$$I_s(\text{p.u.}) = \frac{\sqrt{6}}{\pi} \cdot \sqrt{3} \cdot I_d(\text{p.u.})$$

and if commutation overlap is taken into account, as described in Chapter 3, this equation becomes

$$I_s(\text{p.u.}) = k \frac{3\sqrt{2}}{\pi} \cdot I_d(\text{p.u.}) \qquad (6.3.2)$$

where k is very close to unity. In load-flow studies, equation (6.3.2) can be made sufficiently accurate in most cases by letting:
$$k = 0.995$$

Derivation of equations

The following relationships are derived for the variables defined in Fig. 6.2. The equations are in per unit.

(i) The fundamental current magnitude on the converter side is related to the direct current by the equation

$$I_s = k \frac{3\sqrt{2}}{\pi} \cdot I_d \qquad (6.3.3)$$

(ii) The fundamental current magnitudes on both sides of the lossless transformer are related by the off-nominal tap, i.e.

$$I_p = a \cdot I_s \tag{6.3.4}$$

(iii) The d.c. voltage may be expressed in terms of the a.c. source commutating voltage referred to the transformer secondary, i.e.

$$V_d = \frac{3\sqrt{2}}{\pi} \cdot a \cdot V_{\text{term}} \cos \alpha - \frac{3}{\pi} \cdot I_d \cdot X_c \tag{6.3.5}$$

The converter a.c. source commutating voltage is the busbar voltage on the system side of the converter transformer, V_{term}.

(iv) The d.c. current and voltage are related by the d.c. system configuration,

$$f(V_d, I_d) = 0 \tag{6.3.6}$$

e.g. for a simple rectifier supplying a passive load,

$$V_d - I_d \cdot R_d = 0$$

(v) The assumptions listed at the beginning of this section prevent any real power of harmonic frequencies at the primary and secondary busbars. Therefore, the real power equation relates the d.c. power to the transformer secondary power in terms of fundamental components only, i.e.

$$V_d \cdot I_d = E \cdot I_s \cdot \cos \psi \tag{6.3.7}$$

(vi) As the transformer is lossless, the primary real power may also be equated to the d.c. power, i.e.

$$V_d \cdot I_d = V_{\text{term}} I_p \cdot \cos \phi \tag{6.3.8}$$

(vii) The fundamental component of current flow across the converter transformer can be expressed as

$$I_s = B_t \cdot \sin \psi - B_t \cdot a \cdot V_{\text{term}} \sin \phi \tag{6.3.9}$$

where jB_t is the transformer leakage susceptance.

So far, a total of seven equations have been derived and no other independent equation may be written relating the total set of nine converter variables.

Variables, I_p, I_s, E and ψ can be eliminated as they play no part in defining control specifications.

Thus equations (6.3.3), (6.3.4), (6.3.7) and (6.3.8) can be combined into

$$V_d - k_1 \cdot a \cdot V_{\text{term}} \cos \phi = 0 \tag{6.3.10}$$

where $k_1 = k(3\sqrt{2}/\pi)$.

The final two independent equations required are derived from the specified control mode.

The d.c. model may thus be summarized as follows:

$$\bar{R}(\bar{x}, V_{\text{term}})_k = 0 \tag{6.3.11}$$

130

where

$$R(1) = V_d - k_1 \cdot a \cdot V_{\text{term}} \cdot \cos \phi$$

$$R(2) = V_d - k_1 \cdot a \cdot V_{\text{term}} \cdot \cos \alpha + \frac{3}{\pi} \cdot I_d \cdot X_c$$

$$R(3) = f(V_d, I_d)$$

$$R(4) = \text{control equation}$$

$$R(5) = \text{control equation}$$

and

$$\bar{x} = [V_d, I_d, a, \cos \alpha, \phi]^T$$

V_{term} can either be a specified quantity or an a.c. system variable. The equations for P_{dc} and Q_{dc} may now be written as,

$$Q_{\text{term}}(dc) = V_{\text{term}} \cdot I_p \cdot \sin \phi \qquad (6.3.12)$$

$$= V_{\text{term}} \cdot k_1 \cdot a \cdot I_d \cdot \sin \phi$$

and

$$P_{\text{term}}(dc) = V_{\text{term}} \cdot I_p \cdot \cos \phi \qquad (6.3.13)$$

$$= V_{\text{term}} \cdot k_1 \cdot a \cdot I_d \cdot \cos \phi$$

or

$$P_{\text{term}}(dc) = V_d \cdot I_d \qquad (6.3.14)$$

Incorporation of control equations

Each additional converter in the d.c. system contributes two independent variables to the system and thus two further constraint equations must be derived from the control strategy of the system to define the operating state. For example, a classical two-terminal d.c. link has two converters and therefore requires four control equations. The four equations must be written in terms of the ten d.c. variables (five for each converter).

Any function of the ten d.c. system variables is a valid (mathematically) control equation so long as each equation is independent of all other equations. In practice, there are restrictions limiting the number of alternatives. Some control strategies refer to the characteristics of power transmission (e.g. constant power or constant current), others introduce constraints such as minimum delay or extinction angles.

Examples of valid control specifications are;

(i) Specified converter transformer tap,

$$a - a^{sp} = 0$$

(ii) Specified d.c. voltage

$$V_d - V_d^{sp} = 0$$

(iii) Specified d.c. current

$$I_d - I_d^{sp} = 0$$

(iv) Specified minimum firing angle

$$\cos \alpha - \cos \alpha_{min} = 0$$

(v) Specified d.c. power transmission

$$V_d \cdot I_d - P_{dc}^{sp} = 0$$

These control equations are simple and are easily incorporated into the solution algorithm. In addition to the usual control modes, nonstandard modes such as specified a.c. terminal voltage may also be included as converter control equations (see Section 6.5).

During the iterative solution procedure the uncontrolled converter variables may go outside prespecified limits. When this occurs, the offending variable is usually held to its limit value and an appropriate control variable is freed.[4]

Inverter operation

All the equations presented so far are equally applicable to inverter operation. However, during inversion it is the extinction advance angle (γ) which is the subject of control action and not the firing angle (α). For convenience therefore, equation $R(2)$ of (6.3.11) may be rewritten as

$$V_d - k_1 \cdot a \cdot V_{\text{term}} \cdot \cos(\pi - \gamma) - \frac{3}{\pi} X_c \cdot I_d = 0 \qquad (6.3.15)$$

This equation is valid for rectification or inversion. Under inversion, V_d, as calculated by equation (6.3.15), will be negative.

To specify operation with constant extinction angle the following equation is used

$$\cos(\pi - \gamma) - \cos(\pi - \gamma^{sp}) = 0$$

where γ^{sp} is usually γ minimum for minimum reactive power consumption of the inverter.

6.4. SOLUTION TECHNIQUES

The converter model developed in Section 6.3 is to be incorporated with a fast decoupled a.c. load flow algorithm with the minimum possible modification of the latter in order to retain its computational advantages.

The solution is first discussed with reference to a single converter connected to an a.c. busbar. The extension to multiple or multiterminal d.c. systems is relatively trivial and is discussed in Section 6.6.

Unified solution

The unified method gives recognition to the interdependence of a.c. and d.c. system equations and simultaneously solves the complete system. Referring to equation (6.2.7) the standard Newton–Raphson algorithm involves repeat solutions of the matrix equation:

$$
\begin{bmatrix}
\Delta\bar{P}(\bar{V},\bar{\theta}) \\
\Delta P_{\text{term}}(\bar{V},\bar{\theta},\bar{x}) \\
\Delta\bar{Q}(\bar{V},\bar{\theta}) \\
\Delta Q_{\text{term}}(\bar{V},\bar{\theta},\bar{x}) \\
\bar{R}(V_{\text{term}},\bar{x})
\end{bmatrix}
=
\begin{bmatrix} J \end{bmatrix}
\begin{bmatrix}
\Delta\bar{\theta} \\
\Delta\bar{\theta}_{\text{term}} \\
\Delta\bar{V} \\
\Delta V_{\text{term}} \\
\Delta\bar{x}
\end{bmatrix}
\tag{6.4.1}
$$

where J is the matrix of first order partial derivatives.

$$
\Delta P_{\text{term}} = P^{sp}_{\text{term}} - P_{\text{term}}(ac) - P_{\text{term}}(dc) \tag{6.4.2}
$$

$$
\Delta Q_{\text{term}} = Q^{sp}_{\text{term}} - Q_{\text{term}}(ac) - Q_{\text{term}}(dc) \tag{6.4.3}
$$

and

$$
P_{\text{term}}(dc) = f(V_{\text{term}},\bar{x}) \tag{6.4.4}
$$

$$
Q_{\text{term}}(dc) = f(V_{\text{term}},\bar{x}) \tag{6.4.5}
$$

Applying the a.c. fast decoupled assumptions to all Jacobian elements related to the a.c. system equations, yields:

$\Delta\bar{P}/\bar{V}$		B'				$\Delta\bar{\theta}$
$\Delta P_{\text{term}}/V_{\text{term}}$				DD	AA'	$\Delta\theta_{\text{term}}$
$\Delta\bar{Q}/\bar{V}$	$=$		B''			$\Delta\bar{V}$
$\Delta Q_{\text{term}}/V_{\text{term}}$				B''_{ii}	AA''	ΔV_{term}
$\bar{R}$				BB''	A	$\Delta\bar{x}$

$$(6.4.6)$$

where all matrix elements are zero unless otherwise indicated. The matrices $[B']$ and $[B'']$ are the usual single-phase fast decoupled Jacobians and are constant in value. The other matrices indicated vary at each iteration in the solution process.

A modification is required for the element indicated as B''_{ii} in equation (6.4.6). This element is a function of the system variables and therefore varies at each iteration.

The use of an independent angle reference for the d.c. equations results in

$$
\partial P_{\text{term}}(dc)/\partial\theta_{\text{term}} = 0
$$

i.e. the diagonal Jacobian element for the real power mismatch at the converter terminal busbar depends on the a.c. equations only and is therefore the usual fast decoupled B' element.

In addition,

$$
\partial\bar{R}/\partial\theta_{\text{term}} = 0
$$

which will help the subsequent decoupling of the equation.

In order to maintain the block successive iteration sequence of the usual fast decoupled a.c. load flow, it is necessary to decouple equation (6.4.6). Therefore the Jacobian submatrices must be examined in more detail. The Jacobian submatrices are:

$$DD = \frac{1}{V_{\text{term}}} \partial \Delta P_{\text{term}}/\partial V_{\text{term}}$$

$$= \frac{1}{V_{\text{term}}} \{\partial P_{\text{term}}(ac)/\partial V_{\text{term}}\} + \frac{1}{V_{\text{term}}} \{\partial P_{\text{term}}(dc)/\partial V_{\text{term}}\}$$

Following decoupled load-flow practice

$$DD = 0 + \frac{1}{V_{\text{term}}} \{\partial P_{\text{term}}(dc)/\partial V_{\text{term}}\}$$

and since

$$P_{\text{term}}(dc) = V_d \cdot I_d$$

$$\partial P_{\text{term}}(dc)/\partial V_{\text{term}} = 0$$

therefore: $DD = 0$
Similarly

$$[AA'] = \frac{1}{V_{\text{term}}} [\partial \Delta P_{\text{term}}/\partial \bar{x}]$$

$$= \frac{1}{V_{\text{term}}} [\partial P_{\text{term}}(ac)/\partial \bar{x}] + \frac{1}{V_{\text{term}}} [\partial P_{\text{term}}(dc)/\partial \bar{x}]$$

$$= 0 + \frac{1}{V_{\text{term}}} [\partial P_{\text{term}}(dc)/\partial \bar{x}]$$

$$[AA''] = \frac{1}{V_{\text{term}}} [\partial \Delta Q_{\text{term}}/\partial \bar{x}]$$

$$= \frac{1}{V_{\text{term}}} [\partial Q_{\text{term}}(dc)/\partial \bar{x}]$$

$$[BB''] = \partial \bar{R}/\partial V_{\text{term}}$$

$$[A] = \partial \bar{R}/\partial \bar{x}$$

$$B''_{ii} = \frac{1}{V_{\text{term}}} [\partial \Delta Q_{\text{term}}/\partial V_{\text{term}}]$$

$$= \frac{1}{V_{\text{term}}} \partial Q_{\text{term}}(ac)/\partial V_{\text{term}} + \frac{1}{V_{\text{term}}} [\partial Q_{\text{term}}(dc)/\partial V_{\text{term}}]$$

$$= B''_{ii}(ac) + B''_{ii}(dc)$$

In the above formulation the d.c. variables $\bar{x}$ are coupled to both the real and reactive power a.c. mismatches. However, equation (6.4.6) may be separated to enable a block successive iteration scheme to be used.

The d.c. mismatches and variables can be appended to the two fast decoupled a.c. equations in which case the following two equations result

$$
\begin{bmatrix} \Delta\bar{P}/\bar{V} \\[4pt] \Delta P_{\text{term}}/V_{\text{term}} \\[4pt] \bar{R} \end{bmatrix} =
\begin{bmatrix} \begin{array}{c|c} B' & \\ \hline & AA' \\ \hline A \end{array} \end{bmatrix}
\begin{bmatrix} \Delta\bar{\theta} \\[4pt] \Delta\theta_{\text{term}} \\[4pt] \Delta\bar{x} \end{bmatrix}
\tag{6.4.7}
$$

$$
\begin{bmatrix} \Delta\bar{Q}/\bar{V} \\[4pt] \Delta Q_{\text{term}}/V_{\text{term}} \\[4pt] \bar{R} \end{bmatrix} =
\begin{bmatrix} \begin{array}{c|c|c} B'' & & \\ \hline & B''_{ii} & AA'' \\ \hline & BB'' & A \end{array} \end{bmatrix}
\begin{bmatrix} \Delta\bar{V} \\[4pt] \Delta V_{\text{term}} \\[4pt] \Delta\bar{x} \end{bmatrix}
\tag{6.4.8}
$$

The iteration scheme illustrated in Fig. 6.3 is referred to as—PDC, QDC—and the significance of the mnemonic should be obvious.

The algorithm may be further simplified by recognizing the following physical characteristics of the a.c. and d.c. systems;

— The coupling between d.c. variables and the a.c. terminal voltage is strong.
— There is no coupling between d.c. mismatches and a.c. system angles.
— Under all practical control strategies the d.c. power is well constrained and this implies that the changes in d.c. variables $\bar{x}$ do not greatly affect the real power mismatches at the terminals. This coupling, embodied in matrix AA' of equation (6.4.7) can therefore be justifiably removed.

These features justify the removal of the d.c. equations from equation (6.4.7) to yield a $-P, QDC$-block successive iteration scheme represented by the following two equations,

$$[\Delta\bar{P}/\bar{V}] = [B'][\Delta\bar{\theta}] \tag{6.4.9}$$

$$
\begin{bmatrix} \Delta\bar{Q}/\bar{V} \\[4pt] \Delta Q_{\text{term}}/V_{\text{term}} \\[4pt] \bar{R} \end{bmatrix} =
\begin{bmatrix} \begin{array}{c|c|c} B'' & & \\ \hline & B''_{ii} & AA'' \\ \hline & BB'' & A \end{array} \end{bmatrix}
\begin{bmatrix} \Delta\bar{V} \\[4pt] \Delta V_{\text{term}} \\[4pt] \Delta\bar{x} \end{bmatrix}
\tag{6.4.10}
$$

Programming considerations for the unified algorithms

In order to retain the efficiency of the fast decoupled load flow, the B' and B'' matrices must be factorized only once before the iterative process begins.

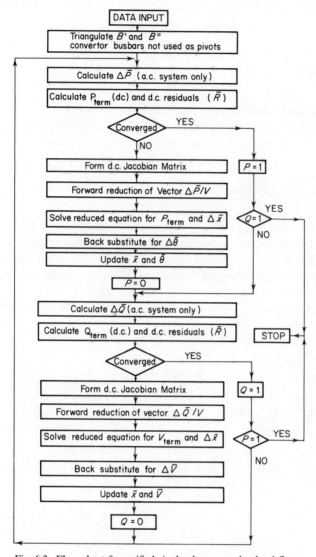

Fig. 6.3. Flow chart for unified single-phase a.c.–d.c. load flow

The Jacobian elements related to the d.c. variables are non-constant and must be re-evaluated at each iteration. It is therefore necessary to separate the constant and nonconstant parts of the equations for the solution routine.

Initially, the a.c. fast decoupled equations are formed with the d.c. link ignored (except for the minor addition of the filter reactance at the appropriate a.c. busbar). The reactive power mismatch equation for the a.c. system is:

$$\begin{bmatrix} \Delta \bar{Q}/\bar{V} \\ \Delta Q'_{\text{term}}/V_{\text{term}} \end{bmatrix} = \begin{bmatrix} B'' \end{bmatrix} \begin{bmatrix} \Delta \bar{V} \\ \Delta V_{\text{term}} \end{bmatrix} \tag{6.4.11}$$

where

$$\Delta Q'_{\text{term}} = Q^{sp}_{\text{term}} - Q_{\text{term}}(ac).$$

is the mismatch calculated in the absence of
the d.c. converter.

and

B'' is the usual constant a.c. fast decoupled Jacobian.

After triangulation down to, but excluding the busbars to which d.c. converters are attached, equation (6.4.11) becomes

$$
\begin{bmatrix} (\Delta \bar{Q}/\bar{V})'' \\ (\Delta Q_{\text{term}}/V_{\text{term}})'' \end{bmatrix} = \begin{bmatrix} & B''' \\ & B'''_{ii} \end{bmatrix} \begin{bmatrix} \Delta \bar{V} \\ \Delta V_{\text{term}} \end{bmatrix} \tag{6.4.12}
$$

where $(\Delta \bar{Q}/V)''$ and $(\Delta Q_{\text{term}}/V_{\text{term}})''$ signify that the left-hand side vector has been processed and matrix B''' is the new matrix B'' after triangulation.

This triangulation (performed before the iterative process) may be achieved simply by inhibiting the terminal busbars being used as pivots during the optimal ordering process.

The processing of $\Delta \bar{Q}$ indicated in the equation is actually performed by the standard forward reduction process used at each iteration.

The d.c. converter equations may then be combined with equation (6.4.12) as follows:

$$
\begin{bmatrix} (\Delta \bar{Q}/\bar{V})'' \\ \left(\dfrac{\Delta Q_{\text{term}}}{V_{\text{term}}}\right)'' + \dfrac{\Delta Q_{\text{term}}(dc)}{V_{\text{term}}} \\ \bar{R} \end{bmatrix} = \begin{bmatrix} 0 & B''' & 0 \\ 0 & B'''_{ii} + B''_{ii}(dc) & AA'' \\ 0 & BB'' & A \end{bmatrix} \begin{bmatrix} \Delta \bar{V} \\ \Delta V_{\text{term}} \\ \Delta \bar{x} \end{bmatrix} \tag{6.4.13}
$$

where

$$B''_{ii}(dc) = \frac{1}{V_{\text{term}}}[\partial Q_{\text{term}}(dc)/\partial V_{\text{term}}]$$

The unprocessed section, i.e.

$$
\begin{bmatrix} \left(\dfrac{\Delta Q_{\text{term}}}{V_{\text{term}}}\right)'' + \dfrac{\Delta Q_{\text{term}}(dc)}{V_{\text{term}}} \\ \bar{R} \end{bmatrix} = \begin{bmatrix} B'''_{ii} + B''_{ii}(dc) & AA'' \\ BB'' & A \end{bmatrix} \begin{bmatrix} \Delta V_{\text{term}} \\ \Delta \bar{x} \end{bmatrix} \tag{6.4.14}
$$

may then be solved by any method suitable for nonsymmetric matrices.

The values of $\Delta\bar{x}$ and ΔV_{term} are obtained from this equation and ΔV_{term} is then used in a back substitution process for the remaining $\Delta\bar{V}$ to be completed, i.e. equation (6.4.12) is solved for $\Delta\bar{V}$.

The most efficient technique for solving equation (6.4.14) depends on the number of converters. For six converters or more, the use of sparsity storage and solution techniques are justified, otherwise all elements should be stored. The method suggested here is a modifed form of gaussian elimination where all elements are stored but only nonzero elements processed.[7]

Sequential method

The sequential method results from a further simplification of the unified method, i.e. the a.c. system equations are solved with the d.c. system modelled simply as a real and reactive power injection at the appropriate terminal busbar. For a d.c.

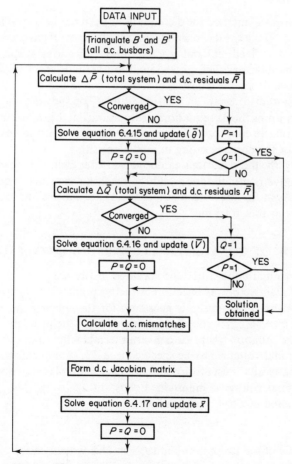

Fig. 6.4. Flow chart for sequential single-phase a.c.–d.c. load flow

solution the a.c. system is modelled simply as a constant voltage at the converter a.c. terminal busbar.

The following three equations are solved iteratively to convergence.

$$[\Delta \bar{P}/\bar{V}] = [B'][\Delta \bar{\theta}] \tag{6.4.15}$$

$$[\Delta \bar{Q}/\bar{V}] = [B''][\Delta \bar{V}] \tag{6.4.16}$$

$$[\bar{R}] = [A][\Delta \bar{x}] \tag{6.4.17}$$

This iteration sequence, referred to as P, Q, DC, is illustrated in the flow chart of Fig. 6.4 and may be summarized as follows;

(i)   Calculate $\Delta \bar{P}/\bar{V}$, solve equation (6.4.15) and update $\bar{\theta}$.
(ii) Calculate $\Delta \bar{Q}/\bar{V}$, solve equation (6.4.16) and update $\bar{V}$.
(iii) Calculate d.c. residuals, $\bar{R}$, solve equation (6.4.17) and update $\bar{x}$.
(iv) Return to (i).

With the sequential method the d.c. equations need not be solved for the entire iterative process. Once the d.c. residuals have converged, the d.c. system may be modelled simply as fixed real and reactive power injections at the appropriate converter terminal busbar. The d.c. residuals must still be checked after each a.c. iteration to ensure that the d.c. system remains converged.

However, in order to establish a direct comparison between the unified and sequential algorithms the d.c. equations are continued to be solved until both a.c. and d.c. systems have converged. This ensures that the sequential technique is an exact parallel of the corresponding unified algorithm.

Alternatively, the d.c. equations can be solved after each real power as well as after each reactive power iteration and the resulting sequence is referred to as P, DC, Q, DC. As in the previous method, the d.c. equations are solved until all mismatches are within tolerance.

6.5. CONTROL OF CONVERTER A.C. TERMINAL VOLTAGE

A converter terminal voltage may be specified in two ways:

(a) By local reactive power injection at the terminal. In this case no reactive power mismatch equation is necessary for that busbar and the relevant variable (i.e. ΔV_{term}) is effectively removed from the problem formulation. This is the situation where the converter terminal busbar is a $P - V$ busbar.

(b) The terminal voltage may be specified as a d.c. system constraint. That is, the d.c. converter must absorb the correct amount of reactive power so that the terminal voltage is maintained constant.

With the unified method, the equation

$$V_{term}^{sp} - V^{term} = 0 \tag{6.5.1}$$

is written as one of the two control equations. This would lead to a zero row in equation (6.4.7) and therefore during the solution of equation (6.4.7) some other variable (e.g. tap ratio) must be specified instead. Although d.c. convergence is marginally slower for the PDC, QDC iteration, the d.c. system is overconverged in this iteration scheme and the overall convergence rate is practically unaffected.

With the sequential method equation (6.5.1) cannot be written. The terminal busbar is specified as a $P - V$ busbar and the control equation

$$Q_{\text{term}}^{sp} (dc) - Q_{\text{term}} (dc) = 0$$

is used, where $Q_{\text{term}}^{sp} (dc)$ is taken as the reactive power required to maintain the voltage constant. The specified reactive power thus varies at each iteration and this discontinuity slows the overall convergence.

6.6. EXTENSION TO MULTIPLE AND/OR MULTITERMINAL D.C. SYSTEMS

The basic algorithm has been developed in previous sections for a single d.c. converter. Each additional converter adds a further five d.c. variables and a corresponding set of five equations. The number of a.c. system Jacobian elements which become modified in the unified solutions is equal to the number of converters.

As an example, consider the system shown in Fig. 6.5. The system represents the North and South Islands of the New Zealand Electricity Division, 220 kV a.c. system. At present converters 1, 2 and 3 are in operation. Converters 1 and 2 form the 600 MW, 500 kV d.c. link between the two islands. Converter 3 represents a 420 MW aluminium smelter. A further three-terminal d.c. interconnection has been added. (Converters 4, 5 and 6) to illustrate the flexibility of the algorithm.

Normally, Converter 4 will operate in the rectifier mode with converters 5 and 6 in the inversion mode.

The reactive power-d.c. Jacobian for the unified method has the following structure,

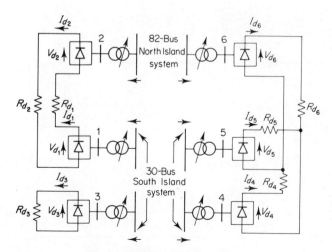

Fig. 6.5. Multiterminal d.c. system

$$
\begin{pmatrix}
\Delta \bar{Q}/\bar{V} \\[2pt]
\dfrac{\Delta Q_{\text{term }1}}{V_{\text{term }1}} \\
\dfrac{\Delta Q_{\text{term }2}}{V_{\text{term }2}} \\
\dfrac{\Delta Q_{\text{term }3}}{V_{\text{term }3}} \\
\dfrac{\Delta Q_{\text{term }4}}{V_{\text{term }4}} \\
\dfrac{\Delta Q_{\text{term }5}}{V_{\text{term }5}} \\
\dfrac{\Delta Q_{\text{term }6}}{V_{\text{term }6}} \\
\Delta \bar{R}_1 \\
\Delta \bar{R}_2 \\
\Delta \bar{R}_3 \\
\Delta \bar{R}_4 \\
\Delta \bar{R}_5 \\
\Delta \bar{R}_6
\end{pmatrix}
=
B''
\begin{pmatrix}
\Delta \bar{V} \\
\Delta V_{\text{term }1} \\
\Delta V_{\text{term }2} \\
\Delta V_{\text{term }3} \\
\Delta V_{\text{term }4} \\
\Delta V_{\text{term }5} \\
\Delta V_{\text{term }6} \\
\Delta \bar{X}_1 \\
\Delta \bar{X}_2 \\
\Delta \bar{X}_3 \\
\Delta \bar{X}_4 \\
\Delta \bar{X}_5 \\
\Delta \bar{X}_6
\end{pmatrix}
$$

B'' matrix blocks:

Upper-left block: B''_{MOD} with diagonal $AA''_{11},\ AA''_{22},\ AA''_{33},\ AA''_{44},\ AA''_{55},\ AA''_{66}$

Lower-left block: diagonal $BB''_{11},\ BB''_{22},\ BB''_{33},\ BB''_{44},\ BB''_{55},\ BB''_{66}$

Lower-right block: A (30×30)

where B''_{MOD} is the part of B'' which becomes modified. Only the diagonal elements become modified by the presence of the converters.

Off diagonal elements will be present in B''_{MOD} if there is any a.c. connection between converter terminal busbars. All off diagonal elements of BB'' and AA'' are zero.

In addition, matrix A is block diagonal in 5×5 blocks with the exception of the d.c. interconnection equations.

Equation R(3) of (6.3.11) in each set of d.c. equations is derived from the d.c. interconnection. For the six-converter system shown in Fig. 6.5 the following equations are applicable.

$$V_{d1} + V_{d2} - I_{d1}(R_{d1} + R_{d2}) = 0$$

$$V_{d3} - I_{d3} \cdot R_{d3} = 0$$

$$I_{d1} - I_{d2} = 0$$

$$V_{d4} + V_{d6} - I_{d4}R_{d4} - I_{d6}R_{d6} = 0$$

$$V_{d5} - V_{d6} - I_{d5} \cdot R_{d5} + I_{d6}R_{d6} = 0$$

$$I_{d4} - I_{d5} - I_{d6} = 0$$

This example indicates the ease of extension to the multiple converter case.

6.7. D.C. CONVERGENCE TOLERANCE

The d.c. p.u. system is based upon the same power base as the a.c. system and on the nominal open circuit a.c. voltage at the converter transformer secondary. The p.u. tolerances for d.c. powers, voltages and currents are therefore comparable with those adopted in the a.c. system.

In general, the control equations are of the form

$$\bar{X} - \bar{X}^{sp} = 0$$

where X may be the tap or cosine of the firing angle, i.e. they are linear and are thus solved in one d.c. iteration. The question of an appropriate tolerance for these mismatches is therefore irrelevant.

An acceptable tolerance for the d.c. residuals which is compatible with the a.c. system tolerance is typically 0.001 p.u. on a 100 MVA base, i.e. the same as that normally adopted for the a.c. system.

6.8. TEST SYSTEM AND RESULTS

The A.E.P. standard 14-bus test system is used to show the convergence properties of the a.c.–d.c. algorithms, with the a.c. transmission line between busbars 5 and 4 replaced by a h.v.d.c. link. As these two buses are not voltage controlled, the interaction between the a.c. and d.c. systems will therefore be considerable.

Table 6.1. Convergence results

Case specification		Number of iterations to convergence (0.1 MW/MVAR)					
	Specified d.c. link constraints m-rectifier end n-inverter end	Unified methods (5 variables)		Sequential methods			
				(5 variables)		(4 variables)	
		$1_{PDC,QDC}$	$2_{P,QDC}$	$1_{P,Q,DC}$	$2_{P,DC,Q,DC}$	$1_{P,Q,DC}$	$2_{P,DC,Q,DC}$
1	$\alpha_m\,P_{dm}\,\gamma_n\,V_{dn}$	4,3	4,3	4,3	4,3	4,4	4,3
2	$\alpha_m\,P_{dm}\,a_n\,V_{dn}$	4,3	4,3	4,4	5,5	4,4	Failed
3	$a_m\,P_{dm}\,a_n\,V_{dn}$	4,3	4,3	4,4	5,5	4,4	Failed
4	$a_m\,P_{dm}\,\gamma_n\,V_{dn}$	4,3	4,3	4,4	4,4	4,4	4,4
5	$a_m\,P_{dm}\,\gamma_n\,a_n$	4,3	4,3	4,4	4,4	4,4	4,4
6	$a_m\,P_{dm}\,\alpha_m\,\gamma_n$	4,3	4,3	4,3	4,3	4,4	4,3
7	$\alpha_m\,I_d\,\gamma_n\,V_{dn}$	4,3	4,3	4,3	4,3	4,4	4,3
8	$a_m\,V_{dm}\,\gamma_n\,P_{dn}$	4,3	4,3	4,4	4,4	4,4	4,4
	Case 1 with initial condition errors						
9	50 per cent error	4,3	4,3	4,4	4,3	4,4	4,3
10	80 per cent error	5,4*	6,5*	7,6*	5,4*	4,4	4,3

Where * indicates a false solution.

Table 6.2. Characteristics of d.c. link

	Converter 1	Converter 2
A.c. Busbar	Bus 5	Bus 4
D.c. voltage base	100 kV	100 kV
Transformer reactance	0.126	0.0728
Commutation reactance	0.126	0.0728
Filter admittance B_f^*	0.478	0.629
D.c. link resistance	0.334 ohms	
Control parameters for Case 1		
D.c. link power	58.6 MW	—
Rectifier firing angle (deg)	7	—
Inverter extinction angle (deg)	—	10
Inverter d.c. voltage	—	−128.87 kV

* Filters are connected to a.c. terminal busbar.

Note: All reactances are in p.u. on a 100 MVA base.

Various control strategies have been applied to the link and the convergence results for the various algorithms are given in Table 6.1. The number of iterations (i,j) should be interpreted as follows:

— i is the number of reactive power-voltage updates required.
— j is the number of real power-angle updates.

Although the number of d.c. iterations varies for the different sequences, this is of secondary importance and may if required be assessed in each case from the number of a.c. iterations. In this respect, a unified QDC iteration is equivalent to a Q iteration and a DC iteration executed separately.

The d.c. link data and specified controls for Case 1 are given in Table 6.2 and the corresponding d.c. link operation is illustrated in Fig. 6.6. The specified conditions for all cases are derived from the results of Case 1. Under those conditions, the a.c. system in isolation, (with each converter terminal modelled as an equivalent a.c. load) requires (4, 3) iterations. The d.c. system in isolation (operating from fixed terminal voltages) requires two iterations under all control strategies.

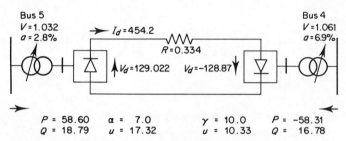

Bus 5
$V = 1.032$
$a = 2.8\%$

$I_d = 454.2$

$R = 0.334$

$V_d = 129.022$ $V_d = -128.87$

Bus 4
$V = 1.061$
$a = 6.9\%$

| $P = 58.60$ | $\alpha = 7.0$ | $\gamma = 10.0$ | $P = -58.31$ |
| $Q = 18.79$ | $u = 17.32$ | $u = 10.33$ | $Q = 16.78$ |

All angles are in degrees. D.C. voltages and current are in kV and Amp respectively. D.C. resistance is in ohms. A.C. powers (P, Q) are in MW and MVARs.

Fig. 6.6. D.c. link operation for Case 1

Unified cases

The results in Table 6.1 show that the unified methods provide fast and reliable convergence in all cases.

For the unified methods 1 and 2 the number of iterations did not exceed the number required for the a.c. system alone.

Sequential cases

The sequential method (P, Q, DC) produces fast and reliable convergence although the reactive power convergence is slower than for the a.c. system alone.

With the removal of the variable ϕ, Q_{term} (dc) converges faster but the convergence pattern is more oscillatory and an overall deterioration of a.c. voltage convergence results.

With the second sequential method, (P, DC, Q, DC) convergence is good in all cases except 2 and 3, i.e. the cases where the transformer tap and d.c. voltage are specified at the inverter end. However, this set of specifications is not likely to occur in practice.

Initial conditions for d.c. system

Initial values for the d.c. variables $\bar{x}$ are assigned from estimates for the d.c. power and d.c. voltage and assuming a power factor of 0.9 at the converter terminal busbar. The terminal busbar voltage is set at 1.0 p.u. unless it is a voltage controlled busbar.

144

This procedure gives adequate initial conditions in all practical cases as good estimates of P_{term} (dc) and V_d are normally obtainable.

With starting values for d.c. real and reactive powers within ± 50 per cent, which are available in all practical situations, all algorithms converged rapidly and reliably (see Case 9).

Effect of a.c. system strength

In order to investigate the performance of the algorithms with a weak a.c. system, the test system described earlier is modified by the addition of two a.c. lines as shown in Fig. 6.7.

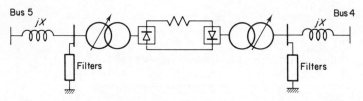

Fig. 6.7. D.c. link operating from weak a.c. system

Table 6.3. Numbers of iterations for weak a.c. systems

Case specification	$x_l = 0.3$			$x_l = 0.4$		
		Sequential P, Q, DC			Sequential P, Q, DC	
m—rectifier n—inverter	Unified P, QDC	(i)	(ii)	Unified P, QDC	(i)	(ii)
11 $\alpha_m P_{dm} \gamma_n V_{dn}$	4, 4	4, 4	4, 4	4, 4	5, 4	4, 4
12 $\alpha_m P_{dm} a_n V_{dn}$	4, 4	9, 8	10, 12	4, 4	> 30	Diverges
13 $a_m P_{dm} a_n V_{dn}$	4, 3	9, 8	10, 12	4, 3	> 30	Diverges
14 $a_m P_{dm} \gamma_n V_{dn}$	4, 3	6, 5	7, 7	4, 3	28, 27	> 30

(i) using the five variable formulation; (ii) using the four variable formulation.

The reactive power compensation of the filters was adjusted to give similar d.c. operating conditions as previously.

The number of iterations to convergence for the most promising algorithms are shown in Table 6.3 for the control specifications corresponding to Cases 1 to 4 in the previous results.

The different nature of the sequential and unified algorithms is clearly demonstrated. The effect of the type of converter control is also shown. For Case 11, both the d.c. real power and the d.c. reactive powers are well constrained by the converter control strategy. Convergence is rapid and reliable for all methods.

In all other cases, where the control angle at one or both converters is free, an oscillatory relationship between converter a.c. terminal voltage and the reactive power of the converter is possible. This leads to poor performance of the sequential algorithms.

To illustrate the nature of the iteration, the convergence pattern of the converter reactive power demand and the a.c. system terminal voltage of the rectifier is plotted in Fig. 6.8.

A measure of the strength of a system in a load-flow sense is the short circuit to converter power ratio (SCR) calculated with all machine reactances set to zero. This short-circuit ratio is invariably much higher than the usual value.

In practice, converter operation has been considered down to a SCR of 3. A survey of existing schemes shows that, almost invariably, with systems of very low SCR, some form of voltage control, often synchronous condensers, is an integral part of the converter installation. These schemes are therefore often very strong in a load-flow sense.

It may therefore be concluded that the sequential integration should converge in all practical situations although the convergence may become slow if the system is weak in a load-flow sense.

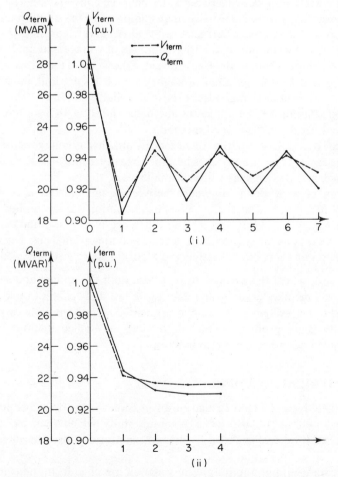

Fig. 6.8. Convergence pattern for a.c.–d.c. load flow with weak a.c. system. (i) Sequential method (P, Q, DC five variable); (ii) Unified method (P, QDC)

Discussion of convergence properties

The overall convergence rate of the a.c.–d.c. algorithms depends on the successful interaction of the two distinct parts. The a.c. system equations are solved using the well-behaved constant tangent fast decoupled algorithm, whereas the d.c. system equations are solved using the more powerful, but somewhat more erratic, full Newton–Raphson approach.

The powerful convergence of the Newton–Raphson process for the d.c. equations can cause overall convergence difficulties. If the first d.c. iteration occurs before the reactive power-voltage update then the d.c. variables are converged to be compatible with the incorrect terminal voltage. This introduces an unnecessary discontinuity which may lead to convergence difficulties in the sequential method. In the unified approach the powerful convergence of the d.c. equations is dampened by the reflection of the a.c. mismatches onto the changes in d.c. variables. This gives faster and better behaved convergence. The solution time of the d.c. equations is normally small compared to the solution time of the a.c. equations. The relative efficiencies of the alternative algorithms may therefore be assessed by comparing the total numbers of voltage and angle updates.

In general, those schemes which acknowledge the fact that the d.c. variables are strongly related to the terminal voltage give the fastest and most reliable performance. The unified methods are the more reliable and the P, QDC solution is the most efficient. Of the sequential methods, the (P, Q, DC) solution is only marginally inferior to the unified method.

When the busbar to which the converters are attached is voltage controlled (as is often the case) the two approaches become virtually identical as the interaction between a.c. and d.c. systems is much smaller.

The only computational difference between a unified and a comparable sequential iteration is that the d.c. Jacobian equation for the unified method (equation 6.4.14) is slightly larger. The difference is one additional row and column for each converter present. In terms of computational cost per iteration the corresponding unified and sequential algorithms are virtually identical.

(i) In cases where the a.c. system is strong, both the unified and sequential algorithms may be programmed to give fast and reliable convergence.

(ii) If the a.c. system is weak the sequential algorithm is susceptible to convergence problems. Thus, in general, the unified method is recommended due to its greater reliability.

6.9. NUMERICAL EXAMPLE

The complete New Zealand Primary transmission system described in Appendix I was used as a basis for a planning study which included an extra multiterminal h.v.d.c. scheme, i.e. involving six converter stations as illustrated in Fig. 6.5.

Representative input and output information obtained from the computer is given in the following pages.

CASE STUDY NUMBER   3

MAXIMUM NUMBER OF ITERATIONS   10

POWER TOLERANCE   .00100

PRINT OUT INDICATOR   000000000

SYSTEM MVA BASE   100.00

D.C. LINK INDICATOR   6

NUMBER OF A.C. SYSTEMS   2

SLACK BUSBARS   80   218

NUMBER OF BUSES   114
NUMBER OF LINES   206
NO OF TRANSFORMERS   19

B U S - D A T A

BUS	NAME	TYPE	VOLTS	LOAD MW	LOAD MVAR	GENERATION MW	GENERATION MVAR	MINIMUM MVAR	MAXIMUM MVAR	SHUNT SUSCEPTANCE
104	AVIEMORE220	1	1.0520	9.00	0.000	220.000	-34.40	-500.000	500.000	0.000
108	BENMORE-220	0	1.0030	329.60	95.80	540.000	46.000	-500.000	500.000	0.000
127	BRLY----220	0	1.0030	0.000	0.000	0.000	0.000	0.000	0.000	0.000
128	CROM1---220	1	1.0520	0.000	0.000	0.000	0.000	0.000	0.000	0.000
129	CLUTHA--220	0	1.0300	0.000	0.000	600.000	0.000	0.000	0.000	0.000
138	GERALDINE20	0	1.0210	0.000	0.000	0.000	0.000	0.000	0.000	0.000
143	HWRS----220	1	1.0270	95.30	80.40	0.000	0.000	0.000	0.000	0.000
167	INVERCARG220	0	1.0050	183.20	40.40	0.000	0.000	0.000	0.000	0.000

L I N E   D A T A

BUS	NAME	BUS	NAME	RESISTANCE	REACTANCE	SUSCEPTANCE
104	AVIEMORE-220	108	BENMORE--220	0.00330	0.01530	0.02298
104	AVIEMORE-220	108	BENMORE--220	0.00330	0.01530	0.02298
108	BENMORE--220	255	WAITAKI--220	0.00370	0.00730	0.01954
108	BENMORE--220	167	TWIZEL---220	0.00210	0.02611	0.05285
118	BRLY----220	104	ISLINGTON220	0.00770	0.00861	0.02751
127	CROM1---220	255	LAND102-220	0.00770	0.04450	0.16746
128	CROM2---220	218	ROXBURGH-220	0.00070	0.09260	0.07251
127	CROM2---220	255	TWIZEL---220	0.00070	0.04450	0.07251

TRANSFORMER - DATA

BUS	NAME	BUS	NAME	RESISTANCE	REACTANCE	TAP	CODE
6	BUNTHORPE110	7	BUNTHORPE220	0.00400	0.09560	1.000	0
6	BUNTHORPE110	7	BUNTHORPE220	0.00400	0.09560	1.000	0
6	BUNTHORPE110	7	BUNTHORPE220	0.00170	0.04590	1.000	0
100	EDGECOMBE110	11	EDGECOMBE220	0.00400	0.09560	1.000	0
11	HAYWARDS-110	22	HAYWARDS-220	0.00170	0.05140	1.000	0
22	HAYWARDS-110	22	HAYWARDS-220	0.00170	0.05120	1.000	0
221	HAYWARDS-110	22	HAYWARDS-220	0.00410	0.10120	1.000	0
223	HENDERSON110	24	HENDERSON220	0.00430	0.10140	1.050	0
39	MARSDEN--110	40	MARSDEN--220	0.00080	0.05550	1.000	0
48	NEWPLYMTH110	49	NEWPLYMTH220	0.00070	0.05240	1.000	0
54	OTAHUHU--110	55	OTAHUHU--220	0.00070	0.04150	1.000	0
54	OTAHUHU--110	55	OTAHUHU--220	0.00070	0.04570	1.000	0
54	OTAHUHU--110	55	OTAHUHU--220	0.00160	0.04570	1.000	0
58	PENROSE--110	59	PENROSE--220	0.00200	0.05290	1.000	0
62	STRATFORD110	63	STRATFORD220	0.00160	0.05290	1.000	0
66	TARUKENGA110	67	TARUKENGA220	0.00080	0.02530	1.000	0

DC SYSTEM EQUATIONS

```
VD1+VD2+VD3+VD4+VD5+VD6+VD7+VD8+VD9+VD10-ID1.RD1-ID2.RD2-ID3.RD3-ID4.RD4-ID5.RD5-ID6.RD6-ID7.RD7-ID8.RD8-ID9.RD9-ID10.RD10=0

 1  1  0  0  0  0  0   25.5600  0.0000  0.0000  0.0000  0.0000  0.0000  0.0000  0.0000  0.0000  0.0000
 0  1  0  1  0  0  0    0.0000  0.0019  0.0000  0.0000  0.0000  0.0000  0.0000  0.0000  0.0000  0.0000
 0  0  0  1  0  1  0    0.0000  0.0000  0.0000 10.0000  0.0000 20.0000  0.0000  0.0000  0.0000  0.0000
 0  0  0  1 -1  0  0    0.0000  0.0000  3.0000-20.0000  0.0000  0.0000  0.0000  0.0000  0.0000  0.0000

ID1+ID2+ID3+ID4+ID5+ID6+ID7+ID8+ID9+ID10=0

 1 -1  0  0  0  0  0

ID1+ID2+ID3+ID4+ID5+ID6+ID7+ID8+ID9+ID10=0

 0  0  1 -1  0  0  0
```

SOLUTION CONVERGED IN 7 P-D AND 6 Q-V ITERATIONS

MW LOAD MVAR		GENERATION MW MVAR		AC LOSSES MW MVAR		MISMATCH MW MVAR		SHUNTS MVAR
4496.80	1518.60	5226.58	791.80	194.91	-306.06	534.87	-182.89	37.85

OPERATING STATE OF CONVERTER 5 WHICH IS ATTACHED TO BUS 167 I.E., ISLINGTON220
===

CONVERTER IS OPERATING IN THE INVERTION MODE
THE CONTROL ANGLE IS THE EXTINCTION ADVANCE ANGLE
DC POWER SUPPLIED TO THE AC SYSTEM = 200.00 MW

CONVERTER AC VOLTAGE (K-VOLTS)	TRANSFORMER TAP (PER CENT)	CONTROL ANGLE (DEGS)	COMMUTATION ANGLE (DEGS)	DC CURRENT (K-AMPS)	DC VOLTAGE (K-VOLTS)
97.08	-5.46	8.00	19.93	0.81	-245.67

POWER TRANSFERS

LINK TERMINAL POWER	=	-200.00 MW	110.89 MVAR
FROM TRANSFORMER TO CONVERTER	=	-200.00 MW	74.63 MVAR
REACTIVE POWER OF FILTERS	=		120.83 MVAR

BUSBAR DATA

BUS	NAME	VOLTS	ANGLE	GENERATION MW	MVAR	LOAD MW	MVAR	SHUNT MVAR
104	AVIEMORE-220	1.052	4.78	220.00	-33.87	0.00	0.00	0.00
108	BENMORE--220	1.052	4.43	540.00	-88.04	97.20	0.00	0.00
118	BRLY----220	0.966	-12.95	0.00	0.00	329.60	95.80	0.00

LINE AND TRANSFORMER DATA

BUS	NAME	MW	MVAR
108	BENMORE--220	41.89	-10.17
108	BENMORE--220	41.89	-10.17
268	WAITAKI--220	136.22	-13.53
	MISMATCH	0.000	0.000
104	AVIEMORE-220	-41.83	7.88
104	AVIEMORE-220	-41.83	7.88
255	TWIZEL---220	-26.47	6.87
	MISMATCH	500.000	-110.672
167	ISLINGTON220	-120.18	-76.16
181	LAND-TO2-220	-20.40	-19.64
	MISMATCH	-0.019	-0.002

	DC CONVERTER NUMBER 1 INPUT DATA	DC CONVERTER NUMBER 6 INPUT DATA
	CONVERTER ATTACHED TO BUS NUMBER108	CONVERTER ATTACHED TO BUS NUMBER 7
NOMINAL DC VOLTAGE	110.00000	99.00000
MAXIMUM DC VOLTAGE	150.00000	140.00000
MINIMUM DC VOLTAGE	0.00000	0.00000
MAXIMUM DC CURRENT	-0.0000	0.0000
COMMUTATION REACTANCE(P.U.)	0.08970	0.07000
TRANSFORMER REACTANCE(P.U.)	0.08970	0.07000
FIRING ANGLE:MINIMUM(DEG)	10.00000	8.00000
MAXIMUM(DEG)	110.00000	150.00000
TRANSFORMER TAP:MINIMUM(P.C.)	0.00000	0.00000
MAXIMUM(P.C.)	0.00000	0.00000
INCREMENT	0.00000	0.00000
FILTER REACTANCE(P.U.)	1.00000	0.70000
NUMBER OF BRIDGES IN SERIES.	4	2
SPECIFIED CONTROLS =============		
CONVERTER POWER FACTOR		
DC LINK VOLTAGE(KV)		-220.00000
CONVERTER CONTROL ANGLE	10.00000	8.00000
TRANSFORMER TAP		
CONVERTER D.C.POWER (MW)	500.00000	
TERMINAL REACTIVE POWER (MVAR)		
AC TERMINAL VOLTAGE (KV)		
CONVERTER DC CURRENT (KA)		

6.10 REFERENCES

1. B. Scott and O. Alsac, 1974. 'Fast decoupled load flow', *Trans. IEEE*, **PAS-93** (3), 859–869.
2. H. Sato and J. Arrillaga, 1969. 'Improved load-flow techniques for integrated a.c.–d.c. systems', *Proc. IEE*, **116** (4), 525–532.
3. J. Reeve, G. Fahmy, and B. Stott, 1976. 'Versatile load-flow method for multi-terminal h.v.d.c. systems', Paper F76-354-1. IEEE PES Summer Meeting, Portland,
4. B. Stott, 'Load flow for a.c. and integrated a.c.–d.c. systems.' Ph.D. Thesis, University of Manchester, 1971.
5. D. A. Braunayal, L. A. Kraft, and J. L. Whysong, 1976. 'Inclusion of the converter and transmission equations directly in a Newton power flow', *IEEE Trans.* **PAS-75** (1), 76–88.
6. J. Arrillaga, B. J. Harker, K. S. and Turner, 1980. 'Clarifying an ambiguity in recent a.c.–d.c. load-flow formulations', *Proc. IEE*, **127**, Pt. C (5), 324–325.
7. P. Bodger, 1977. 'Fast decoupled a.c. and a.c.–d.c. load flows.' Ph.D. Thesis, University of Canterbury, New Zealand.

Three-Phase A.C.–D.C. Load Flow

7.1. INTRODUCTION

Any converter which is operating from an unbalanced a.c. system will itself operate with unbalanced power flows and unsymmetric valve conduction periods. In addition any unbalance present in the converter control equipment or any asymmetry in the converter transformer will introduce additional unbalance.

Considerable interaction exists between the unbalanced operation of the a.c. and d.c. systems. The exact nature of this interaction depends on features such as the converter transformer connection and the converter firing controller.

High-power converters often operate in systems of relatively low short-circuit ratios where unbalance effects are more likely to be significant and require additional consideration. The steady-state unbalance and its effect in converter harmonic current generation may also influence the need for transmission line transpositions and the means of reactive power compensation.

The converter model for unbalanced analysis is considerably more complex than those developed for the balanced case in Chapters 3 and 6. The additional complexity arises from the need to include the effect of the three-phase converter transformer connection and of the different converter firing control modes. Early h.v.d.c. control schemes were based on phase angle control, where the firing of each valve is timed individually with respect to the appropriate crossing of the phase voltages. This control scheme has proved susceptible to harmonic stability problems when operating from weak a.c. systems. An alternative control, based on equidistant firings on the steady state, is generally accepted to provide more stable operation, [1],[2],[3] Under normal steady-state and perfectly-balanced operating conditions, there is no difference between these two basic control strategies. However, their effect on the a.c. system and d.c. voltage and current waveshapes during normal, but not balanced, operation, is quite different. A three-phase converter model must be capable of representing the alternative control strategies.

Similar techniques are available for the integration of the three-phase converter model into the load-flow analysis as were discussed with respect to the balanced single-phase analysis. Based upon the extensive investigations into the behaviour of single-phase a.c.–d.c. load flow described in Chapter 6, the sequential approach is considered the most appropriate for the integration of the d.c. model into the three-phase load flow. The complexity of the unified approach is not considered justified in the three-phase case because, in cases of difficult convergence such as those involving very weak a.c. systems, it is possible to use starting values derived from single-phase analysis.

This chapter describes the development of a model for the unbalanced converter and its sequential integration with the three-phase fast decoupled load flow described in Chapter 5.

7.2. FORMULATION OF THE THREE-PHASE A.C.–D.C. LOAD-FLOW PROBLEM

The operating state of the combined system is defined by:

$$[\bar{V}_{int}, \bar{\theta}_{int}, \bar{V}, \bar{\theta}, \bar{x}]$$

where:

$\bar{V}_{int}/\bar{\theta}_{int}$ are vectors of the balanced internal voltages at the generator internal busbars;

$\bar{V}/\bar{\theta}$ are vectors of the three-phase voltages at every generator terminal busbar and every load busbar;

$\bar{x}$ is a vector of the d.c. variables (as yet, undefined).

The significance of the three-phase a.c. variables was discussed in Chapter 5 and the selection of d.c. variables $\bar{x}$ is discussed in Section 7.3.

To enable a Newton–Raphson based technique to be used, it is necessary to formulate a set of n independent equations in terms of the n variables describing the system. As explained in Chapter 5, the equations which relate to the a.c. system variables are derived from the specified a.c. system operating conditions. The only modification to these equations, which results from the presence of the d.c. system, occurs at the converter terminal busbars. These equations become:

$$\Delta P_{term}^p = (P_{term}^p)^{SP} - P_{term}^p(ac) - P_{term}^p(dc) \tag{7.2.1}$$

$$\Delta Q_{term}^p = (Q_{term}^p)^{SP} - Q_{term}^p(ac) - Q_{term}^p(dc) \tag{7.2.2}$$

where $P_{term}^p(dc)$ and $Q_{term}^p(dc)$ are function of the a.c. terminal conditions and the converter variables, i.e.

$$P_{term}^p(dc) = f(V_{term}^p, \theta_{term}^p, \bar{x}) \tag{7.2.3}$$

$$Q_{term}^p(dc) f(V_{term}^p, \theta_{term}^p, \bar{x}) \tag{7.2.4}$$

The equations for the a.c. system may therefore be summarized as,

$$
\begin{bmatrix}
\Delta \bar{P}(\bar{V}, \bar{\theta}) \\
\Delta \bar{P}_{\text{term}}(\bar{V}, \bar{\theta}, \bar{x}) \\
\Delta \bar{P}_{\text{gen}}(\bar{V}, \bar{\theta}) \\
\Delta \bar{Q}(\bar{V}, \bar{\theta}) \\
\Delta \bar{Q}_{\text{term}}(\bar{V}, \bar{\theta}, \bar{x}) \\
\Delta \bar{V}_{\text{reg}}(\bar{V})
\end{bmatrix} = 0
\tag{7.2.5}
$$

where the mismatches at the converter terminal busbars are indicated separately. Further equations are derived from the d.c. system conditions.

That is, for each converter k a set of equations

$$
\bar{R}(V^p_{\text{term}}, \theta^p_{\text{term}}, \bar{x})_k = 0
\tag{7.2.6}
$$

is derived in terms of the terminal conditions and the converter variables $\bar{x}$.

Equations (7.2.3), (7.2.4) and (7.2.6) form a mathematical model of the d.c. system suitable for inclusion into load-flow analysis.

The three-phase a.c.–d.c. load-flow problem may therefore be formulated as the solution of,

$$
\begin{bmatrix}
\Delta \bar{P}(\bar{V}, \bar{\theta}) \\
\Delta \bar{P}_{\text{term}}(\bar{V}, \bar{\theta}, \bar{x}) \\
\Delta \bar{P}_{\text{gen}}(\bar{V}, \bar{\theta}) \\
\Delta \bar{Q}(\bar{V}, \bar{\theta}) \\
\Delta \bar{Q}_{\text{term}}(\bar{V}, \bar{\theta}, \bar{x}) \\
\Delta \bar{V}_{\text{reg}}(\bar{V}) \\
R(\bar{V}_{\text{term}}, \bar{\theta}_{\text{term}}, \bar{x})
\end{bmatrix}
\tag{7.2.7}
$$

for the set of variables $(\bar{V}, \bar{\theta}, \bar{x})$.

7.3. D.C.-SYSTEM MODELLING

The basic h.v.d.c. interconnection shown in Fig. 7.1 is used as a reference and its extension to other configurations is clarified throughout the development of the model. Under balanced conditions, the converter transformer modifies the source voltages applied to the converter and also affects the phase distribution of current and power. In addition, the a.c. system operation may be influenced (e.g. by a zero-sequence current flow to a star-g–delta transformer) by the transformer connection. Each bridge in Fig. 7.1 will thus operate with a different degree of unbalance, due to the influence of the converter transformer connections, and must be modelled independently. This feature is in contrast to the balanced d.c. model where it is possible to combine bridges in series and in parallel into an

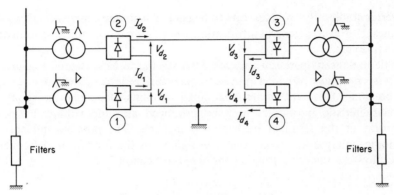

Fig. 7.1. Basic h.v.d.c. interconnection

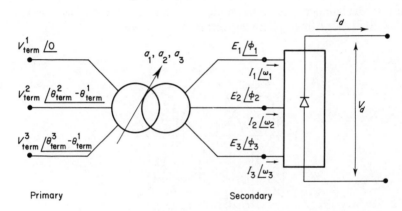

Fig. 7.2. Basic converter unit

equivalent single bridge. The dimensions of the three-phase d.c. model, will, therefore, be much greater than the balanced d.c. model.

All converters, whether rectifying or inverting, are represented by the same model (Fig. 7.2) and their equations are of the same form.

Basic assumptions

To enable the formulation of equation (7.2.6) and to simplify the selection of variables $\bar{x}$ the following assumptions are made:

   (i) The three a.c. phase voltages at the terminal busbar are sinusoidal.
  (ii) The direct voltage and direct current are smooth.
 (iii) The converter transformer is lossless and the magnetizing admittance is ignored.

Assumptions (ii) and (iii) are equally valid for unbalanced three-phase analysis as for single-phase analysis. Assumption (i) is commonly used in unbalanced converter studies [4],[5] and appears to be backed from the experience of existing schemes. However, a general justification will require more critical examination of the problem.

Under balanced operation only characteristic harmonics are produced and, as filtering is normally provided at these frequencies, the level of harmonic voltages will be small. However, under even small amounts of unbalance, significant noncharacteristic harmonics may be produced and the voltage harmonic distortion at the terminal busbars will increase. The possible influence of harmonic voltages at the converter terminal, on the fundamental frequency power flows to the converter, is considered in Section 7.7.

Selection of converter variables

The selection of converter variables was discussed in detail in Chapter 6, with regard to the balanced converter model. The main considerations, also relevant to the unbalanced three-phase converter model, were:

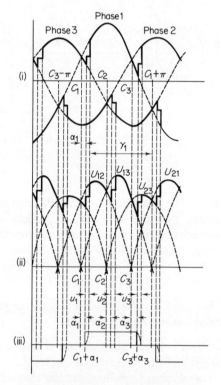

Fig. 7.3. Unbalanced converter voltage and current waveform. (i) Phase voltages; (ii) D.c. voltage waveform; (iii) Assumed current waveshape for Phase 1 (actual waveform is indicated by dotted line)

(i) For computing efficiency, the smallest number of variables should be used. A minimum of six independent variables is required to define the operating state of an unbalanced converter, e.g. the three firing angles and the three transformer tap positions.

(ii) To enable the incorporation of a wide range of control specifications, all variables involved in their formulation should be retained. The following variables, defined with reference to Fig. 7.2 and 7.3, are required in the formulation of the control specifications for unbalanced converter operation:

a_i Off-nominal tap ratios on the primary side.

$U_{13}\underline{/C_1}, U_{23}\underline{/C_2}, U_{21}\underline{/C_3}$
    Phase-to-phase source voltages for the converter referred to the transformer secondary. C_i are therefore the zero crossings for the timing of firing pulses.

α_i Firing delay angle measured from the respective zero crossing.

V_d Total average d.c. voltage from complete bridge.

I_d Average d.c. current.

where $i = 1, 2\,3$ for the three phases involved.

In contrast to the balanced case, the secondary phase to phase source voltages are included among the variables as they depend not only on the transformer taps but also on the transformer connection. Moreover, the zero crossings, C_i are explicitly required in the formulation of the symmetrical firing controller and they are also included.

Although these fourteen variables do not constitute the final d.c. model it is convenient to formulate equation (7.2.6) in terms of these variables at this stage, i.e. vector $\bar{x}$ has the form:

$$[U_i, C_i, \alpha_i, a_i, V_d, I_d]^T$$

The necessary fourteen equations are derived in the following section.

Converter angle reference

In the three-phase a.c. load flow described in Chapter 5 all angles are referred to the slack generators internal busbar. Similarly to the single-phase a.c.–d.c. load flow (Chapter 6) the angle reference for each converter may be arbitrarily assigned. By using one of the converter angles (e.g. θ^1_{term} in Fig. 7.2) as a reference, the mathematical coupling between the a.c. system and converter equations is weakened and the rate of convergence improved.

Per unit system

Similarly to the single-phase case, computational simplicity is achieved by using common power and voltage bases on both sides of the converter.

In the three-phase case, however, the phase-neutral voltage is used as the base parameter and therefore

$$\text{MVA}_{\text{base}} = \text{Base Power per phase}$$
$$V_{\text{base}} = \text{Phase-neutral voltage base}$$

The current base on the a.c. and d.c. sides are also equal. Therefore the p.u. system does not change the form of any of the converter equations.

Converter source voltages

The phase-to-phase source voltages referred to the transformer secondary are found by a consideration of the transformer connection and off-nominal turns ratio. For example, consider the star–star transformer of Fig. 7.4.

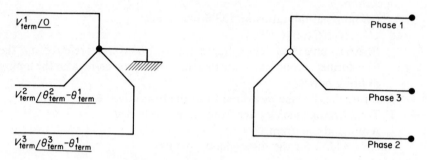

Fig. 7.4. Star–Star transformer connection

The phase-to-phase source voltages referred to the secondary are:

$$U_{13}\underline{/C_1} = \frac{1}{a_1}V^1_{term}\underline{/0} - \frac{1}{a_3}V^3_{term}\underline{/\theta^3_{term} - \theta^1_{term}} \tag{7.3.1}$$

$$U_{23}\underline{/C_2} = \frac{1}{a_2}V^2_{term}\underline{/\theta^2_{term} - \theta^1_{term}} - \frac{1}{a_3}V^3_{term}\underline{/\theta^3_{term} - \theta^1_{term}} \tag{7.3.2}$$

$$U_{21}\underline{/C_3} = \frac{1}{a_2}V^2_{term}\underline{/\theta^2_{term} - \theta^1_{term}} - \frac{1}{a_1}V^1_{term}\underline{/0} \tag{7.3.3}$$

which in terms of real and imaginary parts yield six equations.

D.C. voltage

The d.c. voltage, found by integration of the waveforms in Fig. 7.3(ii), may be expressed in the form:

$$V_d = \frac{\sqrt{2}}{\pi}\{U_{21}[\cos(C_1 + \alpha_1 - C_3 + \pi) - \cos(C_2 + \alpha_2 - C_3 + \pi)]$$

$$+ U_{13}[\cos(C_2 + \alpha_2 - C_1) - \cos(C_3 + \alpha_3 - C_1)]$$

$$+ U_{23}[\cos(C_3 + \alpha_3 - C_2) - \cos(C_1 + \alpha_1 + \pi - C_2)]$$

$$- I_d(X_{c1} + X_{c2} + X_{c3})\} \tag{7.3.4}$$

where X_{ci} is the commutation reactance for phase i.

D.C. interconnection

An equation is derived for each converter, from the d.c. system topology relating the d.c. voltages and currents, i.e.

$$f(V_d, I_d) = 0 \qquad (7.3.5)$$

For example, the system shown in Fig. 7.1 provides the following four equations:

$$Vd_1 + Vd_2 + Vd_3 + Vd_4 - Id_1 \cdot Rd = 0$$

$$Id_1 - Id_2 = 0$$

$$Id_1 - Id_3 = 0$$

$$Id_1 - Id_4 = 0$$

The apparent redundancy in the number of d.c. variables is due to the generality of the d.c. interconnection.

Incorporation of control strategies

Similarly to the single-phase case, any function of the variables is a valid (mathematically) control equation so long as the equation is independent of all the others.

Detailed consideration of the alternative firing controls is of particular interest in this respect.

With reference to symmetrical firing control, one equation results from the specification of minimum firing angle control, i.e.

$$\alpha_i - \alpha_{min} = 0$$

For a six-pulse unit, the interval between firing pulses in specified as 60°. This provides two more equations.

In the equation above, phase (i) is selected during the solution procedure such that the other two phases will have, in the unbalanced case, firing angles greater than α_{min}.

With conventional phase angle control, the firing angle on each phase is specified as being equal to α_{min}, i.e.

$$\alpha_1 - \alpha_{min} = 0 \qquad (7.3.6)$$

$$\alpha_2 - \alpha_{min} = 0 \qquad (7.3.7)$$

$$\alpha_3 - \alpha_{min} = 0 \qquad (7.3.8)$$

The remaining three-control equations required are derived from the operating conditions. Usually, the off-nominal taps are specified as being equal, i.e.

$$a_1 - a_2 = 0 \qquad (7.3.9)$$

$$a_2 - a_3 = 0 \qquad (7.3.10)$$

The final equation will normally relate to the constant current or constant power controller, e.g.

$$I_d - I_d^{sp} = 0 \tag{7.3.11}$$

or

$$V_d \cdot I_d - P_d^{sp} = 0 \tag{7.3.12}$$

Inverter operation with minimum extinction angle

In contrast to the single-phase load flow, for three-phase inverter operation it is necessary to retain the variable α in the formulation, as it is required in the specification of the symmetrical firing controller. Therefore, the restriction upon the extinction advance angle γ, requires the implicit calculation of the commutation angle for each phase.

Using the specification for γ, as defined in Fig. 7.3 the following expression applies:

$$\cos \gamma_1^{sp} + \cos \alpha_1 - I_d \frac{(X_{c1} + X_{c3})}{\sqrt{2} U_{13}} = 0 \tag{7.3.13}$$

Similar equations apply to the other two phases with a cyclic change of suffices.

Enlarged converter model

The three-phase equations so far developed are an exact parallel of the four variable sequential version which was discussed in Chapter 6.

The mathematical model of the converter includes the formulation of equations (7.2.3) and (7.2.4) for the individual phase real and reactive power flows on the primary of the converter transformer. It is in connection with these equations that the three-phase model deviates significantly from the single-phase model developed in Chapter 6.

The calculation of the individual phase, real and reactive powers at the terminal busbar requires the values of both the magnitude and angle of the fundamental components of the individual phase currents flowing into the converter transformer.

In the single-phase analysis, the magnitude of the fundamental current, obtained from the Fourier analysis of the current waveshape on the transformer secondary, was transferred across the converter transformer. This procedure is trivial and the relevant equations were not included in the d.c. solution. The angle of the fundamental component was calculated by simply equating the total real power on the a.c. and d.c. sides of the converter.

A similar procedure may be applied to the three-phase analysis of the unbalanced converter. In this case, however, the transfer of secondary currents to the primary is no longer a trivial procedure due to the influence of the three-phase transformer connection. In addition, the three-phase converter transformer may influence the a.c. system operation, for example, a star-g-delta connection provides a zero sequence path for the a.c. system.

The simplest way of accounting for such influence is to include the converter transformer within the d.c. model.

The three-phase converter transformer is represented by its nodal admittance model, i.e.

$$Y_{\text{node}} = \begin{array}{|c|c|} \hline Y_{pp} & Y_{ps} \\ \hline Y_{sp} & Y_{ss} \\ \hline \end{array}$$

(7.3.14)

where p indicates the primary side
and s the secondary side of the transformer

The 3×3 submatrices (Y_{pp}, etc.) for the various transformer connections, including modelling of the independent phase taps, were derived in Chapter 2.

The inclusion of the converter transformer within the d.c. model requires twelve extra variables, as follows:

$E_i \underline{/\phi_i}$ the fundamental component of the voltage waveshape at the transformer secondary busbar;

$I_i \underline{/\omega_i}$ the fundamental component of the secondary current waveshapes; where $i = 1, 3$ for the three phases.

Thus a total set of twenty-six variables is required for each converter in the d.c. system model, fourteen of which have already been developed in previous sections.

Remaining twelve equations

With reference to Equation (7.3.14), and on the assumption of a lossless transformer (i.e. $Y_{pp} = jb_{pp}$, etc.) the currents at the converter side busbar are expressed as follows:

$$I_i e^{j\omega_i} = - \sum_{k=1}^{3} \left[jb_{ss}^{ik} E_k e^{j\phi k} + jb_{sp}^{ik} V_{\text{term}}^{k} e^{j(\theta^{k}_{\text{term}} - \theta^{1}_{\text{term}})} \right]$$

(7.3.15)

By subtracting θ^{1}_{term} in the above equation, the terminal busbar angles are related to the converter angle reference.

Separating this equation into real and imaginary components, the following six equations result:

$$I_i \cos \omega_i = \sum_{k=1}^{3} \left[b_{ss}^{ik} E_k \sin \phi_k + b_{sp}^{ik} V_{\text{term}}^{k} \sin (\theta^{k}_{\text{term}} - \theta^{1}_{\text{term}}) \right]$$

(7.3.16)

$$I_i \sin \omega_i = \sum_{k=1}^{3} \left[- b_{ss}^{ik} E_k \cos \phi_k - b_{sp}^{ik} V_{\text{term}}^{k} \cos (\theta^{k}_{\text{term}} - \theta^{1}_{\text{term}}) \right]$$

(7.3.17)

Three further equations are derived from approximate expressions for the fundamental r.m.s. components of the line current waveforms as shown in Fig. 7.3, i.e.

$$I_i = 0.995 \frac{4 \cdot I_d}{\sqrt{2}} \sin (T_i/2)$$

(7.3.18)

where T_i is the assumed conduction period for phase i.

The sum of the real powers on the three phases of the transformer secondary may be equated to the total d.c. power, i.e.

$$\sum_{i=1}^{3} E_i \cdot I_i \cdot \cos(\phi_i - \omega_i) - V_d \cdot I_d = 0 \qquad (7.3.19)$$

The derivation of the last two equations is influenced by the position of the fundamental frequency voltage reference for the secondary of the converter transformer.

The voltage reference for the a.c. system is earth, while in d.c. transmission the actual earth is placed on one of the converter d.c. terminals. This point is used as a reference to define the d.c. transmission voltages and the insulation levels of the converter transformer secondary windings.

In load-flow analysis, it is possible to use arbitrary references for each converter unit to simplify the mathematical model. The actual voltages to earth, if required, can then be obtained from knowledge of the particular configuration and earthing arrangements.

With a star-connected secondary winding an obvious reference is the star point itself. If the nodal admittance matrix is formed for a star-g–star-g connection then this reference is implicitly present through the admittance model of the transformer. In this case, however, the converter transformer does not restrict the flow of zero-sequence currents and the following two equations may be written:

$$\sum_{i=1}^{3} I_i \underline{/\omega_i} = 0 \qquad (7.3.20)$$

These two equations (real and imaginary parts) complete the set of twelve independent equations in terms of the twelve additional variables.

However, the above considerations do not apply to delta-connected secondary windings.

To obtain a reference which may be applied to all transformer secondary windings, an artificial reference node is created corresponding to the position of the zero-sequence secondary voltage. This choice of reference results in the following two equations:

$$\sum_{i=1}^{3} E_i \cos \phi_i = 0 \qquad (7.3.21)$$

$$\sum_{i=1}^{3} E_i \sin \phi_i = 0 \qquad (7.3.22)$$

The nodal admittance matrix for the star-connected transformer secondary is now formed for an unearthed star winding. The restriction on the zero-sequence current flowing on the secondary is therefore implicitly included in the transformer model for both star and delta connections.

For a star-connected secondary winding both alternatives yield exactly the same solution to the load-flow problem.

Summary of equations and variables

The twenty-six equations ($\bar{R}$) which define the operation of each converter are:

$$R(1) = \sum_{i=1}^{3} E_i \cos \phi_i = 0$$

$$R(2) = \sum_{i=1}^{3} E_i \sin \phi_i = 0$$

$$R(3) = \sum_{i=1}^{3} E_i I_i \cos (\phi_i - \omega_i) - V_d \cdot I_d$$

$$R(4) = I_1 - \frac{4}{\pi} \cdot \frac{I_d}{\sqrt{2}} \sin (T_1/2)$$

$$R(5) = I_2 - \frac{4}{\pi} \cdot \frac{I_d}{\sqrt{2}} \sin (T_2/2)$$

$$R(6) = I_3 - \frac{4}{\pi} \cdot \frac{I_d}{\sqrt{2}} \sin (T_3/2)$$

$$R(7) = I_1 \cdot \cos \omega_1 - \sum_{k=1}^{3} [b_{ss}^{1k} E_k \sin \phi_k + b_{sp}^{1k} \cdot V_{\text{term}}^k \sin (\theta_{\text{term}}^k - \theta_{\text{term}}^1)]$$

$$R(8) = I_2 \cdot \cos \omega_2 - \sum_{k=1}^{3} [b_{ss}^{2k} E_k \sin \phi_k + b_{sp}^{2k} \cdot V_{\text{term}}^k \sin (\theta_{\text{term}}^k - \theta_{\text{term}}^1)]$$

$$R(9) = I_3 \cdot \cos \omega_3 - \sum_{k=1}^{3} [b_{ss}^{3k} E_k \sin \phi_k + b_{sp}^{3k} \cdot V_{\text{term}}^k \sin (\theta_{\text{term}}^k - \theta_{\text{term}}^1)]$$

$$R(10) = I_1 \cdot \sin \omega_1 + \sum_{k=1}^{3} [b_{ss}^{1k} E_k \cos \phi_k + b_{sp}^{1k} \cdot V_{\text{term}}^k \cos (\theta_{\text{term}}^k - \theta_{\text{term}}^1)]$$

$$R(11) = I_2 \cdot \sin \omega_2 \times \sum_{k=1}^{3} [b_{ss}^{2k} E_k \cos \phi_k + b_{sp}^{2k} \cdot V_{\text{term}}^k \cos (\theta_{\text{term}}^k - \theta_{\text{term}}^1)]$$

$$R(12) = I_3 \cdot \sin \omega_3 + \sum_{k=1}^{3} [b_{ss}^{3k} E_k \cos \phi_k + b_{sp}^{3k} \cdot V_{\text{term}}^k \cos (\theta_{\text{term}}^k - \theta_{\text{term}}^1)]$$

$R(13)$

$\quad \vdots$ depend on transformer connection

$R(18)$

$R(19)$

$\quad \vdots$ depend on the control specifications

$R(24)$

$$R(25) = V_d \cdot \pi - \sqrt{2}U_{21}[\cos(C_1 + \alpha_1 - C_3 + \pi) - \cos(C_2 + \alpha_2 - C_3 + \pi)]$$
$$- \sqrt{2}U_{13}[\cos(C_2 + \alpha_2 - C_1) - \cos(C_3 + \alpha_3 - C_1)]$$
$$- \sqrt{2}U_{23}[\cos(C_3 + \alpha_3 - C_2) - \cos(C_1 + \alpha_1 + \pi - C_2)]$$
$$+ I_d(X_{c1} + X_{c2} + X_{c3})$$

$$R(26) = f(V_{di}, I_{di}) \text{ from d.c. system topology.}$$

The twenty-six variable vector $(\bar{x})$ is:

$$[E_1, E_2, E_3, \phi_1, \phi_2, \phi_3, I_1, I_2, I_3, \omega_1, \omega_2, \omega_3,$$
$$u_{12}, u_{13}, u_{23}, C_1, C_2, C_3, \alpha_1, \alpha_2, \alpha_3, a_1, a_2, a_3,$$
$$V_d, I_d]^T$$

7.4. LOAD-FLOW SOLUTION

A sequential technique, using the three-phase fast-decoupled a.c. algorithm and a full Newton–Raphson algorithm for the d.c. equations, involves the block successive iteration of the following three equations,

$$\begin{bmatrix} \Delta\bar{P}(\bar{V},\bar{\theta})/\bar{V} \\ \Delta\bar{P}_{\text{gen}}/\bar{V}_{\text{int}} \end{bmatrix} = [B'] \begin{bmatrix} \Delta\bar{\theta} \\ \Delta\theta_{\text{int}} \end{bmatrix} \tag{7.4.1}$$

$$\begin{bmatrix} \Delta\bar{Q}(\bar{V},\bar{\theta})/\bar{V} \\ \Delta\bar{V}_{\text{reg}}(\bar{V}) \end{bmatrix} = [B''] \begin{bmatrix} \Delta\bar{V} \\ \Delta\bar{V}_{\text{int}} \end{bmatrix} \tag{7.4.2}$$

$$[\bar{R}(\bar{x})] = [J][\Delta\bar{x}] \tag{7.4.3}$$

where $[B']$ and $[B'']$ are the three-phase fast-decoupled a.c. Jacobian matrices as developed in Chapter 5, and $[J]$ is the d.c. Jacobian of first order partial derivatives.

Equations (7.4.1) and (7.4.2) are the three-phase fast-decoupled algorithmic equations from Chapter 5. For the solution of the equations (7.4.1) and (7.4.2), the d.c. variables $\bar{x}$ are treated as constants and, in effect, the d.c. system is modelled simply by the appropriate real and reactive power injections at the converter terminal busbar.

These power injections are calculated from the latest solution of the d.c. system equations and are used to form the corresponding real and reactive power mismatches. For the d.c. iteration, the a.c. variables at the terminal busbars are considered to be constant.

The iteration sequence for the solution of equations (7.4.1), (7.4.2) and (7.4.3) is illustrated in Fig. 7.5. It is based on the P, Q, DC sequence described in Chapter 6 which proved the most successful sequential technique in the single-phase case.

This sequence acknowledges the fact that the converter operation is strongly related to the magnitude of the terminal voltages and more weakly dependent on their phase angles. Therefore, the converter solution follows the update of the a.c.

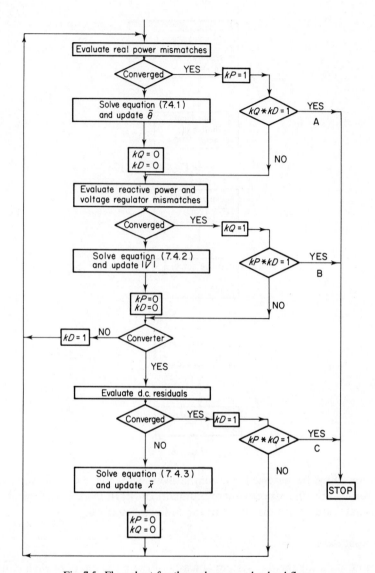

Fig. 7.5. Flow chart for three-phase a.c.–d.c. load flow

terminal voltages. It should be noted, however, that in the three-phase case, final convergence is comparatively slow because the d.c. system behaviour is dependent on the phase-angle unbalance as much as on the voltage unbalance.

7.5. PROGRAM STRUCTURE AND COMPUTATIONAL ASPECTS

The main components of the computer program are illustrated in Fig. 7.6. The additional blocks and increase in size of the a.c.–d.c. program over the purely a.c.

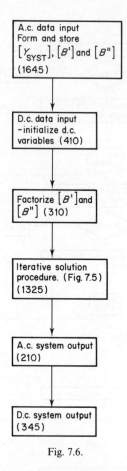

Fig. 7.6.

algorithm may be assessed by comparison with Fig. 5.4. The numbers in parenthesis are the approximate number of FORTRAN statements. The additional features are discussed in the following sections.

D.C. input data

The input data for the d.c. system consists of the parameters of each converter including maximum and minimum variable limits where appropriate. In addition, the d.c. network equations (7.3.5) must be formed from the d.c. system topology. As the d.c. system is relatively small and simple in its interconnection these equations are formed by inspection and effectively input directly by the user.

The d.c. system variables $\bar{x}$ are initialized as the balanced three-phase equivalent of the single-phase converter variables as discussed in Section 6.8.

Programming aspects of the iterative solution

The iterative solution (Fig. 7.5) for the a.c.–d.c. load flow is significantly enlarged over the purely a.c. case (Section 5.5). The basic reason is that the d.c. Jacobian must be reformed and refactorized at each iteration. In addition, because of the nonuniform nature of the d.c. Jacobians and residual equations, each term must be formulated separately in contrast to the a.c. case where compact program loops may be used.

Equations (7.4.1) and (7.4.2) are solved using sparsity techniques and near optimal ordering as described in Chapter 4. Similarly to the single-phase case, the equations for each converter are separate except for those relating to the d.c. interconnection and the solution of equation (7.4.3) is carried out using a modified Gaussian Elimination routine.

This feature may be utilized by appropriate ordering of variables to yield a block sparsity structure for the d.c. Jacobian. With this aim, the d.c. voltage variable is placed last for each block of converter equations and all the d.c. current variables are placed after all converter blocks. The d.c. Jacobian will then have a structure as illustrated in Fig. 7.7.

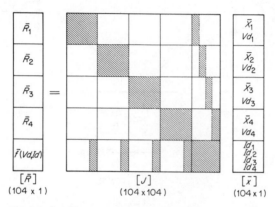

Fig. 7.7. Jacobian structures for four converter d.c. system
(nonzero elements indicated)

By using row pivoting only during the solution procedure, the block sparsity of Fig. 7.7 is preserved. Each block containing nonzero elements is stored in full, but only nonzero elements are processed.

This routine requires less storage than a normal sparsity program for nonsymmetrical matrices and the solution efficiency is improved.

7.6. PERFORMANCE OF THE ALGORITHM

Test system

The performance of the algorithm is discussed with reference to the test system illustrated in Fig. 7.8. The system consists of two a.c. systems interconnected by a 600 kV, 600 MW h.v.d.c. link.

The twenty-bus system is a representation of the 220 kV a.c. network of the South Island of New Zealand. It includes mutually-coupled parallel lines,

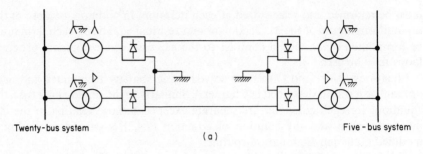

Twenty-bus system　　　　　　　　　　　　　　　　　Five-bus system

(a)

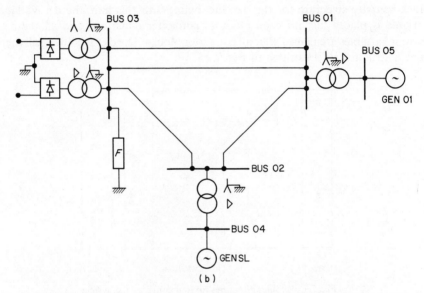

(b)

Fig. 7.8. Three-phase a.c.–d.c. test system. (a) H.V.D.C. interconnection; (b) Five-bus a.c.
system

synchronous generators and condensers, star–star and star–delta connected
transformers and has a total generation in excess of 2000 MW.

At the other end of the link, a fictitious five-bus system represents 800 MW of
remote hydrogeneration connected to a converter terminal and load busbar by
long untransposed high voltage lines.

The small system is used to test the algorithm and to enable detailed discussion
of results. The d.c. link should have considerable influence, as the link power
rating is comparable to the total capacity of the small system. Relevant
parameters for the a.c. system and d.c. link are given in Table 7.1.

Convergence of d.c. model from fixed terminal conditions

Typical convergence patterns for the terminal power flows for the three-phase
model, under both balanced and unbalanced terminal conditions, are shown in
Fig. 7.9. The convergence pattern of the single-phase algorithm developed in

Table 7.1. System data

Data for all lines.

Data for generator transformers

Z_s Series impedance matrix

0.0066 +j0.056	0.0017 +j0.027	0.0012 +j0.021
0.0017 +j0.027	0.0045 +j0.047	0.0014 +j0.022
0.0012 +j0.021	0.0014 +j0.0220	0.0062 +j0.061

Connection	Star–G/Delta
Reactance	$0.0016 + j0.015$
Off-nominal tap	$+2.5\%$ on star

Y_s shunt admittance matrix

$j0.15$	$-j0.03$	$-j0.01$
$-j0.03$	$j0.25$	$-j0.02$
$-j0.01$	$-j0.02$	$j0.125$

Data for all converters

	Phase 1	Phase 2	Phase 3
Transformer reactances	0.0510	0.0510	0.0510
Commutation reactances	0.0537	0.0537	0.0537
Minimum firing angle		7.0 deg	
Minimum extinction angle		10.0 deg	
Nominal voltage		140 kV	

D.C. link resistance = 25.0 ohms.

Generator data

Name	Sequence reactances			Power (MW)	Voltage regulator V^a
	X_0	X_1	X_2		
GEN01	0.02	—	0.004	700.0	1.045
GENSL	0.02	—	0.004	Slack	1.061

Busbar loadings

Pus name	Phase A		Phase B		Phase C	
	P-load	Q-load	P-load	Q-load	P-load	Q-load
BUS01	20.000	10.000	20.000	10.000	20.000	10.000
BUS02	66.667	26.667	66.667	26.667	66.667	26.667
BUS03	0.000	0.000	0.000	0.000	0.000	0.000
BUS04	0.000	0.000	0.000	0.000	0.000	0.000
BUS05	0.000	0.000	0.000	0.000	0.000	0.000

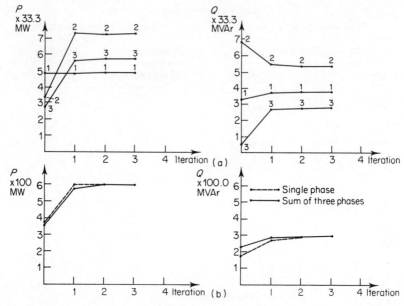

Fig. 7.9. Convergence of terminal powers for three-phase converter model.
(a) Unbalanced; (b) Balanced

Chapter 6, is also illustrated. To enable a comparison to be made, the total three-phase powers are plotted for the balanced case. In all cases the d.c. starting values were selected to give large initial errors in the terminal powers to better illustrate the convergence pattern.

The d.c. equations require two iterations to converge for both the single and three phase models.

Performance of the integrated a.c.–d.c. load flow

With reference to the test system illustrated in Fig. 7.8, the following control specifications are used at the inverting terminal for all test cases

— Symmetrical firing control with the reference phase on minimum extinction angle.
— Off-nominal tap ratios equal on all phases.
— D.C. voltage specified.

A variety of different control strategies are considered at the rectifier terminal and the convergence results are given in Table 7.2. For comparison, the table includes cases with the converters modelled as equivalent a.c. loads.

It should be noted that the iteration scheme illustrated in Fig. 7.5 does not allow for each individual a.c. system to be converged independently, therefore, the number of iterations required is the larger of the two sets given in the table.

It is clear that the integration of the d.c. converter model does not cause any significant deterioration in performance. The only cases where convergence is

Table 7.2. Case descriptions and convergence results

case	Case description and rectifier specifications	Number of iterations to Convergence (0.1 MW/MVAR)	
		20-bus system	5-bus system
a(i)	Converter modelled by equivalent balanced loads*	8,7	6,5
(ii)	Converter modelled by equivalent unbalanced loads*	8,7	6,5
b(i)	Phase-angle control; $\alpha_1 = \alpha_2 = \alpha_3 = \alpha_{min}, a_1 = a_2 = a_3, P_{dc} = P_d^{sp}$	8,7	6,5
(ii)	Symmetrical firing; $\alpha_i = \alpha_{min}$	8,7	6,5
(iii)	Phase-angle control; $\alpha_1 = \alpha_2 = \alpha_3 = \alpha_{min}, a_1 = a_2 = a_3, I_{dc} = I_{dc}^{sp}, V_{d1} = V_{d2}$	8,7	6,5
(iv)	Symmetrical firing; $\alpha_i = \alpha_{min}$	8,7	6,5
(v)	As for case b(I); with poor starting values. $(P_{dc}, Q_{dc}$ in error by 70%)	8,7	8,7
(vi)	As for case b(i); with large unbalanced load at BUS03	8,7	7,6
(vii)	As for case b(ii); with large unbalanced load at BUS03	8,7	7,6
(viii)	As for case b(i); with loss of 1 line BUS01 to BUS03	8,7	9,9
(ix)	Symmetrical firing; $\alpha_i = \alpha_{min}, a_1 = -10\%, a_2 = 0, a_3 = +10\%$	8,7	7,6
(x)	Phase-angle control; $a_1 = a_2 = a_3 = a^{sp}, \alpha_1 = \alpha_2 = \alpha_3, p_{dc} = P_{dc}^{sp}$	8,7	7,6
(xi)	Case (x) loss of 1 line. BUS01 to BUS03	8,8	8,8

* Loading for case a(i) and a(ii) derived from results for case b(i). See Table 7.3.

172

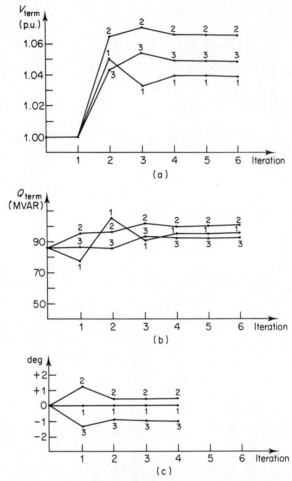

Fig. 7.10. Convergence patterns of terminal conditions for a strong a.c. system (a) A.c. terminal voltages; (b) Terminal reactive power flows; (c) A.c. terminal angle unbalance (deviation from nominal)

slowed are (viii) and (xi) where the system is weakened by the loss of one transmission line. This is to be expected from the discussion of single-phase sequential algorithms given in Section 6.8.

To examine the effect of a weak system in the three-phase case, the convergence patterns for the terminal powers and voltages are shown for case (xi) in Fig. 7.10. The reactive power and voltage unbalance vary considerably over the first few iterations but this initial variation does not cause any convergence problems. With weaker systems, the unbalance increases and the convergence patterns become more oscillatory. The corresponding convergence pattern of the single-phase load flow for case (xi) is shown in Fig. 7.11(b) where a similar oscillatory pattern is observable. Moreover the sum of the three-phase reactive powers and

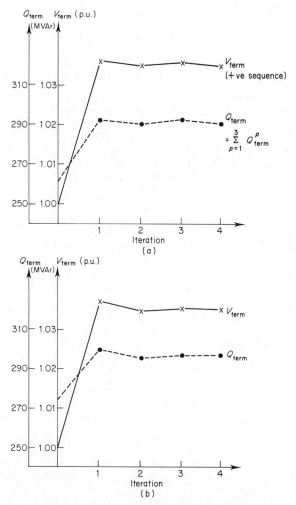

Fig. 7.11. Comparison of single-phase and three-phase positive sequence convergence patterns. (a) Three-phase load flow; (b) Single-phase load flow

the average phase voltage of Fig. 7.10, plotted in Fig. 7.11(a), shows an even closer similarity between the three-phase case and the single-phase behaviour.

Sample results

The operating states of the two converters connected to BUS03 are listed for the most typical cases in Table 7.3. The corresponding a.c. system voltage profiles and generation results are given for cases a(i), b(i) and b(ii) in Table 7.4. The following discussion is with reference to these results.

The results of the realistic three-phase converter model (case b(i)), although

Table 7.3(a) Converter 1 results

					Terminal powers		Dc. conditions	
						Converter 1 (Star–Star)		
Case	Phase	Firing angle α_i (deg)	Tap ratio a_i (%)	Commutation angle u_i (deg)	Real P_i (MW)	Reactive Q_i (MVAr)	Voltage Vd_1 (kv)	Current Id_1 (kA)
b(i)	1	7.00	5.5	29.79	98.1	48.1	292.8	1.0246
	2	7.00	5.5	29.32	101.7	50.8	—	—
	3	7.00	5.5	29.61	100.3	48.3	—	—
b(ii)	1	7.00	5.3	29.78	98.6	49.0	292.8	1.0246
	2	7.20	5.3	29.14	100.9	51.3	—	—
	3	8.43	5.3	28.50	100.6	47.8	—	—
b(vi)	1	7.00	4.8	29.17	95.6	39.5	292.8	1.0246
	2	7.00	4.8	29.16	101.9	50.5	—	—
	3	7.00	4.8	30.43	102.44	57.2	—	—
b(vii)	1	7.00	3.9	29.03	97.6	39.1	292.8	1.0246
	2	11.64	3.9	25.63	101.8	54.7	—	—
	3	9.37	3.9	28.56	100.6	57.7	—	—
b(ix)	1	11.00	−10.0	24.32	104.6	49.4	314.1	0.9483
	2	7.00	0.0	27.76	101.1	45.4	—	—
	3	7.55	10.0	26.08	92.1	44.03	—	—

Table 7.3(b) Converter 2 results

					Terminal powers		Dc. conditions	
						Converter 2 (Star-G–Delta)		
Case	Phase	Firing angle α_i (deg)	Tap ratio a_i (%)	Commutation angle u_i (deg)	Real P_i (MW)	Reactive Q_i (MVAr)	Voltage Vd_2 (kV)	Current Id_2 (kA)
b(i)	1	7.00	5.5	29.80	97.3	49.2	292.8	1.0246
	2	7.00	5.5	29.60	102.6	53.2	—	—
	3	7.00	5.5	29.32	100.14	44.7	—	—
b(ii)	1	8.03	5.2	28.97	96.4	50.0	292.8	1.0246
	2	7.00	5.2	29.57	102.7	52.9	—	—
	3	8.55	5.2	28.08	100.87	45.66	—	—
b(vi)	1	7.00	4.3	30.63	67.9	13.0	292.8	1.0246
	2	7.00	4.3	28.92	95.5	89.4	—	—
	3	7.00	4.3	28.90	136.6	53.7	—	—
b(vii)	1	7.00	3.0	30.48	70.9	17.9	292.8	1.0246
	2	14.95	3.0	23.25	90.1	94.1	—	—
	3	13.41	3.0	24.25	138.9	52.2	—	—
b(ix)	1	8.08	−10.0	25.42	88.9	65.3	314.7	0.9483
	2	8.38	0.0	27.30	122.6	49.9	—	—
	3	7.00	10.0	26.96	86.9	24.2	—	—

Table 7.4. Bus voltages and generation results

Case a(i)

Bus name	Phase A Volt.	Ang.	Phase B Volt.	Ang.	Phase C Volt.	Ang.	Generation Total	
BUS01	1.067	27.294	1.067	−92.891	1.061	147.431	0.000	0.000
BUS02	1.054	25.190	1.065	−94.670	1.057	144.915	0.000	0.000
BUS03	1.038	23.185	1.071	−95.714	1.043	142.567	0.000	0.000
BUS04	1.045	−3.566	1.046	−123.479	1.047	116.436	173.621	74.723
BUS05	1.061	2.683	1.062	−117.367	1.061	122.628	700.000	113.920

Case b(i)

Bus name	Phase A Volt.	Ang.	Phase B Volt.	Ang.	Phase C Volt.	Ang.	Generation Total	
BUS01	1.067	27.362	1.065	−92.955	1.062	147.437	0.000	0.000
BUS02	1.055	25.232	1.064	−94.717	1.057	144.925	0.000	0.000
BUS03	1.038	23.517	1.066	−95.965	1.049	142.543	0.000	0.000
BUS04	1.045	−3.552	1.046	−123.483	1.047	116.438	173.570	74.706
BUS05	1.061	2.690	1.062	−117.369	1.060	122.634	700.000	113.680

Case b(ii)

Bus name	Phase A Volt.	Ang.	Phase B Volt.	Ang.	Phase C Volt.	Ang.	Generation Total	
BUS01	1.066	27.31	1.066	−92.942	1.062	147.421	0.000	0.000
BUS02	1.054	25.238	1.064	−94.705	1.057	144.913	0.000	0.000
BUS03	1.036	23.532	1.066	−95.947	1.049	142.506	0.000	0.000
BUS04	1.045	−3.563	1.046	−123.479	1.047	116.439	173.593	75.949
BUS05	1.061	2.690	1.062	−117.363	1.060	122.635	700.000	115.391

distinguishable from those of the balanced model a(i), are not significantly different as regards the a.c. system operation. They are definitely significant however, as regards converter operation, particularly when consideration is given to the harmonic content.

A comparison of cases b(i) and b(ii) shows an increase in reactive power consumption in case b(ii) due to two phases having greater than minimum firing angles.

The results also show that the transformer connection modifies the converter source voltages and the phase distribution of power flows. Under balanced conditions, a zero-sequence voltage may appear at system busbars. As the converter has no zero-sequence path, zero sequence current will only flow when the converter transformer provides a path, as in the case of the star-G/delta transformer. A typical example is illustrated in Fig. 7.12 where the zero-sequence voltages and currents are shown for case b(i). Accurate converter transformer models must therefore be included in the converter modelling.

Conclusions on performance of the algorithm

The fast-decoupled three-phase a.c.–d.c. load flow behaves in a very similar

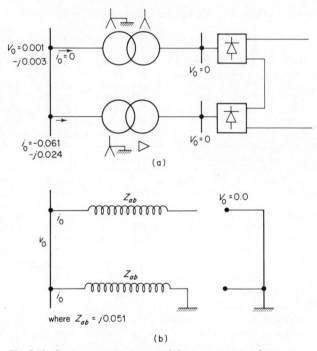

Fig. 7.12. Sequence components and the converter transformer connection. (a) Zero sequence potentials for case b(i); (b) Zero sequence network for converter transformers. (Note: Transformer secondary zero sequence reference is provided by equations (7.3.21) and (7.3.22))

manner to the corresponding single-phase version. The following general conclusions can be made on its performance:

— the number of iterations to convergence is not significantly increased by the presence of the d.c. converters;
— d.c. convergence is not dependent on the specific control specifications applied to each converter;
— wide errors in initial conditions may be tolerated;
— for very weak a.c. systems the interaction of the converter with the a.c. system is increased and the convergence is slowed. Successful convergence can, however, be expected in all practical cases;
— the algorithm exibits good reliability even under extreme unbalance.

7.7. EFFECT OF HARMONIC VOLTAGES ON THE FUNDAMENTAL FREQUENCY OPERATING STATE OF D.C. CONVERTERS

The three-phase model of the d.c. converter has been formulated under the assumption of perfectly sinusoidal voltages at the converter terminal busbar. Without this assumption the steady-state load-flow formulation would be very complex due to the difficulty of calculating accurately the level of harmonic voltages which may be present.

However, the presence of harmonic voltages at the converter terminal will alter the fundamental frequency power flows primarily because the actual zero crossings of the phase-to-phase voltages will be shifted from those calculated from the fundamentals alone. Other effects (e.g. modification of average d.c. voltages and commutation time) are of secondary importance and may be ignored.

The effect of harmonic voltages is also influenced by the control system in operation. However, the order of magnitude of any possible errors may be assessed by assuming that the control system, whether phase-angle control or symmetrical firing, is perfect (i.e. there is no firing angle error) and is presented with the actual zero crossings of the commutating voltages (i.e. there are no control system filters).

Harmonic voltage limits

In order to reduce the undesirable effects caused by the harmonic voltages and currents the supply authorities specify maximum permissible values for the harmonics.

Typical recommended limits for harmonic voltages in percentage of the fundamental are as follows:

Supply voltage	Odd	Even
415 V	4	2
33 k V	2	1
110/132 k V	1	0.5

In addition to limits on individual harmonics the maximum total harmonic distortion (THD) is also specified.

The total harmonic distortion is defined by:

$$THD = \sqrt{\sum_{n=2}^{\infty} V_n^2}$$

and a typical limit for a 110/132-kV network is 1.5 percent.

Shift in zero crossing of phase–phase voltages

There is considerable difficulty in assigning a realistic worst case for the investigation of the influence of harmonic voltages. The following considerations apply:

— Under balanced conditions the characteristic harmonic orders do not influence the intervals between firing pulses, i.e. all firings are shifted by equal amounts. Therefore these voltages will have very little effect on the magnitude of the fundamental currents.

— Harmonic limits are usually applied at the point of common coupling to the

supply network which may or may not be the converter terminal busbar.
— The majority of the allowed triplen harmonics will consist of the usual zero sequence components and these have no influence on the position of phase to phase crossings.

The order of the nonzero sequence triplen harmonics, which may be considered to cause the most significant shifts, is not important as, under balanced conditions with the subsequent crossings 120 degrees apart, a small magnitude of any lower order triplen harmonic will result in comparable shifts for all three phases.

As a consequence of these features, it is considered reasonable to investigate the worst case effect of 2 percent of nonzero sequence third harmonic voltage.

The worst case for a shift in any zero crossing is for all harmonics to add to the fundamental voltage on one phase and to subtract from the other phase at the position of actual zero crossing.

With consequent zero crossings 60 degrees apart, i.e. one-half rotation of the third harmonic phasors, then one subsequent shift will be approximately half the initial worst case, and the third shift will be equal to the worst case initial shift except in the opposite direction.

For 2 percent nonzero sequence third harmonic, the worst case shift is approximately one degree.

The order of magnitude of the resultant errors can be assessed on the basis of nominally balanced operation and by interpreting shifts in zero crossings as firing angle errors determined in accordance with the firing controller in operation.

For a converter operating with fixed tap ratios and specified d.c. current, any firing-angle errors are reflected into the converter operation through errors in the calculated d.c. voltage and in the calculated magnitudes for the fundamental component of the phase current waveforms. Both errors occur to some extent with both firing controllers.

Effect of crossing error on the average d.c. voltage

The largest error occurs with symmetrical firing control where any shift in the reference is reflected as a change in firing angle on all phases. The d.c. voltage error depends upon the nominal firing angle. Typical errors in d.c. voltage under symmetrical firing control are illustrated in Table 7.5 on the assumption of balanced fundamental voltages. Although the errors are small at normal firing angles, they are significant at large firing angles.

Table 7.5. Percentage error in calculated d.c. voltage (symmetrical firing)

α (degrees)	Firing angle errors due to shift of zero crossings			
	$0.5°$	$1°$	$2°$	$3°$
0	0.5°	1°	2°	3°
10	0.2	0.3	0.7	1
20	0.3	0.6	1.3	2
40	0.6	1.1	2.3	3.5
70	0.8	1.6	3.3	5.0

Errors in calculated phase current magnitudes

Appreciable errors occur with the case of phase-angle control; errors with symmetrical firing are limited to the effects of commutation angle unbalance.

The effect of up to two degrees modulation has been investigated on the basis of the current waveform of a converter with commutation angle of ten degrees and balanced voltages. Ignoring the shifts in commutation angle which will occur with an alteration of firing angle, the effect of alteration in the period of conduction has been investigated using a Fourier Transform algorithm. The results are shown in Table 7.6. In addition to the fundamental, the percentages of

Table 7.6. Harmonic currents

	120°	121°	122°		120°	121°	122°
0	0			11	4.88	4.57	4.25
1	100	100.50	100.99	12	—	—	—
2	—	—	—	13	3.09	3.26	3.41
3	0.01	0.95	1.92	14	—	—	—
4	—	—	—	15	0.01	0.29	0.56
5	17.85	17.3	16.89	16	—	—	—
6	—	—	—	17	1.05	0.94	0.82
7	11.32	11.64	12.04	18	—	—	—
8	—	—	—	19	0.75	0.81	0.85
9	0.01	0.68	1.35	20	—	—	—
10	—	—	—				

the harmonics are also given. Note that no even harmonics are present as the waveform was assumed to be symmetrical.

It is important to note that the percentage errors presented in the Table are the maximum that can occur; in all practical cases there will be an alteration in the commutation angle which will inhibit the change in the waveform. This effect will be most noticeable at small firing angles. In general therefore, the errors will be less than those indicated in the table.

The change in fundamental is not, therefore, expected to exceed 1 percent of the value calculated by the load flow.

7.8. REFERENCES

1. J. D. Ainsworth, 1967. 'Harmonic instability between controlled static converters and a.c. networks', *Proc. IEE*, **114** (7), 949–957.
2. J. D. Ainsworth, 1968. 'The phase-locked oscillator—A new control system for controlled static converters', *IEEE, PAS*, **PAS-87** (3), 859–865.
3. J. Kauferle, R. Mey and Y. Rogowsky, 1970. 'H.V.D.C. stations connected to weak a.c. systems', *IEEE, PAS*, **PAS-89** (7), 1610–1617.
4. A. G. Phadke and J. H. Harlow, 1966. 'Unbalanced Converter Operation', *IEEE PAS*, **PAS-85**, 233–239.
5. J. Arrillaga and A. E. Efthymiadis, 1968. 'Simulation of converter performance under unbalanced conditions', *Proc. IEE*, **115** (12), 1809–1818.

Faulted System Studies

8.1. INTRODUCTION

The main object of Fault Analysis is the calculation of fault currents and voltages for the determination of circuit-breaker capacity and protective relays performance.

Early methods used in the calculation of fault levels involved the following approximations:

— All voltage sources assumed a one per unit magnitude and zero relative phase, which is equivalent to neglecting the prefault load current contribution.
— Transmission plant components included only inductive parameters.
— Transmission line shunt capacitance and transformer magnetizing impedance were ignored.

Based on the above assumptions, simple equivalent sequence impedance networks were calculated and these were interconnected according to the fault specification. Conventional circuit analysis was then used to calculate the sequence voltage and currents and with them, by means of the inverse sequence component transformation, the phase components.

Although the basic procedure of the computer solution is still the same, the need for the various approximations has disappeared.

The three-phase models of transmission plant developed in Chapter 2, which included interphase and parallel line mutual effects, could be easily combined to produce the faulted system matrix admittance or matrix impedance and hence provide an accurate model for the analysis of a.c. system faults.

However, the main reasons given for the use of the phase frame of reference in load flows are less relevant here. Extra losses and harmonic content are less of a problem in the short period of time prior to fault clearance. Fault studies are normally performed on systems reasonably well balanced either at the operational or planning stage; in the latter case only after prospective system configurations have been proved acceptable through load-flow studies.

Moreover, unbalanced faults involving large converter plant, while in need of a three-phase representation, cannot be assessed by steady-state models as explained in Chapter 11.

180

Finally, faulted system studies constitute an integral part of multi-machine transient stability programs, the complexity of which will not normally permit the three-phase approach.

A single-phase representation, achieved with the help of the symmetrical components transformation[1] is used in this chapter as a basis for the development of a fault-study program.[2],[3],[4],[5]

8.2. ANALYSIS OF THREE-PHASE FAULTS

A preliminary stage to the analysis is the collection of appropriate data specifying the system to be analysed in terms of prefault voltage, loading and generating conditions. Such data is then processed to form a nodal equivalent network constituted by admittances and injected currents.

The equivalent circuits of loads, lines and transformers discussed in Chapter 2 are directly applicable here. The generators can be represented by a constant voltage E^M behind an approximate machine admittance y^M which value depends on the time of the calculation from the instant of fault inception. This is illustrated in Fig. 8.1(a).

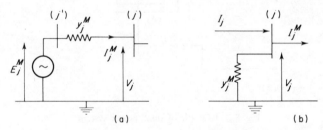

Fig. 8.1. Generator representation

When analysing the first two or three cycles following the fault, the subtransient admittance of the machine is normally used, whilst for longer times, it is more appropriate to use the transient admittance. The machine model, illustrated in Fig. 8.1(a), is then converted to a nodal equivalent by means of Norton's Theorem which changes the voltage source into a current source injected at the bus j as shown in Fig. 8.1(b). This is most effective as otherwise a further node at j' is necessary to define the machine admittance y^M.

The injected nodal current is given by

$$I_j = y_j^M E_j^M \tag{8.2.1}$$

where

$$E_j^M = V_j + \frac{I_j^M}{y_j^M} \tag{8.2.2}$$

so that

$$I_j = y_j^M V_j + I_j^M \tag{8.2.3}$$

182

I_j^M is the current required at the voltage V_j to produce the machine power $P_j^M + jQ_j^M$, so

$$(I_j^M)^* V_j = P_j^M + j \cdot Q_j^M \tag{8.2.4}$$

Thus from the load-flow data of P^M, Q^M and V^M we may calculate the injected nodal current I_j, as

$$I_j = y_j^M \cdot V_j + \frac{P_j^M - jQ_j^M}{V_j^*} \tag{8.2.5}$$

Admittance matrix equation

Let us take as a reference the small system of Fig. 8.2. Each element is converted to its nodal equivalent. These are connected together as shown in Fig. 8.3 and finally simplified to the equivalent circuit of Fig. 8.4.

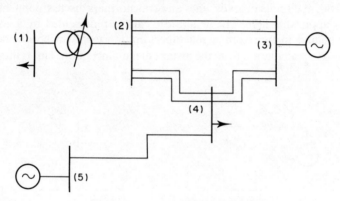

Fig. 8.2. Example of small power system

The following equations may then be written for the network of Fig. 8.4.

$$I_1 = y_{11}V_1 + y_{12}(V_1 - V_2) \tag{8.2.6}$$

$$I_2 = y_{12}(V_2 - V_1) + y_{22}V_2 + y_{23}(V_2 - V_3) + y_{24}(V_2 - V_4) \tag{8.2.7}$$

$$I_3 = y_{23}(V_3 - V_2) + y_{33}V_3 + y_{34}(V_3 - V_4) \tag{8.2.8}$$

$$I_4 = y_{24}(V_4 - V_2) + y_{34}(V_4 - V_3) + y_{44}V_4 + y_{45}(V_4 - V_5) \tag{8.2.9}$$

$$I_5 = y_{45}(V_5 - V_4) + y_{55}V_5 \tag{8.2.10}$$

or in matrix form after grouping together the terms common to each voltage:

$$\begin{bmatrix} I_1 \\ I_2 \\ I_3 \\ I_4 \\ I_5 \end{bmatrix} = \begin{bmatrix} Y_{11} & Y_{21} & Y_{31} & Y_{41} & Y_{51} \\ Y_{12} & Y_{22} & Y_{32} & Y_{42} & Y_{52} \\ Y_{13} & Y_{23} & Y_{33} & Y_{43} & Y_{53} \\ Y_{14} & Y_{24} & Y_{34} & Y_{44} & Y_{54} \\ Y_{15} & Y_{25} & Y_{35} & Y_{45} & Y_{55} \end{bmatrix} \begin{bmatrix} V_1 \\ V_2 \\ V_3 \\ V_4 \\ V_5 \end{bmatrix} \tag{8.2.11}$$

183

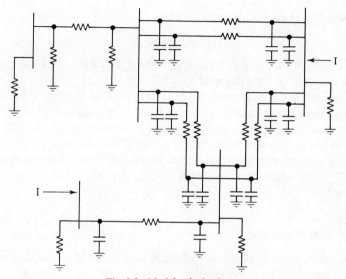

Fig. 8.3. Model substitution

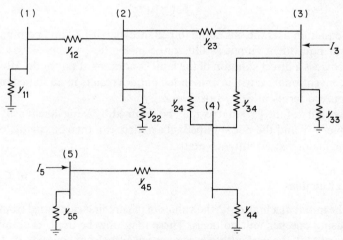

Fig. 8.4. Final equivalent

where

$$Y_{ii} = \sum_j y_{ij}$$

$$Y_{ij} = -y_{ij} \qquad i \neq j$$

Equation (8.2.11) is usually written as

$$[I] = [Y] \cdot [V] \tag{8.2.12}$$

where $[I]$ and $[V]$ are the current and voltage vectors and $[Y]$ is the nodal admittance matrix of the system of Fig. 8.2.

It can be seen from equations (8.2.6) to (8.2.10) that nonzero elements only occur where branches exist between nodes. Since each node or busbar is normally connected to less than four other nodes, there are usually quite a number of zero elements in any system with more than ten busbars. Such sparsity is exploited by only storing and processing the nonzero elements. Moreover, the symmetry of the matrix $(Y_{ij} = Y_{ji})$ permits using only the upper right hand terms in the calculations.

Impedance matrix equation

The nodal admittance equation is inefficient as it requires a complete iterative solution for each fault type and location. Instead, equation (8.2.12) can be written as

$$[V] = [Y]^{-1} \cdot [I] \tag{8.2.13}$$

$$= [Z] \cdot [I]$$

This equation uses the bus nodal impedance matrix $[Z]$ and permits using the Thevenin Equivalent circuit as illustrated in Fig. 8.5, which, as will be shown later, provides a direct solution of the fault conditions at any node. However, the use of conventional matrix inversion techniques results in an impedance matrix with nonzero terms in every position Z_{ij}.

The sparsity of the $[Y]$ matrix may be retained by using an efficient inversion technique[6],[7] and the nodal impedance matrix can then be calculated directly from the factorized admittance matrix.

Fault calculations

From the initial machine data, the values of $[I]$ are first calculated from equation (8.2.5) using one per unit voltages. These may now be used to obtain a better estimate of $[V]$, the prefault voltage at every node from equation (8.2.13). If the initial data is supplied from a load flow, this calculation will not make any difference.

The program now has sufficient information to calculate the voltages and currents during a fault.

The voltage at the fault bus k is from Fig. 8.6.

$$V_k^f = Z^f \cdot I^f \qquad (8.2.14)$$

where

k is the bus to be faulted,
Z^f is the fault impedance and
I^f is the fault current.

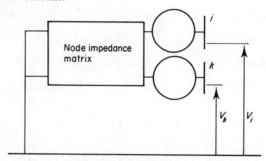

Fig. 8.5. Thevenin equivalent of prefault system

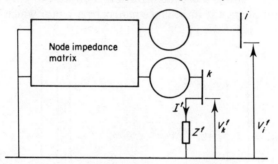

Fig. 8.6. Thevenin equivalent of faulted system

Equation (8.2.13) may be expanded to yield

$$
\begin{bmatrix} V_1 \\ V_2 \\ \cdot \\ V_k \\ \cdot \\ V_n \end{bmatrix} =
\begin{bmatrix}
Z_{11} & Z_{12} & \cdot & Z_{1k} & \cdot & Z_{1n} \\
Z_{21} & Z_{22} & \cdot & Z_{2k} & \cdot & Z_{2n} \\
\cdot & \cdot & \cdot & \cdot & \cdot & \cdot \\
Z_{k1} & Z_{k2} & \cdot & Z_{kk} & \cdot & Z_{kn} \\
\cdot & \cdot & \cdot & \cdot & \cdot & \cdot \\
Z_{n1} & Z_{n2} & \cdot & Z_{nk} & \cdot & Z_{nn}
\end{bmatrix}
\begin{bmatrix} I_1 \\ I_2 \\ \cdot \\ I_k \\ \cdot \\ I_n \end{bmatrix}
\qquad (8.2.15)
$$

Selecting row k and expanding:

$$V_k = Z_{k1}I_1 + Z_{k2}I_2 + \cdots + Z_{kk}I_k + \cdots + Z_{kn}I_n \qquad (8.2.16)$$

This equation describes the voltage at bus k prior to the fault. During a fault a large fault current I^f flows out of bus k. Including this current in equation (8.2.16) and using equation (8.2.14)

$$V_k^f = Z^f I^f = Z_{k1}I_1 + \cdots + Z_{kk}I_k + \cdots + Z_{kn}I_n - Z_{kk}I^f \qquad (8.2.17)$$

or

$$Z^f I^f = V_k - Z_{kk} I^f \tag{8.2.18}$$

and so the fault current is given directly by

$$I^f = \frac{V_k}{Z_{kk} + Z^f} \tag{8.2.19}$$

Also from equation (8.2.15) the prefault voltage at any other bus j is

$$V_j = Z_{j1} I_1 + Z_{j2} I_2 + \cdots + Z_{jk} I_k + \cdots + Z_{jn} I_n \tag{8.2.20}$$

and during the fault

$$V_j^f = Z_{j1} I_1 + Z_{j2} I_2 + \cdots + Z_{jk} I_k + \cdots + Z_{jn} I_n - Z_{jk} I^f \tag{8.2.21}$$

or

$$V_j^f = V_j - Z_{jk} \cdot I^f \tag{8.2.22}$$

From equations (8.2.19) and (8.2.22) the fault voltages at every bus in the system may be calculated, each calculation requiring only one column of the impedance matrix. Using the bifactorization routines the kth column can be obtained by multiplying the impedance matrix by a vector which has a '1' in the kth row and '0''s elsewhere, i.e.

$$
\begin{bmatrix} Z_{1k} \\ Z_{2k} \\ \cdot \\ Z_{kk} \\ \cdot \\ Z_{nk} \end{bmatrix}
=
\begin{bmatrix}
Z_{11} & Z_{12} & \cdot & Z_{1k} & \cdot & Z_{1n} \\
Z_{21} & Z_{22} & \cdot & Z_{2k} & \cdot & Z_{2n} \\
\cdot & & \cdot & & \cdot & \\
Z_{k1} & Z_{k2} & \cdot & Z_{kk} & \cdot & Z_{kn} \\
\cdot & & & & & \\
Z_{n1} & Z_{n2} & \cdot & Z_{nk} & \cdot & Z_{nn}
\end{bmatrix}
\begin{bmatrix} 0 \\ 0 \\ \cdot \\ 1 \\ \cdot \\ 0 \end{bmatrix}
\cdot \tag{8.2.23}
$$

Once Z_{kk} is known then I^f is calculated from equation (8.2.19). I^f is then subtracted from the initial prefault nodal currents to form a new vector $[I^f]$ defined by

$$I_j^f = I_j \qquad \text{for } j \neq k, \qquad j = 1 \text{ to } n$$

$$I_k^f = I_k - I^f \quad \text{for } j = k$$

The voltages during the fault are given by the product of the impedance matrix and this new vector $[I^f]$, i.e.

$$[V^f] = [Z] \cdot [I^f] \tag{8.2.24}$$

Equation (8.2.24) is equivalent to (8.2.22) because of the following expansion.

$$[I^f] = [I] - [0, 0, 0, \ldots I^f, \ldots 0]^T$$

from which equation (8.2.24) expands as

$$[V^f] = [Z]\{[I] - [0, 0, 0, \ldots I^f, \ldots 0]^T\}$$

or

$$[V^f] = [V] - [Z] \cdot [0, 0, \ldots I^f, \ldots 0]^T$$

which is equivalent to equation (8.2.22).

Once the fault voltages are known the branch currents between buses can be calculated from the original branch admittances, i.e.

$$I_{ij}^f = y_{ij}\{V_i^f - V_j^f\} \tag{8.2.25}$$

A correction is necessary for the sending end current of a tapped transformer, i.e.

$$I_{ij}^f = y_{ij}\{(1 - \tau)V_i^f - V_j^f\} \tag{8.2.26}$$

With reference to Fig. 8.7, a machine fault current contribution is

$$I_i^{Mf} = (E_i^M - V_i^f) \cdot y_i^M$$

or substituting $I_i = y_i^M \cdot E_i^M$ (from equation (8.2.1)).

$$I_i^{Mf} = I_i^M - V_i^f y_i^M \tag{8.2.27}$$

8.3. ANALYSIS OF UNBALANCED FAULTS

If the network is unsymmetrically faulted or loaded, neither the phase currents nor the phase voltage will possess three-phase symmetry. The analysis can no longer be limited to one phase and the admittance of each element will consist of a 3×3 matrix which on the assumption of a reasonably balanced tranmission system, will be symmetrical, i.e.

$$\begin{bmatrix} ^{aa}Y & ^{ab}Y & ^{ac}Y \\ & ^{aa}Y & ^{ab}Y \\ & & ^{aa}Y \end{bmatrix} \tag{8.3.1}$$

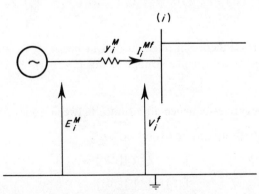

Fig. 8.7. Machine representation showing fault current contribution

Matrix (8.3.1) can be diagonalized by the symmetrical components transformation $(T^*)^t Y T$ into its sequence component equivalent, i.e.

$$\begin{bmatrix} {}^0Y & & \\ & {}^1Y & \\ & & {}^2Y \end{bmatrix} \tag{8.3.2}$$

where

$$
\begin{aligned}
{}^0Y &= {}^{aa}Y + {}^{ab}Y + {}^{ac}Y \\
{}^1Y &= {}^{aa}Y + a({}^{ab}Y) + a^2({}^{ac}Y) \\
{}^2Y &= {}^{aa}Y + a^2({}^{ab}Y) + a({}^{ac}Y) \\
a &= e^{j2\pi/3}
\end{aligned}
\tag{8.3.3}
$$

Moreover, for stationary balanced system elements the admittances ${}^{ab}Y$ and ${}^{ac}Y$ are equal and equations (8.3.3) show that the corresponding positive and negative sequence admittances are also equal. Further, the simplifying assumption is often made that the positive and negative sequence admittances of rotating machines are equal. This assumption is only reasonable when the subtransient admittances are being used and in such case the storage required by the program can be substantially reduced by deleting the negative sequence matrices.

Admittance matrices

The data specifying each element of the system is then used to form the following three nodal equations:

$$
{}^0I_i = {}^0V_i\, {}^0y_{ii} + ({}^0V_i + {}^0V_1)\, {}^0y_{ii} + \ldots + ({}^0V_i - {}^0V_n)\, {}^0y_{ni} \tag{8.3.4}
$$

$$
{}^1I_i = {}^1V_i\, {}^1y_{ii} + ({}^1V_i - {}^1V_1)\, {}^1y_{ii} + \ldots + ({}^1V_i - {}^1V_n)\, {}^1y_{ni} \tag{8.3.5}
$$

$$
{}^2I_i = {}^2V_i\, {}^2y_{ii} + ({}^2V_i - {}^2V_1)\, {}^2y_{ii} + \ldots + ({}^2V_i - {}^2V_n)\, {}^2y_{ni} \tag{8.3.6}
$$

where

${}^0I_i$ is the zero-sequence injected current at bus i.
${}^1V_i$ is the positive-sequence voltage at bus i.

and

${}^2y_{ni}$ is the negative-sequence admittance between nodes n and i.

The above equations can be expressed as:

$$[{}^0I] = [{}^0Y][{}^0V] \tag{8.3.7}$$

$$[{}^1I] = [{}^1Y][{}^1V] \tag{8.3.8}$$

$$[{}^2I] = [{}^2Y][{}^2V] \tag{8.3.9}$$

where

$$^\gamma Y_{ij} = -\,^\gamma Y_{ij} \quad \text{for } i = 1, n; j = 1, n \quad j \neq i \quad \text{and} \quad \gamma = 0, 1, \text{ or } 2$$

and

$$^\gamma Y_{ii} = \sum_{k=1}^{n} \,^\gamma y_{ik} \quad \text{for } i = 1, n; \gamma = 0, 1, \text{ or } 2$$

The sequence-admittance matrices can now be triangularized by the bifactorization method. Since the three admittance matrices have identical structure, this can be made more efficient by triangularizing them simultaneously, i.e. in programming terms, only one set of vectors is needed to form pointers to the three arrays as they are stored by the bifactorization routine.

Fault calculations

As already explained for the three-phase fault, the nodal impedance matrices may now be calculated directly from the reduced admittance matrices and the following sequence impedance matrix equations result.

$$[^0V] = [^0Z][^0I] \tag{8.3.10}$$

$$[^1V] = [^1Z][^1I] \tag{8.3.11}$$

$$[^2V] = [^2Z][^2I] \tag{8.3.12}$$

Because the system is assumed to be balanced prior to the fault, the vectors of negative and zero-sequence currents are zero, i.e. there are no prefault negative or zero-sequence voltages.

The positive-sequence network then models the prefault network condition and equation (8.3.11) is used to calculate the prefault voltages. If the original voltages used in the machine models were obtained from a load-flow calculation, then the use of equation (8.3.11) will make no difference to those results; however, if the voltages were assumed at one p.u. with zero angle then this calculation will provide more accurate prefault voltages.

The single-phase equivalent circuit is then set up by linking the three sequence networks together according to the type of fault to be analysed.[8]

Short-circuit faults

A convenient way of simulating the fault location F for the analysis of short-circuit faults is illustrated in Fig. 8.8.

It includes three fault impedances $^aZ, ^bZ$ and ^cZ and three injected currents $^aI^f, ^bI^f$, and $^cI^f$

For each type of fault, it is possible to write 'boundary conditions' for the currents and voltages at the fault location. For example, Fig. 8.9 shows the case of a line to ground fault at bus k.

190

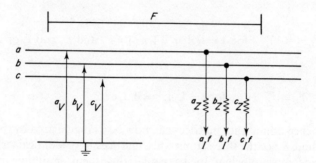

Fig. 8.8. The fault location

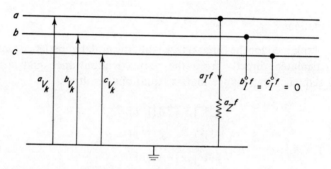

Fig. 8.9. Single line to ground fault

The boundary conditions are

$$^bI_k^f = {}^cI_k^f = 0 \tag{8.3.13}$$

and

$$^aV_k^f = {}^aZ^f \cdot {}^aI^f \tag{8.3.14}$$

Using equations (8.3.13) and (8.3.14) with the sequence components transformation the following relationships result.

$$^0I^f = {}^1I^f = {}^2I^f = {}^aI^f/3 \tag{8.3.15}$$

and

$$^0V_k^f + {}^1V_k^f + {}^2V_k^f = {}^aZ^f \cdot {}^aI^f = 3 \cdot (Z^f)^1I^f \tag{8.3.16}$$

Also, the sequence voltages at the fault location may be described by the equations

$$^0V_k^f = -{}^0Z_{kk} \cdot {}^0I^f \tag{8.3.17}$$

$$^1V_k^f = {}^1V_k - {}^1Z_{kk} \cdot {}^1I^f \tag{8.3.18}$$

$$^2V_k^f = -{}^2Z_{kk} \cdot {}^2I^f \tag{8.3.19}$$

From equations (8.3.15) to (8.3.19), the following relationships are obtained.

$$^0I^f = {}^1I^f = {}^2I^f = \frac{{}^1V_k}{{}^0Z_{kk} + {}^1Z_{kk} + {}^2Z_{kk} + 3Z^f} \tag{8.3.20}$$

Similar considerations yield the fault currents for other types of short-circuit fault. The results for line-to-ground, line-to-line, line-to-line-to-ground, and line-to-line-to-line faults are illustrated in Table 8.1.

Table 8.1. Fault currents for short-circuit faults

Fault	$^1I^f$	$^2I^f$	$^0I^f$
L–G	$\dfrac{V_i}{{}^1Z_{ii} + {}^2Z_{ii} + {}^0Z_{ii} + 3.Z^f}$	$^1I^f$	$^1I^f$
L–L	$\dfrac{V_i}{{}^1Z_{ii} + {}^2Z_{ii} + Z^f}$	$-{}^1I^f$	0
L–L–G	$\dfrac{V_i}{({}^2Z'_{ii}\cdot{}^0Z'_{ii})/({}^2Z'_{ii} + {}^0Z'_{ii}) + {}^1Z'_{ii}}$	$\dfrac{-{}^0Z'_{ii}\cdot{}^1I^f}{{}^2Z'_{ii} + {}^0Z'_{ii}}$	$\dfrac{-{}^2Z'_{ii}\cdot{}^1I^f}{{}^2Z'_{ii} + {}^0Z'_{ii}}$
L–L–L–G	$\dfrac{V_i}{{}^1Z_{ii} + Z^f}$	0	0

where

$$^1Z'_{ii} = {}^1Z_{ii} + 0.5\,Z^f$$

$$^2Z'_{ii} = {}^2Z_{ii} + 0.5\,Z^f$$

$$^0Z'_{ii} = {}^0Z_{ii} + 0.5\,Z^f$$

These fault currents at the fault location are then added to the current vectors $[^0I]$, $[^1I]$ and $[^2I]$ to produce the fault current vectors $[^0I^f]$, $[^1I^f]$ and $[^2I^f]$. For a fault at bus k these are

$$^0I^f_i = \begin{cases} 0 & \text{for } i \neq k \\ -{}^0I^f & \text{for } i = k \end{cases} \tag{8.3.21}$$

$$^1I^f_i = \begin{cases} {}^1I_i & \text{for } i \neq k \\ {}^1I_k - {}^1I^f & \text{for } i = k \end{cases} \tag{8.3.22}$$

$$^2I^f_i = \begin{cases} 0 & \text{for } i \neq k \\ -{}^2I^f & \text{for } i = k \end{cases} \tag{8.3.23}$$

The fault voltages are then obtained from equations (8.3.10) to (8.3.12) by substituting the fault current vector for the prefault current vector, i.e.

$$[^0V^f] = [^0Z][^0I^f] \tag{8.3.24}$$

$$[^1V^f] = [^1Z][^1I^f] \tag{8.3.25}$$

$$[^2V^f] = [^2Z][^2I^f] \tag{8.3.26}$$

Open-circuit faults

The system is now represented by a two-port network across which the faulty line is connected as shown in Fig. 8.10. In this case, the prefault voltages have to be obtained from a load-flow study.

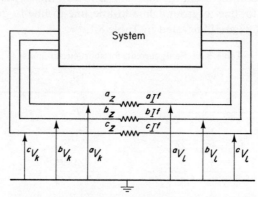

Fig. 8.10. Two-part network with faulty line where ^aZ, ^bZ or ^cZ may be on open circuit

For an open-circuit fault on phases 'b' and 'c' the boundary conditions are

$$I_b = I_c = 0$$
$$(^aV_l - {}^aV_k) = {}^aZ^aI$$

Using these equations with the sequence transformation, the following relationships result

$$^0I^f = {}^1I^f = {}^2I^f = {}^aI^f/3 \tag{8.3.27}$$

and

$$^0V_{kl}^f + {}^1V_{kl}^f + {}^2V_{kl}^f = {}^aZ \cdot {}^aI^f = 3 \cdot ({}^aZ)^1I^f \tag{8.3.28}$$

where

$$V_{kl} = V_l - V_k$$

Equations (8.3.27) and (8.3.28) define the connection of the Thevenin equivalent sequence networks at the fault location to solve for the fault currents.

The equivalent Thevenin impedances are the sequence impedances of the system between the two buses k and l, i.e.

$$Z_{eqv} = Z_{kk} + Z_{ll} - Z_{lk} - Z_{kl} \tag{8.3.29}$$

and the equivalent Thevenin voltage is given by the difference between the voltage at buses l and k with the faulted line disconnected.

During the fault the sequence voltages $^0V_{kl}^f$, $^1V_{kl}^f$ and $^2V_{kl}^f$ have the same expressions as equations (8.3.17) (8.3.18) (8.3.19). Thus similar considerations, as in the case of the line-to-ground short-circuit, lead to the following expression for the fault currents.

$$^0I^f = {}^1I^f = {}^2I^f = \frac{^1V_{kl}}{^0Z_{eqv} + {}^1Z_{eqv} + {}^2Z_{eqv} + 3({}^aZ)} \tag{8.3.30}$$

The case of a single open-circuit fault can be analysed in a similar manner and the final relevant equations are shown in Table 8.2.

The fault current vector is formed as follows:

$$^0I_i^f = \begin{cases} 0 & \text{for } i = 1, n \quad i \neq k \text{ or } l \\ -^0I^f & \text{for } i = k \\ ^0I^f & i = l \end{cases}$$

$$^1I_i^f = \begin{cases} ^1I_i & \text{for } i = 1, n \quad i \neq k \text{ or } l \\ ^1I_k - ^1I^f & i = k \\ ^1I_l + ^1I^f & i = l \end{cases}$$

$$^2I_i^f = \begin{cases} 0 & i = 1, n \quad i \neq k \text{ or } l \\ -^2I^f & i = k \\ ^2I^f & i = l \end{cases}$$

and the voltage vector is given by equations (8.3.24) to (8.3.26).

Table 8.2. Fault currents for open–circuit faults

Fault	$^1I^f$	$^2I^f$	$^0I^f$
ONE $0 - C$	$\dfrac{V_l - V_k}{(^2Z' \cdot ^0Z')/(^2Z' + ^0Z') + ^1Z'}$	$\dfrac{^0Z' \cdot ^1I^f}{^2Z' + ^0Z'}$	$\dfrac{^2Z' \cdot ^1I^f}{^2Z' + ^0Z'}$
TWO $0 - C$	$\dfrac{V_l - V_k}{^1Z + ^2Z + ^0Z + Z^f}$	$^1I^f$	$^1I^f$

where

$$^1Z = {^1Z_{kk}} + {^1Z_{ll}} - {^1Z_{kl}} - {^1Z_{lk}}$$

$$^0Z = {^0Z_{kk}} + {^0Z_{ll}} - {^0Z_{kl}} - {^0Z_{lk}}$$

$$^2Z = {^2Z_{kk}} + {^2Z_{ll}} - {^2Z_{kl}} - {^2Z_{lk}}$$

Z^f is the sum of the positive, negative and zero sequence impedances of the faulty circuit, i.e.

$$Z^f = {^1Z^f} + {^2Z^f} + {^0Z^f}$$

$$^1Z' = {^1Z} + {^1Z^f}$$

$$^0Z' = {^0Z} + {^0Z^f}$$

$$^2Z' = {^2Z} + {^2Z^f}$$

From the fault voltages the branch currents are obtained as follows:

$$^0I_{ij}^f = {^0y_{ij}}(^0V_i^f - {^0V_j^f}) \tag{8.3.31}$$

$$^1I_{ij}^f = {^1y_{ij}}(^1V_i^f - {^1V_j^f}) \tag{8.3.32}$$

$$^2I_{ij}^f = {^2y_{ij}}(^2V_i^f - {^2V_j^f}) \tag{8.3.33}$$

Where necessary the corrections for taps on the positive and negative sequence networks are

$$^1I_{ij}^f = {}^1y_{ij}\{{}^1V_i^f(1-\tau_{ij}) - {}^1V_j^f\}$$ (8.3.34)

$$^2I_{ij}^f = {}^2y_{ij}\{{}^2V_i^f(1-\tau_{ij}) - {}^2V_j^f\}$$ (8.3.35)

Finally, the machine contributions may be calculated, i.e.

$$^0I_i^{Mf} = -\,{}^0y_i^M \cdot {}^0V_i^f$$

$$^1I_i^{Mf} = I_i - {}^1y_i^M \cdot {}^1V_i^f$$

$$^2I_i^{Mf} = -\,{}^2y_i^M \cdot {}^2V_i^f$$

8.4. PROGRAM DESCRIPTION AND TYPICAL SOLUTIONS

A fault analysis program must be capable of analysing the following a.c. system faults.

(1) line-to-ground short circuit;
(2) line-to-line short circuit;
(3) line-to-line-to-ground short circuit;
(4) line-to-line-to-line to ground short circuit;
(5) single open-circuit line;
(6) double open-circuit line.

However, the conventional a.c. program cannot in general be used for the solution of similar faults in the presence of h.v.d.c. links. An alternative program for this purpose is described in Chapter 11.

Basic to the fault study program is the determination of the impedance matrix of the system, the elements of which can be used, along with the conditions imposed by the type of fault, to directly solve for the fault currents and voltages.

The main steps of a general-purpose fault program are indicated in Fig. 8.11. Prefault information and typical outputs for balanced and unbalanced fault conditions are illustrated in the following computer printouts. The printouts relate to studies carried out for the New Zealand South Island a.c. system.

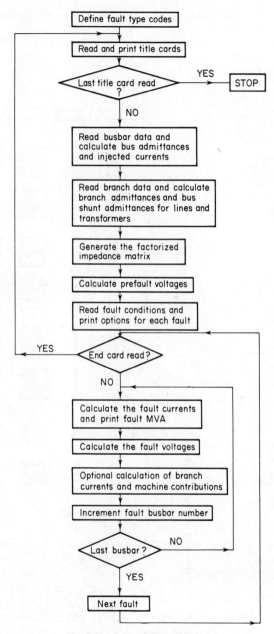

Fig. 8.11. General flow diagram

220KV NEW ZEALAND SYSTEM   SOUTH ISLAND
————————————————————————

XXXXXXXXXXXXXXXXXXXXXXXXXXXXXXXXXXXXXXXXXXXXXXXXXXXXXXXXXXXXXXXXXXXXXXXX

0 BUSES   100. MVA BASE   0 LINES

ALL AC SYSTEM DATA WILL BE PRINTED

B U S – D A T A

NAME	LOAD MW	MVAR	M/C GENERATION MW	MVAR	POS R	X	NEG R	X	ZERO R	X	BUS VOLTAGE (KV)
AVIEMORE			220.0	-6.9	0.00182	0.09900	0.00182	0.07980	0.00182	0.04500	0.0000
BENMORE	597.2	180.0	540.0	163.2	0.00066	0.07800	0.00066	0.05396	0.00066	0.02268	0.0000
BROMLEY	129.6	38.0									0.0000
HALFWAYB	195.3	40.4									0.0000
INVERCAR	183.2	20.0	0.0	142.8	0.00490	0.33500	0.00490	0.32200	0.00490	0.13830	0.0000
ISLINGTO	594.1	124.3									0.0000
LIVINGTO	59.2	9.2									0.0000
LIVINGST											0.0000
MANAPOUR			400.0	42.9	0.00050	0.03930	0.00050	0.02630	0.00050	0.01310	0.0000
OHAU-A			264.0	-4.9	0.00153	0.00936	0.00153	0.00634	0.00153	0.00345	0.0000
OHAU-B			175.0	-7.4	0.00165	0.00950	0.00165	0.00640	0.00165	0.00375	0.0000
ROXBURGH			174.3	10.3	0.00169	0.08550	0.00169	0.06690	0.00169	0.03190	0.0000
SOUTHDUN	34.2	-18.3									0.0000
STOKE	53.2	-20.3									0.0000
TEKAPO-B			160.0	-10.7	0.00112	0.09900	0.00112	0.11250	0.00112	0.05621	0.0000
TIWAI.76	288.0	105.7									0.0000
TWIZEL											
WAITAKI	30.0			-6.6	0.00555	0.19200	0.00555	0.12840	0.05550	0.64200	0.0

END OF BUSBAR DATA

New Zealand South Island 220KV (Results)

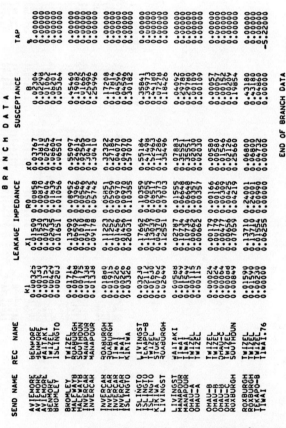

New Zealand South Island 220KV (Results)

198

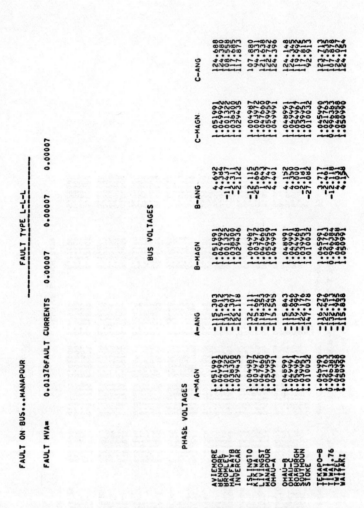

New Zealand South Island 220KV (Results)

BRANCH CURRENTS

PHASE CURRENTS

FROM	TO	MAGN	PHASE A ANG	MAGN	B ANG	MAGN	C ANG
AVIEMORE	BENMORE	0.388349	122.788	0.388349	-12.793	0.388349	117.203
AVIEMORE	WEITAKI	0.388944	-77.740	0.388944	162.776	0.388944	117.224
BENMORE	TWIZEL	0.330732	120.655	0.330732	-12.546	0.330732	-112.546
BROMLEY	ISLINGTO	0.748692	107.605	0.748892	12.351	0.748892	123.438
BROMLEY	TWIZEL	.985947	53.577	.985948	173.503	.985947	-66.431
HALFWAYB	ROXBURGH	.088263	34.623	.088263	153.169	.682963	-76.081
INVERCAR	SOUTHDUN	.011743	34.646	.011743	152.641	.620753	-87.363
INVERCAR	MANAPOUR	1.011743	49.646	1.011743	169.641	1.011743	70.363
INVERCAR	MANAPOUR	.406253	.932	.406253	154.927	.406253	-85.077
INVERCAR	ROXBURGH	.023022	35.086	.023022	155.000	.023022	-84.928
INVERCAR	ROXBURGH	.067663	126.790	.067663	144.200	.637663	74.809
ISLINGTO	KIKIWA	1.067063	126.701	1.067063	120.706	1.067063	176.290
ISLINGTO	LIVINGST	1.389375	55.354	1.389375	175.449	1.389375	-64.655
ISLINGTO	TEKAPO-B	.974797	.867	.974797	174.463	.974797	-64.503
KIKIWA	STOKEL	0.224208	55.602	0.224208	174.501	0.257838	-105.507
LIVINGST	ROXBURGH	0.224208	-9.854	0.224208	28.301	0.224209	146.133
LIVINGST	WAITAKI	1.621177	.684	1.621177	163.321	1.621177	-49.325
MANAPOUR	TIMAI	.940191	-33.613	.940191	-79.622	.974917	-106.374
OHAU-A	TWIZEL	1.019275	144.598	1.019275	55.397	1.019275	165.393
OHAU-B	TWIZEL	1.289990	109.008	1.289990	10.886	1.289990	130.882
OHAU-B	TWIZEL	1.079608	56.430	1.079608	176.890	1.079608	63.557
OHAU-B	OHAUEC	0.598291	134.898	0.598291	-14.795	0.598291	105.093
ROXBURGH	SOUTHDUN	0.538277	73.271	0.538277	-66.733	0.538277	46.778
ROXBURGH	TWIZEL	0.414163	73.963	0.414163	104.959	0.414163	4.738
ROXBURGH	TWIZEL	3.249168	-152.272	3.249168	-132.276	3.249168	-87.732
TEKAPO-B	TIMAI.76						

MACHINE CONTRIBUTIONS

PHASE CURRENTS

BUS NAME	PHASE A MAGN	ANG	B MAGN	ANG	C MAGN	ANG
AVIEMORE	2.092327	-113.512	2.092328	6.484	2.092327	126.480
BENOKE	5.372743	-132.400	5.372765	-12.434	5.372743	107.560
BROMLEY B	.0000	.000	.0000	.000	.0000	.000
INVERCAR	.0000	.000	.0000	.000	.0000	.000
ISLINGTO	1.421128	137.904	1.421128	-102.088	1.421128	17.908
MIKIWAI	.000094	-.000	.000094	.000	.000094	103.003
LIVINGST	.000971	-.000	.000971	.000250	.000971	119.796
MANAPOUR	2.038571	-114.566	2.038571	5.423	2.038571	125.425
OMAU-A						
OMAU-B	1.714448	102.545	1.714448	17.450	1.714448	137.446
OMAU-CC	1.671500	-6.800	1.671500	-113.636	1.671500	111.375
SOUBURGH	-.100	-.100	.100	-.8700	.100	-.100
STOKE	.0000	.000	.0000		.0000	.000
TEKAPO-B	1.533137	-113.463	1.533138	7.530	1.533137	127.529
TIWAI-6	.000000	.00000	.000000	7.00000	.000000	.00000
TIWAL.76	.000000	.00422	.000000	.00004	.000000	.00000
WAITAKI	.292333	-101.422	.292333	16.574	.292333	136.569

FAULT ON BUS...MANAPOUR   FAULT TYPE L-G

FAULT MVA= 3085.79068FAULT CURRENTS 50.42398   0.00000   0.00000

BUS VOLTAGES

SEQUENCE VOLTAGES

BUS NAME	PSEQ-REAL	PSEQ-IMAG	ZSEQ-REAL	ZSEQ-IMAG	NSEQ-REAL	NSEQ-IMAG
AVIEMORE						
BENMORE						
BROMLEY						
HALFWAYB						
INVERCAR						
ISLINGTO						
KIKIWA						
LIVINGST						
MANAPOUR						
OHAU-A						
OHAU-B						
OHAU-C						
ROXBURGH						
SOUTHDUN						
STOKE						
TEKAPO-B						
TIWAI.76						
TIWAI.76						
WAITAKI						

PHASE VOLTAGES

BUS NAME	A-MAGN	A-ANG	B-MAGN	B-ANG	C-MAGN	C-ANG
AVIEMORE						
BENMORE						
BROMLEY						
HALFWAYB						
INVERCAR						
ISLINGTO						
KIKIWA						
LIVINGST						
MANAPOUR						
OHAU-A						
OHAU-B						
OHAU-C						
ROXBURGH						
SOUTHDUN						
STOKE						
TEKAPO-B						
TIWAI.76						
TIWAI.76						
WAITAKI						

New Zealand South Island 220KV (Results)

FAULT ON BUS...MANAPUR   FAULT TYPE L-L

FAULT MVA= 4452.54917 FAULT CURRENTS   0.00000   36.37889   36.37889

BUS VOLTAGES

SEQUENCE VOLTAGES

BUS NAME	PSEQ-REAL	PSEQ-IMAG	ZSEQ-REAL	ZSEQ-IMAG	NSEQ-REAL	NSEQ-IMAG
AVIEMORE	1.012641	0.087434	0.000000	0.000000	0.018771	0.000512
BENMORE	1.124447	0.082077	0.000000	0.000000	0.017496	0.000197
BROMLEY	0.945049	0.089711	0.000000	0.000000	0.012043	0.004298
INVERCAR	0.556222	0.016651	0.000000	0.000000	0.333317	0.006637
ISLINGTO	0.946069	0.201219	0.000000	0.000000	0.020949	0.004392
KIKIWA	0.874721	0.148133	0.000000	0.000000	0.037116	0.010008
LIVINGST	0.844481	0.133168	0.000000	0.000000	0.037116	0.010008
MANAPOUR	0.445532	0.028149	0.000000	0.000000	0.425516	0.020038
OHAU-A	1.005502		0.000000	0.000000	0.022276	0.000038
OHAU-B	1.004390	0.077727	0.000000	0.000000	0.023010	0.000225
OHAU-C	0.993690	0.081560	0.000000	0.000000	0.025266	0.000657
ROXBURGH	0.873990	0.055450	0.000000	0.000000	0.112580	0.003872
SOUTHDUN	0.861344	0.027605	0.000000	0.000000	0.018330	0.000972
STOKE						
TEKAPO-B	1.007006	0.069498	0.000000	0.000000	0.020687	0.000078
TIWAI	0.555696	0.097544	0.000000	0.000000	0.342991	0.007995
TIWAI.76	0.223731	0.108054	0.000000	0.000000	0.283451	0.006313
WAITAKI	1.008047	0.077572	0.000000	0.000000	0.021654	0.000543

PHASE VOLTAGES

BUS NAME	A-MAGN	A-ANG	B-MAGN	B-ANG	C-MAGN	C-ANG
AVIEMORE	1.006201	-115.953	1.035154	4.874	1.008131	125.885
BENMORE	1.099091	-132.424	1.043124	4.571	1.018718	125.827
BROMLEY	0.927720	-132.459	0.974819	-11.321	0.992853	109.807
HALFWAY B				-11.956		125.357
INVERCAR	0.496413	157.459	0.899829	-1.454	0.489836	154.208
ISLINGTO	0.956762	-133.116	0.986636	-25.005	0.956653	109.096
KIKIWA	0.865266	-179.052	0.897728	-25.574	0.867662	123.892
LIVINGST	0.869406	-174.320	0.902511	-3.856	0.846406	-176.330
MANAPOUR	0.997030	168.390	1.031481	4.587	0.999914	125.811
OHAU-A	0.994754	-116.671	1.036354	4.339	0.997479	125.608
OHAU-B	0.993456	-116.945	1.036354	4.534	0.994747	125.600
OHAU-C	0.820335	-119.223	1.043254	-0.326	0.820511	127.467
ROXBURGH	0.869659	148.083	0.993776	-0.826	0.808532	125.487
SOUTHDUN	0.961565		0.993399	-26.971	0.961456	94.129
TEKAPO-B	0.998063	-117.035	1.030045	3.873	1.000177	125.006
TIWAI	0.504669	150.174	0.886053	-3.745	0.603651	-149.971
TIWAI.76	0.221816	155.811	0.643772	-14.216	0.290564	-151.513
WAITAKI	0.993390	-118.633	1.033708	4.344	1.001360	125.508

FAULT ON BUS...MANAPOUR   FAULT TYPE L-L-G

FAULT MVA= 5298.3094#FAULT CURRENTS   0.00000   42.80370   43.77439

BUS VOLTAGES

SEQUENCE VOLTAGES

BUS NAME	PSEQ-REAL	PSEQ-IMAG	ZSEQ-REAL	ZSEQ-IMAG	NSEQ-REAL	NSEQ-IMAG
AVIEMORE	0.999767	0.087942	0.000924	0.000152	0.009495	0.000510
BENMORE	1.009731	0.088476	0.000907	0.000257	0.008854	0.000339
BROMLEY	0.933387	-0.085588	0.000902	0.000163	-0.004040	-0.000385
HALFWAYB	0.740024	-0.085524	0.125906	0.003352	0.168811	0.001231
ISLINGTO	0.932942	-0.197725	0.000891	0.000110	0.010663	0.001943
KIKIWA	0.957790	-0.10367	0.000478	0.000309	0.009104	0.003785
LIVINGST	0.906733	-0.032638	0.025244	0.000207	0.021184	0.002089
MANAPOUR	0.225144	0.208231	0.205455	0.001215	0.011471	0.000469
OHAU-A	0.991171					
OHAU-B	0.989352	0.078378	0.000925	0.001491	0.011645	0.000422
OHAU-C	0.991401	0.081853	0.007742	0.001257	0.010277	0.000447
ROXBURGH	0.809969	-0.077492	0.020144	0.000200	0.009796	0.003949
SOUTHDUN	0.797158	-0.072925	0.001207	0.000260	0.008349	0.000654
STOKE	0.854505	0.034410	0.000661	0.000417		
TEKAPO-B	0.993787	0.070117	0.000533	0.000076	0.010471	0.000316
TIWAI	0.274475	-0.010952	0.165510	0.000020	0.039728	0.002868
TIWAI.76	0.358679	-0.071062	0.000730	0.001624	0.161106	0.002420
TWIZEL	0.388682	0.076269	0.000073		0.161099	0.000565
WAITAKI	0.993609	0.078214	0.001769	0.000654	0.011075	

PHASE VOLTAGES

BUS NAME	A-MAGN	A-ANG	B-MAGN	B-ANG	C-MAGN	C-ANG
AVIEMORE	0.998094	-115.387	1.010064	5.013	0.998778	125.456
BENMORE	0.998947	-115.868	1.013569	5.744	1.000024	125.147
BROMLEY	0.943765	-131.814	0.943204	-11.234	0.941945	109.439
HALFWAYB	0.244552	-124.909	0.877504	-0.893	0.234771	121.608
ISLINGTO	0.947880	-132.503	0.965348	-25.931	0.947844	108.533
KIKIWA	0.957248	-118.047	0.966863	-25.948	0.966460	94.998
LIVINGST	0.909000	-118.660	0.678650	5.130	0.950159	122.780
MANAPOUR	0.987199	-115.767	0.678743	4.810	0.989994	125.270
OHAU-B	0.984537	-115.026	1.005132	4.580	0.987781	125.033
OHAU-C	0.987337	-115.620	0.893491	4.759	0.990178	123.228
ROXBURGH	0.770761	-123.103	0.893137	-0.459	0.776315	123.022
SOUTHDUN	0.762482	-147.463	0.829177	-0.843	0.052504	93.582
STOKE	0.952992		0.970119	-26.920	0.952335	
TEKAPO-B	0.989677	-116.465	0.997318	4.001	0.991852	124.514
TIWAI	0.283391	-153.065	0.535825	-11.741	0.245018	136.705
TIWAI.76	0.389182		0.547600	-4.479	0.368808	
WAITAKI	0.990152	-115.945	1.009435	4.479	0.990025	124.950

FAULT ON BUS...MANAPOUR   FAULT TYPE L-L-L

FAULT MVA= 6660.5B621FAULT CURRENTS   36.27956   36.27954   36.27954   36.27954

BUS VOLTAGES

SEQUENCE VOLTAGES

BUS NAME	PSEQ-REAL	PSEQ-IMAG	ZSEQ-REAL	ZSEQ-IMAG	NSEQ-REAL	NSEQ-IMAG
AVIEMORE	0.986561	0.087756	0.000000	0.000000	0.0000000	0.0000000
BENMORE	0.988194	0.082998	0.000000	0.000000	0.0000000	0.0000000
BROMLEY	0.917656	0.182949	0.000000	0.000000	0.0000000	0.0000000
HALEWAYB	0.730208	0.023518	0.000000	0.000000	0.0000000	0.0000000
INVERCAR	0.229464	0.003930	0.000000	0.000000	0.0000000	0.0000000
ISLINGTO	0.913313	0.194867	0.000000	0.000000	0.0000000	0.0000000
KIKIWA	0.847223	0.041095	0.000000	0.000000	0.0000000	0.0000000
LIVINGST	0.936790	0.024109	0.000000	0.000000	0.0000000	0.0000000
MANAPOUR	0.000000	0.002519	0.000000	0.000000	0.0000000	0.0000000
OHAU-A	0.976061		0.000000	0.000000	0.0000000	0.0000000
OHAU-B	0.973924	0.078220	0.000000	0.000000	0.0000000	0.0000000
OHAU-C	0.976332	0.081658	0.000000	0.000000	0.0000000	0.0000000
ROXBURGH	0.742290	0.005997	0.000000	0.000000	0.0000000	0.0000000
SOUTHDUN	0.731421	0.021891	0.000000	0.000000	0.0000000	0.0000000
STOKE	0.842014	0.427798	0.000000	0.000000	0.0000000	0.0000000
TEKAPO-B	0.980223	0.070026	0.000000	0.000000	0.0000000	0.0000000
TIWAI	0.199147	0.001307	0.000000	0.000000	0.0000000	0.0000000
TIWAI.76	0.199147	0.004352	0.000000	0.000000	0.0000000	0.0000000
WAITAKI	0.988301	0.076977	0.000000	0.000000	0.0000000	0.0000000

PHASE VOLTAGES

BUS NAME	A-MAGN	A-ANG	B-MAGN	B-ANG	C-MAGN	C-ANG
AVIEMORE	0.990456	-115.912	0.990457	5.084	0.990456	125.079
BENMORE	0.991666	-115.260	0.991686	-0.977	0.991666	124.793
BROMLEY	0.935886	-131.240	0.935886	-11.265	0.935886	120.774
HALEWAYB	0.730586	-118.564	0.730586	-1.845	0.730586	118.480
INVERCAR	0.229662	-122.374	0.229662	-2.378	0.229662	117.617
ISLINGTO	0.934739	-131.964	0.934739	-11.969	0.934739	108.027
KIKIWA	0.938790	-145.514	0.938790	-25.508	0.938790	94.477
LIVINGST	0.934941	-117.512	0.934941	-2.564	0.934910	122.779
MANAPOUR	0.000040	-117.003	0.000003	4.803	0.000010	125.779
OHAU-A	0.979543	-115.163	0.979543	4.828	0.979542	124.828
OHAU-B	0.977060	-115.403	0.977060	4.592	0.977060	124.588
OHAU-C	0.979772	-115.330	0.979772	4.663	0.979772	124.677
ROXBURGH	0.742148	-119.687	0.742114	-0.214	0.742148	114.291
SOUTHDUN	0.731748	-117.814	0.731748	1.148	0.731748	93.060
STOKE	0.944457	-146.931	0.944457	-26.936	0.944457	93.060
TEKAPO-B	0.980251	-115.209	0.980251	4.087	0.982021	124.087
TIWAI	0.200035	-133.231	0.200035	-13.276	0.200035	116.776
TIWAI.76	0.199027	-132.234	0.197601	-12.475	0.200067	107.524
WAITAKI	0.981901	-115.520	0.981902	4.475	0.981901	124.551
WAITAKI	0.981901	-115.440	0.981902	4.555	0.981901	124.551

END OF FAULT STUDY

New Zealand South Island 220KV (Results)

8.5. REFERENCES

1. Wagner and Evans, 1933. *Symmetrical Components*. McGraw-Hill, New York.
2. A. Brameller, 1972. 'Three-phase short-circuit calculation using digital computer.' Internal Report, University of Manchester Institute of Science and Technology.
3. A. Brameller, 1972. 'User manual for unbalanced fault analysis.' Internal Report, University of Manchester Institute of Science and Technology.
4. J. Preece, 1975. 'Symmetrical components fault studies.' M.Sc. Dissertation, University of Manchester Institute of Science and Technology.
5. C. B. Lake, 1978. 'A computer program for a.c. fault studies in a.c.–d.c. systems.' M.E. thesis, University of Canterbury, New Zealand.
6. B. Stott and E. Hobson, 1971. 'Solution of large power-system networks by ordered elimination: a comparison of ordering schemes', *Proc. IEE*, **118** (1), 125–134.
7. K. Zollenkopf, 1970. 'Bifactorization—basic computational algorithm and programming techniques.' Conference on Large Sets of Sparse Linear Equations, Oxford, 75–96.
8. P. M. Anderson, 1973. *Analysis of Faulted Power Systems*. Iowa State University Press.

Power System Stability—
Basic Model

9.1. INTRODUCTION

The stability of a power system following some predetermined operating condition is a dynamic problem and requires more elaborate plant component models than the ones discussed in previous chapters. It is normally assumed that prior to the dynamic analysis, the system is operating in the steady state and that a load-flow solution is available.

Two types of stability studies are normally carried out. The subsequent recovery from a sudden large disturbance is referred to as 'Transient Stability' and the solution is obtained in the time domain. The period under investigation can vary from a fraction of a second, when first swing stability is being determined, to over ten seconds when multiple swing stability must be examined.

The term 'Dynamic Stability' is used to describe the long-time response of a system to small disturbances or badly set automatic controls. The problem can be solved either in the time domain or in the frequency domain. In this book, dynamic stability is treated as an extension of transient stability and is thus solved in the time domain. Such extension normally requires modification of some plant component models and often the introduction of new models, but because of the smaller perturbations and longer study duration the small time constant effects can be ignored.

Consideration is given in this chapter to the dynamic modelling of a power system containing synchronous machines and basic loads. More advanced synchronous machine models as well as other power-system components, such as induction motors and a.c.–d.c. converters, are considered in Chapter 10.

The form of the equations

To a greater or lesser extent, all system variables require time to respond to any change in operating conditions and a large set of differential equations can be written to determine this response. This is impractical, however, and many assumptions must be made to simplify the system model. The assumptions made depend on the problem being investigated and no clear definitive model exists.

A major problem with a time domain solution is the 'stiffness' of the system (Appendix I). That is, the time constants associated with the system variables vary enormously. When only synchronous machines are being considered, rotor swing stability is the principal concern. The main time constants associated with the rotor are of the order of 1 to 10 seconds. The form of the solution is dominated by time constants of this order and smaller or greater time constants have less significance.

The whole of the a.c. transmission network responds rapidly to configurational changes as well as loading changes. The time constants associated with the network variables are extremely small and can be considered to be zero without significant loss of accuracy. Similarly the synchronous machine stator time constants may be taken as zero. The relevant differential equations for these rapidly changing variables are transformed into algebraic equations.

When the time constant is large or the disturbance is such that the variable will not change greatly, the time constant may be regarded as infinite, that is the variable becomes a constant. Excitation voltage or mechanical power to the synchronous machine may often be treated as constant in short duration studies without appreciable loss of accuracy. Depending on how the computer program is written variables which become constant may be treated by either:

(1) retaining the differential equation but assigning a very large value to the relevant time constant;
(2) removing the differential equation.

For flexibility both methods are usually incorporated in a program.

A system which after these initial simplifications contains α differential variables, contained in the vector Y_α, and β algebraic variables, contained in the vector X_β, may be described by the matrix equations:

$$p Y_\alpha = F_\alpha(Y_\alpha, X_\beta) \tag{9.1.1}$$

$$0 = G_\beta(Y_\alpha, X_\beta) \tag{9.1.2}$$

where p denotes the differential operator $\dfrac{d}{dt}$.

Frames of reference

The choice of axes, or frame of reference, in which the system equations are formulated is of great importance, as it influences the analysis.

For synchronous machines, the most appropriate frame of reference is one which is attached to the rotor, i.e. it rotates at the same speed as the rotor. The main advantage of this choice is that the coefficients of the equations developed for the synchronous machine are not time-dependent. The major axis of this frame of reference is taken as the rotor pole or 'direct axis'. The second axis lies 90 degree (electrical) from each pole and is referred to as the 'quadrature axis'.

In the dynamic state, each synchronous machine is rotating independently from each other and transforming between synchronous machine frames through the network is difficult. This is overcome by choosing an independent frame of reference for the network and transforming between this frame and the synchronous machine frames at the machine terminals. The most obvious choice for the network is a frame of reference which rotates at synchronous speed. The two axes are obtained from the initial steady-state load-flow slack busbar. Although the network frame is rotating synchronously, this does not stop each nodal voltage or branch current from having an independent frequency during the dynamic analysis.

9.2. SYNCHRONOUS MACHINES—BASIC MODELS

Mechanical equations

The mechanical equations of a synchronous machine are very well established[1],[2] and need be only briefly outlined. Three basic assumptions are made in deriving the equations:

(1) Machine rotor speed does not vary greatly from synchronous speed (1.0 p.u.).
(2) Machine rotational power losses due to windage and friction are ignored.
(3) Mechanical shaft power is smooth, that is the shaft power is constant except for the results of speed governor action.

Assumption 1 allows per unit power to be equated with per unit torque. From Assumption 2, the accelerating power of the machine (Pa) is the difference between the shaft power (Pm) as supplied by the prime mover or absorbed by the load and the electrical power (Pe). The acceleration (α) is thus:

$$\alpha = \frac{Pa}{Mg} = \frac{(Pm - Pe)}{Mg}$$ (9.2.1)

where Mg is the angular momentum.

The acceleration is independent of any constant speed frame of reference and it is convenient to choose a synchronously rotating frame to define the rotor angle (δ). Thus:

$$\frac{d^2\delta}{dt^2} = \frac{(Pm - Pe)}{Mg}$$ (9.2.2)

The angular momentum may be further defined by the inertia constant Hg (measured in MWS/MVA) which is relatively constant regardless of the size of the machine, i.e.:

$$Mg = \frac{Hg}{\pi fo}$$ (9.2.3)

where fo is the system base frequency.

Eddy currents induced in the rotor iron or in the damping windings produce torques which oppose the motion of the rotor relative to the synchronous speed.

A deceleration power can be introduced into the mechanical equations to account for this damping, giving

$$\frac{d^2\delta}{dt^2} = \frac{1}{Mg}\left(Pm - Pe - Da\frac{d\delta}{dt}\right) \tag{9.2.4}$$

The damping coefficient (Da), measured in watts/rad/sec, has been largely superseded by a synchronous machine model which includes the subtransient effect of the damper windings in the electrical equations, but it is still used in some programs.

Two single-order ordinary differential equations may now be written to describe the mechanical motion of the synchronous machine, i.e.

$$p\omega = \frac{1}{Mg}(Pm - Pe - Da(\omega - 2\pi fo)) \tag{9.2.5}$$

$$p\delta = \omega - 2\pi fo \tag{9.2.6}$$

Electrical equations

The derivation of equations to account for flux changes in a synchronous machine has been given by Concordia[3] and Kimbark.[4] A brief outline only will be given in this section, so that various electrical quantities may be defined and phasor diagrams constructed. The approximations made in the derivation are as follows:

(1) The rotor speed is always sufficiently near 1.0 p.u. that it may be considered a constant.
(2) All inductances defined in this section are independent of current. The effects due to saturation of iron are considered in Chapter 10.
(3) Machine winding inductances can be represented as constants plus sinusoidal harmonics of rotor angle.
(4) Distributed windings may be represented as concentrated windings.
(5) The machine may be represented by a voltage behind an impedance.
(6) There are no hysteresis losses in the iron, and eddy currents are only accounted for by equivalent windings on the rotor.
(7) Leakage reactance only exists in the stator.

Using these assumptions, classical theory permits the construction of a model for the synchronous machine in the steady-state, transient and subtransient states.

The per unit system adopted is normalized to eliminate factors of $\sqrt{2}, \sqrt{3}, \pi$ and turns ratio, although the term 'proportional' should be used instead of 'equal' when comparing quantities. Note that one p.u. field voltage produces 1.0 p.u. field current and 1.0 p.u. open-circuit terminal voltage at rated speed.

STEADY STATE EQUATIONS

Figure 9.1 shows the flux and voltage phasor diagram for a cylindrical rotor synchronous machine in which all saturation effects are ignored. The flux Ff

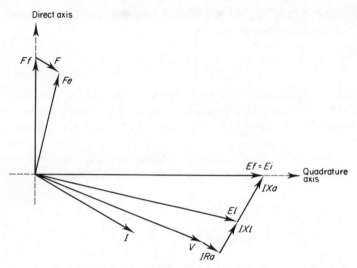

Fig. 9.1. Phasor diagram of a cylindrical rotor synchronous machine in the steady state

is proportional to the field current If and the applied field voltage and it acts in the direct axis of the machine. The stator open-circuit terminal voltage Ei is proportional to Ff but lies on the quadrature axis. The voltage Ei is also proportional to the applied field voltage and may be referred to as Ef.

When the synchronous machine is loaded, a flux F proportional to and in phase with the stator current I is produced which, when added vectorially to the field flux Ff, gives an effective flux Fe. The effective internal stator voltage El is due to Fe and lags it by 90 degrees. The terminal voltage V is found from this voltage El by considering the volt drops due to the leakage reactance Xl and armature resistance Ra. By similar triangles, the difference between Ef and El is in phase with the IXl volt drop and is proportional to I. Therefore the voltage difference may be treated as a volt drop across an armature reactance Xa. The sum of Xa and Xl is termed the synchronous reactance.

For the salient pole synchronous machine the phasor diagram is more complex. Because the rotor is symmetrical about both the d and q axes it is convenient to resolve many phasor quantities into components in these axes. The stator current may be treated in this manner. Although F_d will be proportional to I_d and F_q will be proportional to I_q, because the iron paths in the two axes are different, the total armature reaction flux F will not be proportional to I nor necessarily be in phase with it. Retaining our earlier normalizing assumptions, it may be assumed that the proportionality between I_d and F_d is unity but the proportionality between I_q and F_q is less than unity and is a function of the saliency.

In Fig. 9.2 the phasor diagram of the salient pole synchronous machine is shown. Note that the d and q axes armature reactances have been developed as in the cylindrical rotor case. From these, direct and quadrature synchronous

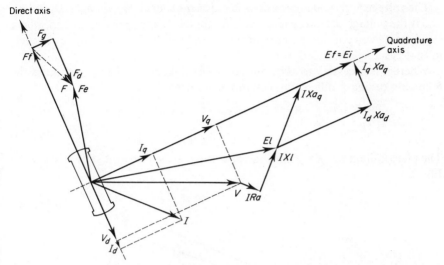

Fig. 9.2. Phasor diagram of a salient pole synchronous machine in the steady state

reactances (X_d and X_q) can be established, i.e.

$$X_d = Xl + Xa_d \tag{9.2.7}$$

$$X_q = Xl + Xa_q \tag{9.2.8}$$

$$Ei - V_q = RaI_q - X_dI_d \tag{9.2.9}$$

$$-V_d = RaI_d + X_qI_q \tag{9.2.10}$$

where V_d and V_q are the axial components of the terminal voltage V.

In steady-state conditions it is quite acceptable to use as the machine model, the field voltage Ef or the voltage equivalent to field current Ei behind the synchronous reactances. In these circumstances the rotor position (quadrature axis) with respect to the synchronously rotating frame of reference is given by the angular position of Ef.

Only the salient pole machine will now be considered, as the cylindrical rotor model may be regarded as a special case of a salient machine ($X_d = X_q$).

TRANSIENT EQUATIONS

For faster changes in the conditions external to the synchronous machine, the above model is no longer suitable. Due to the 'inertia' of the flux linkages these changes can not be reflected throughout the whole of the model immediately. It is therefore necessary to create new fictitious voltages E_d' and E_q' which represent the flux linkages of the rotor windings. These transient voltages can be shown to exist behind the transient reactances X_d' and X_q'.

$$E_q' - V_q = RaI_q - X_d'I_d \tag{9.2.11}$$

$$E_d' - V_a = RaI_d + X_q'I_q \tag{9.2.12}$$

The voltage E_i should now be considered as the sum of two voltages, E_d and E_q, and is the voltage behind synchronous reactance. In the previous section, where steady state was considered, current flowed only in the field winding and, hence, in that case, $E_d = 0$ and $E_q = E_i$.

Where it is necessary to allow the rotor flux linkages to change with time, the following ordinary differential equations are used:

$$pE'_q = (Ef - E_q)/T'_{d0} = (Ef + (X_d - X'_d)I_d - E'_q)/T'_{d0} \qquad (9.2.13)$$

$$pE'_d = -E_d/T'_{q0} = (-(X_q - X'_q)I_q - E'_d)/T'_{q0} \qquad (9.2.14)$$

The phasor diagram of the machine operating in the transient state is shown in Fig. 9.3.

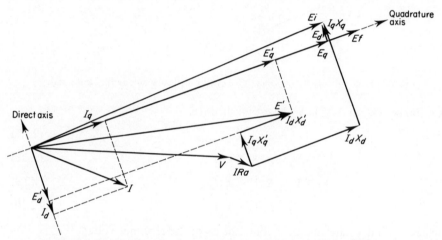

Fig. 9.3. Phasor diagram of a synchronous machine in the transient state

SUBTRANSIENT EQUATIONS

Either deliberately, as in the case of damper windings, or unavoidably, other circuits exist in the rotor. These circuits are taken into account if a more exact model is required. The reactances and time constants involved are small and can often be justifiably ignored. When required, the development of these equations is identical to that for transients and yields:

$$E''_q - V_q = RaI_q - X''_d I_d \qquad (9.2.15)$$

$$E''_d - V_d = RaId + X''_q I_q \qquad (9.2.16)$$

$$pE''_q = (E'_q + (X'_d - X''_d)I_d - E''_q)/T''_{d0} \qquad (9.2.17)$$

$$pE''_d = (E'_d - (X'_q - X''_q)I_q - E''_d)/T''_{q0} \qquad (9.2.18)$$

The equations are developed assuming that the transient time constants are large compared with the subtransient time constants. A phasor diagram of the synchronous machine operating in the subtransient state is shown in Fig. 9.4. It

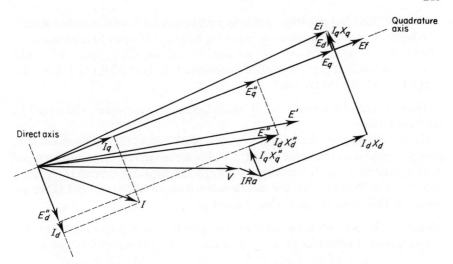

Fig. 9.4. Phasor diagram of a synchronous machine in the subtransient state

should be noted that equations (9.2.11) and (9.2.12) are now true only in the steady-state mode of operation, although once subtransient effects have decayed, the error will be small.

MACHINE MODELS

It is possible to extend the model beyond subtransient level but this is seldom done in multi-machine programs. Investigations[5] using a generator model with up to seven rotor windings, have shown that using the standard machine data the more complex models do not necessarily give more accurate results. However, improved results can be obtained if the data, especially the time constants are suitably modified.

The most convenient method of treating synchronous machines of differing complexity is to allow each machine the maximum possible number of equations and then let the actual model used be determined automatically according to the data presented.

Five models are thus possible for a four-winding rotor.

Model 1—constant voltage magnitude behind d-axis transient reactance (X_d') requiring no differential equations. Only the algebraic equations (9.2.11) and (9.2.12) are used.

Model 2—D-axis transient effects requiring one differential equation (pE_q'). Equations (9.2.11), (9.2.12) and (9.2.13) are used.

Model 3—D- and q-axis transient effects requiring two differential equations (pE_q' and pE_d'). Equations (9.2.11), (9.2.12), (9.2.13) and (9.2.14) are used.

Model 4—D- and q-axis subtransient effects requiring three differential equations (pE_q', pE_q'' and pE_d''). Equations (9.2.13), (9.2.15), (9.2.16), (9.2.17) and:

$$pE_d'' = \frac{(-(X_q - X_q'')I_q - E_d'')}{T_{q0}''}$$

(9.2.19)

are used. This last equation is merely equation (9.2.14) with modified primes. Whether it is a subtransient or transient equation is open to argument.

Model 5—D- and q-axis subtransient effects requiring four differential equations (pE'_q, pE'_d, pE''_q and pE''_d). Equations (9.2.13), (9.2.14), (9.2.15), (9.2.16), (9.2.17) and (9.2.18) are used.

The mechanical equations (9.2.5) and (9.2.6) are also required to be solved for all these models.

Groups of synchronous machines or parts of the system may be represented by a single synchronous machine model. An infinite busbar, representing a large stiff system, may be similarly modelled as a single machine represented by model 1, with the simplification that the mechanical equations (9.2.5) and (9.2.6) are not required. This sixth model is thus defined as:

Model 0—Infinite machine-constant voltage (phase and magnitude) behind d-axis transient reactance (X'_d). Only equations (9.2.11) and (9.2.12) are used.

9.3. SYNCHRONOUS MACHINE AUTOMATIC CONTROLLERS

For dynamic power system simulations of 1 second or longer duration, it is necessary to include the effects of the machine controllers, at least for the machine most affected by the disturbance. Moreover, controller representation is becoming necessary, even for first swing stability, with systems being operated at their limits with near critical fault clearing times.

The two principal controllers of a turbine-generator set are the automatic-voltage regulator (AVR) and the speed governor. The AVR model consists of voltage sensing equipment, comparators and amplifiers controlling a synchronous machine which can be generating or motoring. The speed governor may be considered to have similar equipment but in addition it is necessary to take the turbine into account.

Automatic voltage regulators

Many different AVR models have been developed to represent the various types used in a power system. The application of such models is difficult and a better approach is to develop a single general purpose AVR model, on a similar basis to the synchronous machine model. The model can then revert to any desired type by using the correct data. The IEEE defined several AVR types,[6] the main two of which (Type 1 and Type 2) are shown in Fig. 9.5.

A composite model of these two AVR types can be constructed. This model may also include a secondary signal which can be taken from any source, but usually either machine rotor speed deviation from synchronous speed or rate of change of machine output power. This model is shown in Fig. 9.6 and has been found to be satisfactory for all the systems studied so far. It is acknowledged that other AVR models may be necessary for specific studies.

In many systems studied, the amount of data available for an AVR model is

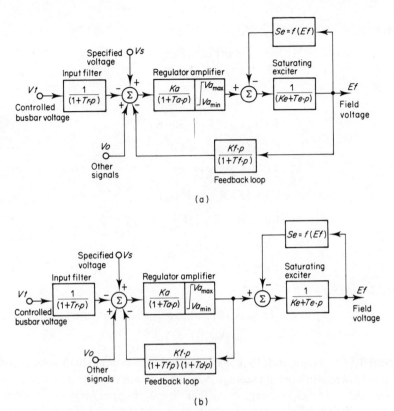

Fig. 9.5. Block diagrams for two commonly used AVR models. (6) (a) IEEE Type 1
AVR model; (b) IEEE Type 2 AVR model. (© 1982 IEEE)

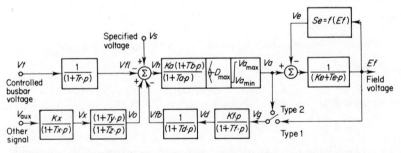

Fig. 9.6. Block diagram of a composite automatic voltage regulator model

quite small. The composite model can degenerate into a very simple model easily
by defaulting time constants to zero and gains to either zero, unity or an
extremely large value depending on their position.

The equations for the AVR model shown in Fig. 9.6 are as follows:

$$pVfl = (Vt - Vfl)/Tr \tag{9.3.1}$$

$$pVa = (Ka(1 + T_b \cdot p)Vh - Va)/Ta \tag{9.3.2}$$

subject to

$$|pVa| \leq D_{max}$$

and

$$Va_{max} \geq Va \geq Va_{min}$$

$$pEf = (Va - Ve - Ke \cdot Ef)/Te \tag{9.3.3}$$

$$pVd = (Kf \cdot pVg - Vd)/Tf \tag{9.3.4}$$

$$pVfb = (Vd - Vfb)/T_d \tag{9.3.5}$$

$$pVx = (KxV_{aux} - Vx)/Tx \tag{9.3.6}$$

$$pVo = ((1 + Ty \cdot p)Vx - Vo)/Tz \tag{9.3.7}$$

$$Vh = Vs - Vfb - Vfl + Vo \tag{9.3.8}$$

$$Ve = SeEf \tag{9.3.9}$$

where $Se = f(Ef)$

$$Vg = Ef \text{ [unless IEEE Type 2 when } Vg = Va] \tag{9.3.10}$$

$$V_{aux} = \text{A predefined signal}$$

The IEEE[6] recommends that Se be specified at maximum field voltage (Se_{max}) and at 0.75 of maximum field voltage ($Se_{0.75max}$). From this Se may be determined for any value of field voltage by either linear interpolation or by fitting a quadratic. Where linear interpolation is used, equation (9.3.9) may by transformed to:

$$Ve = (k_1 \cdot Ef - k_2)Ef \tag{9.3.11}$$

where

$$\left. \begin{aligned} k_1 &= (4 \cdot Se_{0.75max})/(3 \cdot Ef_{max}) \\ k_2 &= 0 \end{aligned} \right\} \qquad \text{if } Ef \leq 0.75 Ef_{max}$$

or

$$\left. \begin{aligned} k_1 &= 4(Se_{max} - Se_{0.75max})/Ef_{max} \\ k_2 &= 4 \cdot Se_{0.75max} - 3 \cdot Se_{max} \end{aligned} \right\} \qquad \text{if } Ef > 0.75 Ef_{max}$$

A means of modelling lead-lag circuits such as those in the regulator amplifier, the stabilizing loop and the auxiliary signal circuits is given at the end of this section.

Despite the advantages of one composite AVR model, if there are a great many AVRs to be modelled most of which have simple characteristics then it is better to make two models. One model, which contains only the commonly used parts of the composite model can then be dimensioned for all AVRs. The other model, which contains only the less commonly used parts of the composite model can be quite small dimensionally. A connection vector is all that is necessary to interconnect the two models whenever necessary.

Speed governors

For speed governors, as with AVRs, a composite model which can be reduced to any desired level is the most satisfactory. The speed governor models recommended by the IEEE[7] are shown in Fig. 9.7. It will be noticed that if limits are not exceeded, the two models are identical. The difference is due to the

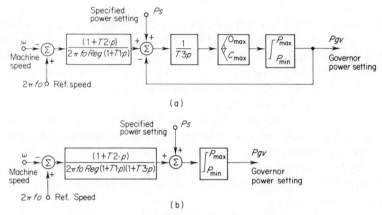

Fig. 9.7. Typical models of speed governors and valves. (7) (a) Thermal governor and valve; (b) Hydro governor and valve. (© 1982 IEEE)

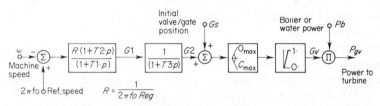

Fig. 9.8. Generalized model of a speed governor and valve

assumption that, in a hydro governor, gate servo and gate positions are the same. One model can be used for the governors of both turbines provided that the limits are either internal or external to the second transfer function block of Fig. 9.8. Also, very little extra effort is required to divorce the governor from the actual turbine power and keep it instead as a function of valve position.

The equations of the speed governor shown in Fig. 9.8 are:

$$pG1 = [R(1 + T2p) \cdot (2\pi f0 - \omega) - G1]/T1 \tag{9.3.12}$$

$$pG2 = (G1 - G2)/T3 \tag{9.3.13}$$

$$Gv = G2 + Gs \tag{9.3.14}$$

The valve/gate position setting (Gv) is subject to opening and closing rate limits

(o_{max} and c_{max} respectively) and to physical travel limits so that:

$$-c_{max} < pGv < o_{max}$$

$$0 \cdot < Gv < 1 \tag{9.3.15}$$

The valve equation is:

$$Pgv = Gv \cdot Pb \tag{9.3.16}$$

For thermal turbines, where a boiler is modelled, in the steady state, Pb will be the actual power delivered and Gs will be unity, i.e. the valve will be fully open. If a boiler is not modelled or a hydro turbine is being controlled then, in the steady state, Pb will be the maximum output from the boiler or water system (i.e. maximum turbine mechanical power output) and Gv, and hence Gs, will be such that Pgv is the actual mechanical power output of the turbine.

This method of modelling a valve has the advantage that nonlinearities between valve position and power can be easily included and also the operation of the governor and valve can be readily interpreted.

For a hydro governor where the limits are external, the model is as given in equations (9.3.12) to (9.3.16) but for a thermal governor, $G2$ is reset after the valve limits are applied to be:

$$G2_{(lim)} = Gv - Gs \quad \text{(thermal governor only)} \tag{9.3.17}$$

Hydro and thermal turbines

This section is restricted to the modelling of simple turbines only. Compound thermal turbines may require a detailed model, as given in Chapter 10, but, for stability studies of only 1 or 2 seconds duration, the effect of all but the high pressure (HP) turbine can usually be ignored. The time constant associated with the steam entrained between the HP turbine outlet and the IP or LP turbine inlet is usually very large (greater than 5 sec.) and the output from all turbines other than the HP turbine may be treated as constant.

Simple linear models of hydro and thermal turbines are shown in Fig. 9.9. The hydro-turbine model includes the penstock which gives the characteristic lead-

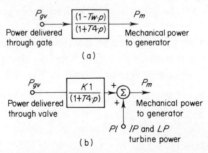

Fig. 9.9. Simple linear models of turbines.
(a) Hydro turbine; (b) Thermal turbine

lag response of this type of turbine. The model is generally sufficient for all hydro turbines and, from Fig. 9.9, the differential equation for the mechanical power output (Pm) of the turbine is:

$$pPm = ((1 - Tw \cdot p)Pgv - Pm)/T4 \qquad (9.3.18)$$

with $T4 = 0.5Tw$ as a further close approximation.

For the thermal turbine using Fig. 9.9(b) this equation is:

$$pPm = (K1 \cdot Pgv - Pm)/T4 \qquad (9.3.19)$$

with $K1$ representing the fraction of power delivered by the HP turbine. For simple turbines $K1$ is thus unity. For compound turbines, the power (Pl) from the IP and LP turbines is obtained from:

$$Pl = (1 - K1)Pm_0 \qquad (9.3.20)$$

here Pm_0 is the initial steady-state mechanical power. Note that for this simple model, the initial value of Pgv is $Pm/K1$, even though all the steam passes through the valve.

Provided that the HP valve does not close fully, then, rather than inject the power from the IP and LP turbines as shown, it is easier to treat it as a simple turbine ($Pl = 0$) but with the speed regulation modified by:

$$Reg_{(mod)} = \frac{Reg}{K1} \qquad (9.3.21)$$

Modelling lead-lag circuits

Lead-lag circuits may present a problem depending on the integration scheme adopted. Where the differential equations are not used directly and the derivatives are not explicitly calculated, the following can be used to convert the model into a more acceptable form.

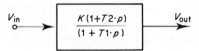

Fig. 9.10. Typical lead-lag circuit block diagram

For the circuit shown in block diagram form in Fig. 9.10, the equation is:

$$V_{out} = \frac{K(1 + T2p)}{(1 + T1p)}V_{in} \qquad (9.3.22)$$

This can be transformed to:

$$V_{out} = \frac{K \cdot T2}{T1}\left(\frac{(T1/T2) + T1 \cdot p}{1 + T1 \cdot p}\right)V_{in}$$

and then to:

$$V_{\text{out}} = \left(\frac{K \cdot T2}{T1} + \frac{K(1 - T2/T1)}{(1 + T1 \cdot p)} \right) V_{\text{in}} \qquad (9.3.23)$$

which can be represented by the block diagram in Fig. 9.11, and is a lag circuit in parallel with a gain.

It is important to remember that the time constant $T1$, must be nonzero even if the integration method can accommodate zero-time constants.

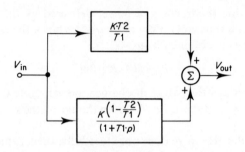

Fig. 9.11. Modified block diagram of a lead-lag circuit

9.4. LOADS

Early transient-stability studies were concerned primarily with generator stability, and little importance was attached to loads. In the two-machine problem for example, the remainder of the system, generators and loads, were represented by an infinite busbar. A great deal of attention has been given to load modelling since then.

Much of the domestic load and some industrial load consist of heating and lighting, especially in the winter, and in early load models these were considered as constant impedances. Rotating equipment was often modelled as a simple form of synchronous machine and composite loads were simulated by a mixture of these two types of load.

A lot of work has gone into the development of more accurate load models. These include some complex models of specific large loads which are considered in the next chapter. Most loads, however, consist of a large quantity of diverse equipment of varying levels and composition and some equivalent model is necessary.

A general load characteristic[8] may be adopted such that the MVA loading at a particular busbar is a function of voltage (V) and frequency (f):

$$P = Kp(V)^{pv} \cdot (f)^{pf} \qquad (9.4.1)$$

$$Q = Kq(V)^{qv} \cdot (f)^{qf} \qquad (9.4.2)$$

where Kp and Kq are constants which depend upon the nominal value of the variables P and Q.

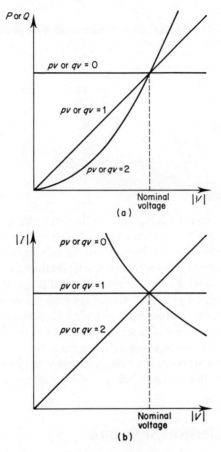

Fig. 9.12. Characteristic of different load models. (a) Active and reactive power against voltage; (b) Current against voltage

Static loads are relatively unaffected by frequency changes, i.e. $pf = qf = 0$, and with constant impedance loads $pv = qv = 2$.

The importance of accurate load models has been demonstrated by Dandeno and Kundur[8] when considering voltage sensitive loads. Figure 9.12 demonstrates the power and current characteristics of constant power, constant current and constant impedance loads. Berg[9] has identified the characteristic load parameters for various homogeneous loads and these are given in Table 9.1. These characteristics may be combined to give the overall load characteristic at a busbar. For example, a group of n homogeneous loads, each with a characteristic of pv_j and a nominal power of P_j may be combined to give an overall characteristic of:

$$pv_{(\text{overall})} = \sum_{j=1}^{n} (pv_j \cdot P_j) \bigg/ \sum_{j=1}^{n} (P_j) \qquad (9.4.3)$$

The other three overall characteristics may be similarly determined.

222

Table 9.1. Typical values of characteristic load parameters[9]

Load	pv	qv	pf	qf
Filament lamp	1.6	0	0	0
Fluorescent lamp	1.2	3.0	− 1.0	− 2.8
Heater	2.0	0	0	0
Induction motor half load	0.2	1.6	1.5	− 0.3
Induction motor full load	0.1	0.6	2.8	1.8
Reduction furnace	1.9	2.1	− 0.5	0
Aluminium plant	1.8	2.2	− 0.3	0.6

Low-voltage problems

When the load parameters pv and qv are less than or equal to unity, a problem can occur when the voltage drops to a low value. As the voltage magnitude decreases, the current magnitude does not decrease. In the limiting case with zero-voltage magnitude, a load current flows which is clearly irrational, given the nondynamic nature of the load model. From a purely practical point of view, then the load characteristics are only valid for a small voltage deviation from nominal. Further, if the voltage is small, small errors in magnitude and phase produce large errors in current magnitude and phase. This results in loss of accuracy and with iterative solution methods poor convergence or divergence.

These effects can be overcome by using a constant impedance characteristic to represent loads where the voltage is below some predefined value, for example 0.8 p.u.

9.5. THE TRANSMISSION NETWORK

It is usual to represent the static equipment which constitute the transmission system by lumped 'equivalent pi' parameters independent of the changes occurring in the generating and load equipment. This representation is used for multi-machine stability programs because the inclusion of time varying parameters would cause enormous computational problems. Moreover, frequency, which is the most obvious variable in the network, usually varies by only a small amount and thus, the errors involved are small. Also, the rates of change of network variables are assumed to be infinite which avoids the introduction of differential equations into the network solution.

The transmission network can thus be represented in the same manner as in the load-flow or short-circuit programs, that is by a square complex admittance matrix.

The behaviour of the network is described by the matrix equation:

$$[I_{inj}] = [Y][V] \qquad (9.5.1)$$

where $[I_{inj}]$ is the vector of injected currents into the network due to generators and loads and $[V]$ is the vector of nodal voltages.

Any loads represented by constant impedances may be directly included in the network admittance matrix with the injected currents due to these loads set to zero. Their effect is thus accounted for directly by the network solution.

9.6. OVERALL SYSTEM REPRESENTATION

Two alternative solution methods are possible. The preferred method uses the nodal matrix approach, while the alternative is the mesh matrix method.

Matrix reduction techniques can be used with both methods if specific network information is not required, but this gives little advantage as the sparsity of the reduced matrix is usually very much less.

Mesh matrix method

In this method, the system-loading components are treated as Thevenin equivalents of voltages behind impedances. The network is increased in size to include these impedances and the mesh impedance matrix of the increased network is created. This is then inverted or the factorized form of the inverse determined.

The solution process is as follows:

(1) Calculate the Thevenin voltages of the system loading components by solving the relevant differential and algebraic equations.
(2) Determine the network currents using the Y matrix or factors. As the network current around a mesh containing the Thevenin voltage is the loading current this may affect the Thevenin voltage in which case an iterative process will be required.

Nodal matrix method

In this method, all network loading components are converted into Norton equivalents of injected currents in parallel with admittance. The admittances can be included in the network admittance matrix to form a modified admittance matrix which is then inverted, or preferably factorized by some technique so that solution at each stage is straightforward.

The following solution process applies:

(1) For each network loading component, determine the injected currents into the modified admittance matrix by solving the relevant differential and algebraic equations.
(2) Determine network voltages from the injected currents using the Z-matrix or factors.

As the network voltages affect the loading components, an iterative process is often required, although good approximations[8] can be used to avoid this.

With the Nodal Matrix Method, busbar voltages are available directly and branch currents can be calculated if necessary while with the mesh matrix

224

method, mesh currents are available directly and busbar voltages and branch currents must be calculated if necessary.

Although much work has been spent on the systematic construction of the mesh impedance matrix, the nodal admittance matrix is easier to construct and has gained wide acceptance in load-flow and fault analysis. For this reason, the remainder of this section will consider the nodal matrix method.

Synchronous machine representation in the network

The equations representing a synchronous machine, as defined in Section 9.2, are given in the form of Thevenin voltages behind impedances. This must be modified to a current in parallel with an admittance by use of Norton's theorem. The admittance of the machine thus formed may be added to the shunt admittance of the machine busbar and treated as a network parameter. The vector $[I_{inj}]$ in equation (9.5.1) thus contains the Norton equivalent currents of the synchronous machines.

The synchronous machine equations are written in a frame of reference rotating with its own rotor. The real and imaginary components of the network equations, as given in Fig. 9.13, are obtained from the following transformation:

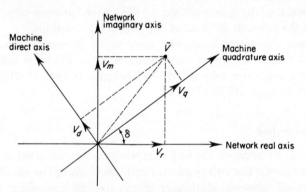

Fig. 9.13. Synchronous machine and network frames of reference

$$
\begin{array}{|c|} V_r \\ \hline V_m \end{array}
=
\begin{array}{|c|c|} \cos \delta & -\sin \delta \\ \hline \sin \delta & \cos \delta \end{array}
\cdot
\begin{array}{|c|} V_q \\ \hline V_d \end{array}
\qquad (9.6.1)
$$

This transformation is equally valid for currents as is the reverse transformation:

$$
\begin{array}{|c|} V_g \\ \hline V_d \end{array}
=
\begin{array}{|c|c|} \cos \delta & \sin \delta \\ \hline -\sin \delta & \cos \delta \end{array}
\cdot
\begin{array}{|c|} V_r \\ \hline V_m \end{array}
\qquad (9.6.2)
$$

When saliency exists, the values of X_d'' and X_q'' used in equations (9.2.15) and (9.2.16) and/or X_d' and X_q' used in equations (9.2.11) and (9.2.12) are different. Therefore, the Norton shunt admittance will have a different value in each axis and when transformed into the network frame of reference, will have time varying components. However, a constant admittance can be used, provided that the injected current is suitably modified to retain the accuracy of the Norton equivalent.[10] This approach can be justified by comparing the two circuits of Fig. 9.14 in which $\bar{Y}_t$ is a time varying admittance, whereas $\bar{Y}_0$ is fixed.

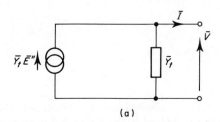

(a)

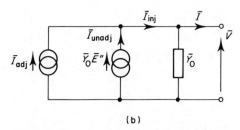

(b)

Fig. 9.14. Method of representing synchronous machines in the network. (a) Norton equivalent circuit; (b) Modified equivalent circuit

At any time t, the Norton equivalent of the machine is illustrated in Fig. 9.14(a), but the use of a fixed admittance results in the modified circuit of Fig. 9.14(b). The machine current is:

$$\bar{I} = \bar{Y}_t(\bar{E}'' - \bar{V}) = \bar{Y}_0(\bar{E}'' - \bar{V}) + \bar{I}_{adj}$$

and hence

$$\bar{I}_{adj} = (\bar{Y}_t - \bar{Y}_0)(\bar{E}'' - \bar{V}) \qquad (9.6.3)$$

where $\bar{I}_{adj}$ accounts for the fact that the apparent current source is not accurate in this case.

The injected current into the network which includes Y_0 is given by:

$$\bar{I}_{inj} = \bar{I}_{unadj} + \bar{I}_{adj} \qquad (9.6.4)$$

where

$$\bar{I}_{unadj} = \bar{Y}_0 \bar{E}''$$

A suitable value for $\bar{Y}_0$ is found by using the mean of direct and quadrature admittances, i.e.

$$\bar{Y}_0 = \frac{(Ra - jX_{dq})}{(Ra^2 + X_d'' \cdot X_q'')} \tag{9.6.5}$$

where

$$X_{dq} = \tfrac{1}{2}(X_d'' + X_q'')$$

The unadjusted value of current injected into the busbar is:

$$\left| \begin{array}{c} I_{\text{unadj}_r} \\ \hline I_{\text{unadj}_m} \end{array} \right| = \frac{1}{(Ra^2 + X_d'' \cdot X_q'')} \left| \begin{array}{cc} Ra & X_{dq} \\ \hline -X_{dq} & Ra \end{array} \right| \cdot \left| \begin{array}{c} E_r'' \\ \hline E_m'' \end{array} \right|$$

$$\tag{9.6.6}$$

The adjusting current is not affected by rotor position in the machine frame of reference but it is when considered in the network frame. From equation (9.6.3) and also equations (9.2.15) and (9.2.16):

$$\left| \begin{array}{c} I_{\text{adj}_q} \\ \hline I_{\text{adj}_d} \end{array} \right| = \frac{\tfrac{1}{2}(X_d'' - X_q'')}{(Ra^2 + X_d'' X_q'')} \left| \begin{array}{cc} 0 & 1 \\ \hline 1 & 0 \end{array} \right| \cdot \left| \begin{array}{c} E_q'' - V_q \\ \hline E_d'' - V_d \end{array} \right| \tag{9.6.7}$$

and transforming

$$\left| \begin{array}{c} I_{\text{adj}_r} \\ \hline I_{\text{adj}_m} \end{array} \right| = \frac{\tfrac{1}{2}(X_d'' - X_q'')}{(Ra^2 + X_d'' X_q'')} \left| \begin{array}{cc} -\sin 2\delta & \cos 2\delta \\ \hline \cos 2\delta & \sin 2\delta \end{array} \right| \cdot \left| \begin{array}{c} E_r'' - V_r \\ \hline E_m'' - V_m \end{array} \right| \tag{9.6.8}$$

The total nodal injected current is therefore

$$\left| \begin{array}{c} I_{\text{inj}_r} \\ \hline I_{\text{inj}_m} \end{array} \right| = \left| \begin{array}{c} I_{\text{unadj}_r} \\ \hline I_{\text{unadj}_m} \end{array} \right| + \left| \begin{array}{c} I_{\text{adj}_r} \\ \hline I_{\text{adj}_m} \end{array} \right| \tag{9.6.9}$$

Load representation in the network

To be suitable for representation in the overall solution method, loads must be transformed into injected currents into the transmission network from which the terminal voltages can be calculated. A Norton equivalent model of each load must therefore be created. In a similar way to that adopted for synchronous machines, the Norton admittance may be included directly in the network admittance matrix.

A constant impedance load is therefore totally included in the network admittance matrix and its injected current is zero. This representation is extremely simple to implement, causes no computational problems and improves the accuracy of the network solution by strengthening the diagonal elements in the admittance matrix.

Nonimpedance loads may be treated similarly. In this case, the steady-state values of voltage and complex power obtained from the load flow are used to obtain a steady-state equivalent admittance ($\bar{Y}_0$) which is included in the network admittance matrix $[Y]$. During the stability run, each load is solved sequentially along with the generators, etc. to obtain a new admittance ($\bar{Y}$), i.e.:

$$\bar{Y} = \frac{\bar{S}^*}{|V|^2} \tag{9.6.10}$$

The current injected into the network thus represents the deviation of the load characteristic from an impedance characteristic.

$$\bar{I}_{inj} = (\bar{Y}_0 - \bar{Y})\bar{V} \tag{9.6.11}$$

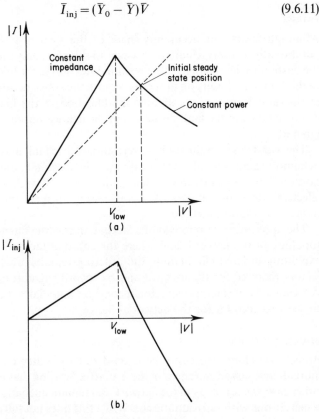

Fig. 9.15. Load and injected currents for a constant power type load with low-voltage adjustment. (a) Load current; (b) Injected current

By converting the load characteristic to that of a constant impedance, when the voltage drops below some predetermined value (V_{low}), as described in Section 9.4, the injected current is kept relatively small. An example of a load characteristic and its corresponding injected current is shown in Fig. 9.15.

In an alternative model the low-voltage impedance is added to the network and the injected current compensates for the deviation from the actual characteristic. In this case, there is a nonzero injected current in the initial steady-state operating condition.

System faults and switching

In general most power-system disturbances to be studied will be caused by changes in the network. These changes will normally be caused by faults and subsequent switching action but occasionally the effect of branch or machine switching will be considered.

FAULTS

Although faults can occur anywhere in the system, it is much easier computationally to apply a fault to a busbar. In this case, only the shunt admittance at the busbar need be changed, that is, a modification to the relevant self-admittance of the Y matrix. Faults on branches require the construction of a dummy busbar at the fault location and suitable modification of the branch data unless the distance between the fault position and the nearest busbar is small enough to be ignored.

The worst case is a three-phase zero-impedance fault and this involves placing in infinite admittance in parallel with the existing shunt admittance. In practice, a nonzero but sufficiently low-fault impedance is used so that the busbar voltage is effectively brought to zero. This is necessary to meet the requirements of the numerical solution method.

The application or removal of a fault at an existing busbar does not affect the topology of the network and where the solution method is based on sparsity exploiting ordered elimination, the ordering remains unchanged and only the factors required for the forward and backward substitution need be modified. Alternatively the factors can remain constant and diakoptical techniques[11] can be used to account for the network change.

BRANCH SWITCHING

Branch switching can easily be carried out by either modifying the relevant mutual- and self-admittances of the Y matrix or using diakoptical techniques. In either case, the topology of the network can remain unchanged as an open branch is merely one with zero admittance. While this does not fully exploit sparsity, the gain in computation time by not reordering exceeds the loss by retaining zero elements, in almost all cases.

The only exception is the case of a branch switched into a network where no interconnection existed prior to that event. In this case, either diakoptical or reordering techniques become necessary. To avoid this problem, a dummy branch may be included with the steady-state data of sufficiently high impedance that the power flow is negligible under all conditions, or alternatively, the branch resistance may be set negative to represent an initial open circuit. A negative branch reactance should not be used as this is a valid parameter where a branch contains series capacitors.

Where a fault occurs on a branch but very close to a busbar, nonunit protection at the near busbar will normally operate before that at the remote end. Therefore, there will be a period when the fault is still being supplied from the remote end. There are two methods of accounting for this type of fault.

The simplest method only requires data manipulation. The fault is initially assumed to exist at the local busbar rather than on the branch. When the specified time for the protection and local circuit bareaker to operate has elapsed, the fault is removed and the branch on which the fault is assumed to exist is opened. Simultaneously, the fault is applied at the remote busbar, but in this case, with the fault impedance increased by the faulted branch impedance, similarly the fault is maintained until the time specified for the protection and remote circuit breaker to operate has elapsed.

The second method is generally more involved but it is better when protection schemes are modelled. In this case, a dummy busbar is located at the fault position, even though it is close to the local busbar and a branch with a very small impedance is inserted between the dummy busbar and the local busbar. The faulted branch then connects the dummy busbar to the remote busbar and the branch shunt susceptance originally associated with the local busbar is transferred to the dummy busbar. This may all be done computationally at the time when the fault is being specified. The two branches can now be controlled independently by suitable protection systems. An advantage of this scheme is that the fault duration need not be specified as part of the input data. Opening both branches effectively isolates the fault, which can remain permanently attached to the dummy busbar, or if auto-reclosing is required, it can be removed automatically after a suitable deionization period.

The second method will give problems if the network is not being solved by a direct method. During the iterative solution of the network, slight voltage errors will cause large currents to flow through a branch with a very small impedance. This will slow convergence and in extreme cases will cause divergence. With a direct method, based on ordered elimination, an exact solution of the busbar voltages is obtained for the injected currents specified at that particular iteration. Thus, provided that the impedance is not so small that numerical problems occur when calculating the admittance, and the subsequent factors for the forward and backward substitution, then convergence of the overall solution between machines and network will be unaffected. The value of the low-impedance branch between the dummy and local busbars may be set at a fraction of the total branch impedance, subject to a minimum value. If this fraction is under 1/100, the change in branch impedance is very small compared to the accuracy of the network data input and it is unnecessary to modify the impedance of the branch from the remote to the dummy busbar.

MACHINE SWITCHING

Machine switching may be considered, either as a network or as a machine operation. It is a network operation if a dummy busbar is created to which the machine is connected. The dummy busbar is then connected to the original machine busbar by a low-impedance branch.

Alternatively, it may be treated as a machine operation by retaining the original network topology. When a machine is switched out, it is necessary to remove its injected current from the network solution. Also, any shunt admittance included in the network Y matrix, which is due to the machine, must be removed.

Although a disconnected machine can play no direct part in system stability, its response should still be calculated as before, with the machine stator current set to zero. Thus machine speed, terminal voltage, etc., can be observed even when disconnected from the system and in the event of reconnection, sensible results are obtained.

Where an industrial system is being studied many machines may be disconnected and reconnected at different times as the voltage level changes. This process will require many recalculations of the factors involved in the forward and backward substitution solution method of the network. However, these can be avoided by using the method adopted earlier to account for synchronous machine saliency. That is, an appropriate current is injected at the relevant busbar, which cancels out the effect of the shunt admittance.

9.7. INTEGRATION

Many integration methods have been applied to the power-system transient stability problem and the principal methods are discussed in Appendix I. Of these, only three are considered in this section. They are simple and easily applied methods which have gained wide acceptance. The purpose of the third method is not to provide another alternative but to clarify the differences between the other two methods.

Explicit Runge–Kutta methods have been used extensively in transient stability studies. They have the advantage that a 'packaged' integration method is usually available or quite readily constructed and the differential equations are incorporated with the method explicitly. It has only been with the introduction of more detailed system component models with very small time constants that the problems of stability have caused interest in other methods.

Fourth-order methods ($p = 4$) have probably been the most popular and among these the Runge–Kutta Gill method has the advantage that round-off error is minimized. With reference to equations (I.4.1) to (I.4.3), for this method the number of function substitutions is four ($v = 4$) and:

$$w_1 = 1/6$$

$$w_2 = (2 - \sqrt{2})/6$$

$$w_3 = (2 + \sqrt{2})/6 \tag{9.7.1}$$

$$w_4 = 1/6$$

$$k_1 = hf(t_n, y_n)$$

$$k_2 = hf(t_n + h/2, y_n + k_1/2)$$

$$k_3 = hf(t_n + h/2, y_n + (\sqrt{2} - 1)k_1/2 + (2 - \sqrt{2})k_2/2) \tag{9.7.2}$$

$$k_4 = hf(t_n + h, y_n - \sqrt{2}k_2/2 + (2 + \sqrt{2})k_3/2)$$

The characteristic root of this fourth-order method, when applied to equation (I.3.3), is:

$$z_1 = 1 + h\lambda + \tfrac{1}{2}h^2\lambda^2 + \tfrac{1}{6}h^3\lambda^3 + \tfrac{1}{24}h^4\lambda^4 \tag{9.7.3}$$

and to ensure stability, the step length h must be sufficiently small that z_1 is less than unity.

The basic trapezoidal method is very well known, having been established as a useful method of integration before digital computers made hand calculation redundant.

More recently an implicit trapezoidal integration method has been developed for solving the multi-machine transient-stability problem[10], and has gained recognition as being very powerful, having great advantages over the more traditional methods.

The method is derived from the general multistep equation given by equation (I.3.2) with k equal to unity and is thus a single-step method. The solution at the end of $n+1$ steps is given by:

$$y_{n+1} = y_n + \frac{h_{n+1}}{2}(py_{n+1} + py_n) \tag{9.7.4}$$

It has second-order accuracy with the major term in the truncation error being $-\tfrac{1}{12}h^3$.

The characteristic root when applied to equation (I.3.3) is:

$$z_1 = 1 - 2b_{n+1} \tag{9.7.5}$$

where

$$b_{n+1} = \frac{h_{n+1}}{(h_{n+1} - 2/\lambda)} \tag{9.7.6}$$

If $\text{Re}(\lambda) < 0$ then $0 \le b_{n+1} \le 1.0$ and $|z_1| \le 1.0$. The trapezoidal method is therefore A-stable, a property which is shown in Appendix I to be more important in the solution process than accuracy. The trapezoidal method is linear and thus in a multivariable problem, like power-system stability, the method is Σ-stable.

It can be shown that an A-stable linear multistep method cannot have an order of accuracy greater than two, and that the smallest truncation error is achieved by

232

the trapezoidal method. The trapezoidal method is thus the most accurate $\sum$-stable finite difference method possible.

The method, as expressed by equation (9.7.4), is implicit and requires an iterative solution. However, the solution can be made direct by incorporating the differential equations into equation (9.7.4). Rearranging forms algebraic equations as described in Appendix I.

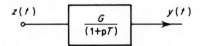

Fig. 9.16. Simple transfer function

For example, consider the trivial transfer function shown in Fig. 9.16. The differential equation for this system is given by:

$$py(t) = (G \cdot z(t) - y(t))/T \qquad (9.7.7)$$

with the input variable being denoted by 'z' to indicate that it may be either integrable or nonintegrable.

The algebraic form of equation (9.7.7) has a solution at the end of the $(n + 1)$th step of:

$$y_{n+1} = c_{n+1} + m_{n+1} \cdot z_{n+1} \qquad (9.7.8)$$

where

$$c_{n+1} = (1 - 2b_{n+1})y_n + b_{n+1} \cdot G \cdot z_n \qquad (9.7.9)$$
$$m_{n+1} = b_{n+1} \cdot G \qquad (9.7.10)$$

and

$$b_{n+1} = h_{n+1}/(2T + h_{n+1}) \qquad (9.7.11)$$

Provided that the step length h remains constant it is unnecessary to re-evaluate b or m at each step. That is:

$$\left. \begin{array}{l} b_{n+1} = b_n \\ m_{n+1} = m_n \end{array} \right\} \quad \text{if } h_{n+1} = h_n \qquad (9.7.12)$$

There is little to be gained by this, however, as it is a simple process and it is often desirable to change h during a study.

A comparison between the Runge–Kutta Gill and the trapezoidal methods when used to solve two power system transient stability problems is given in Tables 9.2 and 9.3. The comparison is made in terms of maximum error (based on results using very small step lengths) and central processor unit (CPU) execution time.

The advantages of the Σ-stable trapezoidal method are apparent from both tables, but the results are sufficiently different to show that an absolute comparison between methods cannot be made. The nonlinearity of the equations in any system also effect the errors obtained. CPU time using the Runge–Kutta Gill method is a function of the step length but this is not so with the trapezoidal method. For very small step lengths, only one iteration per step is needed using the trapezoidal method but as the step length increases so does the number of iterations. The relationship between step length and iterations is nonlinear, with the result that there is an optimum step length in which the iterations per step are small but greater than one.

For comparison, the backward Euler method is also included. This is a first-order method with the solution given by:

Table 9.2

Step length (ms)	Runge-Kutta Gill		Trapezoidal		Backward Euler	
	Max. error (degs)	CPU time (s)	Max. error (degs)	CPU time (s)	Max. error (degs)	CPU time (s)
100.0	—	—	2.2	0.26	—	—
50.0	—	—	0.7	0.27	—	—
25.0	21.0	0.43	0.1	0.29	5.7	0.41
10.0	13.0	0.72	—	0.49	2.4	0.47
5.0	7.8	1.18	—	0.69	1.3	0.67
2.0	3.7	2.57	—	1.34	.5	1.31
1.0	1.9	4.88	—	2.42	.2	2.35
0.5	1.0	9.52	—	4.60	—	4.42
0.2	0.4	24.19	—	—	—	10.58
0.1	0.2	47.95	—	—	—	—

Table 9.3

Step length (ms)	Runge-kutta Gill		Trapezoidal		Backward Euler	
	Max. error (degs)	CPU time (s)	Max. error (degs)	CPU time (s)	Max. error (degs)	CPU time (s)
10.0	8.6	1.67	0.5	2.37	—	—
5.0	4.4	3.06	0.1	2.31	8.5	2.76
2.0	1.7	7.24	—	3.74	3.8	3.64
1.0	1.2	14.19	—	7.12	1.8	6.80
0.5	0.9	28.00	—	13.88	0.6	13.24

$$y_{n+1} = y_n + h \cdot p y_{n+1} \qquad (9.7.13)$$

and the characteristic root when applied to equation (I.3.3) is:

$$z_1 = 1/(1 - h\lambda) \qquad (9.7.14)$$

Despite the three orders of accuracy difference between it and the Runge–Kutta Gill, the backward Euler method compares well.

The results for the trapezoidal and backward Euler methods were obtained using linear extrapolation of the nonintegrable variables at the beginning of each step. This required the storing of machine terminal voltages and currents together with other nonintegrable variables obtained at the end of the previous step.

Problems with the trapezoidal method

Although the trapezoidal method is Σ-stable and the step length is not constrained by the largest negative eigenvalue, the accuracy of the solution corresponding to the largest negative eigenvalues will be poor if a reasonable step length is not chosen.

With the backward Euler method, the larger the step length the smaller the characteristic root

$$z_{1(BE)} \to 0 \quad \text{as } h\lambda \to -\infty \tag{9.7.15}$$

whereas for the trapezoidal method

$$z_{1(TRAP)} \to -1 \quad \text{as } h\lambda \to -\infty \tag{9.7.16}$$

For small step lengths the characteristic roots of both methods tend towards, but never exceed, unity (positive).

$$z_{1(BE)} \text{ and } z_{1(TRAP)} \to +1 \quad \text{as } h\lambda \to 0 \tag{9.7.17}$$

The effect of too large a step length can be shown in a trivial but extreme example. The system in Fig. 9.16 and equation (9.7.7) with a zero time constant T, and unity gain G, is such an example.

If the input $z(t)$ is a unit step function from an initial value of zero, then with a zero-time constant, the output $y(t)$ should follow the input exactly, that is a constant output of unity. In fact, the output oscillates with $y_1 = 2, y_2 = 0, y_3 = 2$, etc.

Table 9.4 shows the effect of different step lengths on this simple system with a nonzero-time constant T. This table shows that oscillations occur when $h\lambda$ is smaller than -2, that is, when the characteristic root z_1 is negative. The oscillations decay with a rate dependent on $h\lambda$, that is, the rate is dependent on the

Table 9.4. The effect of different step lengths on the solution of a simple system (Fig. 9.16) by the trapezoidal method

	$h\lambda = -0.5$		$h\lambda = -2.0$		$h\lambda = -8.0$		$h\lambda = -32.0$	
Step No.	Trap method	Exact soln.	Trap method	Exact soln.	Trap method	Exact soln.	Trap method	Exact soln.
0	0	0	0	0	0	0	0	0
1	0.4000	0.3935	1.0000	0.8647	1.6000	0.9997	1.8824	1.0000
2	0.6400	0.6321		0.9817	0.6400	1.0000	0.2215	
3	0.7840	0.7769		0.9975	1.2160		1.6870	
4	0.8704	0.8647		0.9997	0.8704		0.3938	
5	0.9222	0.9179	1.0000	1.0000	1.0778	1.0000	1.5349	1.0000

magnitude of z_1. It can also be seen that accuracy is good provided that $h\lambda$ is greater than or equal to -0.5.

Oscillations are only initiated at a discontinuity. Provided that there is no step function input, the output of a transfer function with zero-time constant duplicates the input.

The example given is an extreme case and for the power-system stability problem this usually only occurs in the input circuit of the AVR.

For the mechanical equation of the synchronous machine, the speed is given by:

$$p\omega = \frac{1}{Mg}(Pa) \qquad (9.7.18)$$

where Pa is the accelerating power given by $Pa = P_m - Pe$, and the damping factor Da is zero. Therefore, in this case

$$\omega_{n+1} = c_{n+1} + m_{m+1}Pa_{n+1} \qquad (9.7.19)$$

where

$$c_{n+1} = \omega_n + m_{n+1}Pa_n \qquad (9.7.20)$$

and

$$m_{n+1} = h_{n+1}/2Mg \qquad (9.7.21)$$

and oscillations do not occur. Da, when it does exist, is usually very small and any oscillations will similarly be very small.

For the electrical equations of the synchronous machine, only the current can change instantaneously, and the effect is not as pronounced as for a unit step function.

Techniques[12] are available to remove the oscillations but they require a lot of storage and it is simpler to reduce the step length.

Programming the trapezoidal method

There is no means of estimating the value of errors in the trapezoidal method but the number of iterations required to converge at each step may be used as a very good indication of the errors. As previously mentioned, the number of iterations increases more rapidly than the step length and thus the number of iterations is a good reference for the control of the step length. It is suggested[13] to double the step length if the number of iterations per step is less than 3 and to halve it if the number of iterations per step exceed 12. The resulting bandwidth (3 to 12) is necessary to stop constant changes in the step length.

To avoid problems of step length chattering a factor of about 1.5 (instead of 2) may be used. Unfortunately it is difficult to maintain a regular printout interval if a noninteger factor is used.

Even using a step length changing factor of 2, it is difficult to maintain a regular printout interval. Step halving can be carried out at any time but indiscriminate step doubling may mean that there is no solution at the desired printout time. Doubling the step length thus should only be done immediately after a printout and the step length should not be allowed to exceed the printout interval.

On rare occasions, it is possible that the number of iterations at a particular step greatly exceeds the upper desired limit. It can be shown that the convergence pattern is geometric and usually oscillatory [13] after the first five or six iterations. Even when diverging, the geometric and oscillatory pattern can be observed. Schemes can thus be devised which estimate the correct solution. However, these schemes are relatively costly to implement in terms of programming, storage and execution and a more practical method is to stop iterating after a fixed number of iterations and start again with a half-step. It is not necessary to store all the information obtained at the end of the previous step, in anticipation of a restart, as this information is already available for the nonintegrable variables if an extrapolation method is being used at the beginning of each step. Further, much information is available in the C and M constants of the algebraic form of the integration method.

For example, with the two-variable problem given by equation (9.7.7), if $z(t)$ is a nonintegrable variable, then its value at the end of the nth step z_n is stored. The value of the integrable variable $y(t)$ at the end of the (n)th step y_n can be re-evaluated from equations (9.7.9) and (9.7.10) to be:

$$y_n = \frac{(c_{n+1} - m_{n+1}z_n)}{(1 - 2m_{n+1}/G)} \tag{9.7.22}$$

In only a few cases where the differential equation is complex need the value of y_n be stored at the beginning of each step. While the method requires programming effort it is very economical on storage and the few instances where it is used do not affect the overall execution time appreciably.

Linear extrapolation of nonintegrable variables at the beginning of each step is a very worthwhile addition to the trapezoidal method. Although not essential, the number of iterations per step is reduced and the storage is not prohibitive. Higher orders of extrapolation give very little extra improvement and as they are not effective until some steps after a discontinuity their value is further reduced.

It is only at the first step after a discontinuity that linear extrapolation can not be used. As this often coincides with a large rate of change of integrable variables, the number of iterations to convergence can be excessive. This is overcome by automatic step length reduction after a discontinuity. Two half-step lengths, before returning to the normal step length, has been found to be satisfactory in almost all cases.[13]

Application of the trapezoidal method

The differential equations developed in this chapter have all been associated with the synchronous machine and its controllers. These equations can be transformed into the algebraic form of the trapezoidal method given by equation (9.7.8). While these algebraic equations can be combined to make a matrix equation this has little merit and makes discontinuities such as regulator limits more difficult to apply.

In order to simplify the following equations, the subscripts on the variables have been removed. It is rarely necessary to retain old values of variables and, where this is necessary, it is noted. The variable values are thus overwritten by new information as soon as they are available. The constants C and M associated with the algebraic form are evaluated at the beginning of a new integration step and hence use the information obtained at the end of the previous step.

SYNCHRONOUS MACHINE

The two mechanical differential equations are given by equations (9.2.5) and (9.2.6) and the algebraic form is:

$$\omega = C_\omega + M_\omega (Pm - Pe) \tag{9.7.23}$$

where

$$C_\omega = (1 - 2 \cdot M_\omega \cdot Da)\omega + M_\omega (Pm - Pe + 4\pi \cdot fo \cdot Da)$$
$$M_\omega = h/(2Mg + hDa)$$

and also

$$\delta = C_\delta + M_\delta(\omega) \tag{9.7.24}$$

where

$$C_\delta = \delta + M_\delta(\omega - 4\pi \cdot fo)$$
$$M_\delta = 0.5h$$

It would be possible to combine these equations to form a single simultaneous solution of the form

$$\delta = C'_\delta + M'_\delta (Pm - Pe) \tag{9.7.25}$$

where

$$C'_\delta = C_\delta + M_\delta \cdot C_\omega$$
$$M'_\delta = M_\delta \cdot M_\omega$$

but machine speed ω is a useful piece of information and would still require evaluation in most problems.

It is also more convenient to retain the electrical power (Pe) as a variable rather than attempt to reduce it to its constituent parts:

$$Pe = I_d \cdot V_d + I_q \cdot V_q + (I_d^2 + I_q^2)Ra \tag{9.7.26}$$

Thus Pe is extrapolated after C_ω and M_ω have been evaluated.

The mechanical power Pm is an integrable variable which, in the absence of a speed governor model for the machine, is constant.

There are four electrical equations associated with the change in flux in the

238

synchronous machine and these are given by equations (9.2.13), (9.2.14), (9.2.17) and (9.2.18). The algebraic form of these equations are:

$$E'_q = C_q + M_q(Ef + (X_d - X'_d)I_d) \qquad (9.7.27)$$

where

$$C_q = (1 - 2M_q)E'_q + M_q(Ef + (X_d - X'_d)I_d)$$
$$M_q = h/(2T'_{d0} + h)$$

also

$$E'_d = C_d - M_d(X_q - X'_q)(I_d) \qquad (9.7.28)$$

where

$$C_d = (1 - 2M_d)E'_d - M_d(X_q - X'_q)I_q$$
$$M_d = h/(2T'_{q0} + h)$$

also

$$E''_q = C_{qq} + M_{qq}(E'_q + (X'_d - X''_d)I_d) \qquad (9.7.29)$$

where

$$C_{qq} = (1 - 2M_{qq})E''_q + M_{qq}(E'_q + (X'_d - X''_d)I_d)$$
$$M_{qq} = h/(2T''_{d0} + h)$$

and also

$$E''_d = C_{dd} + M_d(E'_d - (X'_q - X''_q)I_q) \qquad (9.7.30)$$

where

$$C_{dd} = (1 - 2M_{dd})E''_d + M_{dd}(E'_d - (X'_q - X''_q)I_q)$$
$$M_{dd} = h/(2T''_{q0} + h)$$

SYNCHRONOUS MACHINE CONTROLLER LIMITS

There are usually limits associated with AVRs and speed governors and these require special consideration when applying the algebraic form of the trapezoidal rule. It is best to ignore the limits at first and develop the whole set of 'limitless' equations. Rather than confuse this discussion, it is easier to consider a simple AVR system as shown in Fig. 9.17, for which can be written:

$$pV_{out} = (G1 \cdot (V_{in} - Vfb) - V_{out})/T1 \qquad (9.7.31)$$

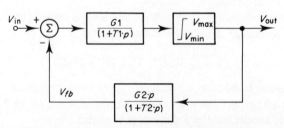

Fig. 9.17. Block diagram of a simple controller

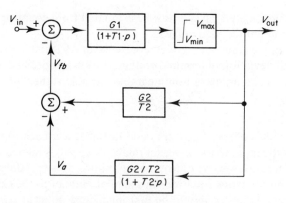

Fig. 9.18. Modified block diagram of a simple controller of Fig. 9.17

subject to

$$V_{max} \geq V_{out} \geq V_{min}$$

$$pVfb = (G2 \cdot pV_{out} - Vfb)/T2 \qquad (9.7.32)$$

The feedback loop can be rearranged to avoid the derivative of V_{out} being explicitly required as described in Section 9.3 and this is shown in Fig. 9.18. Equation (9.7.32) is now replaced by:

$$Vfb = \frac{G2}{T2}V_{out} - Va \qquad (9.7.33)$$

$$pVa = \left(\frac{G2}{T2}V_{out} - Va\right)\Big/T2 \qquad (9.7.34)$$

Equations (9.7.31) and (9.7.34) can be transformed into the algebraic form:

$$V_{out} = C_1 + M_1(V_{in} - Vfb) \qquad (9.7.35)$$

$$V_a = C_2 + M_2(V_{out}) \qquad (9.7.36)$$

where C_1, C_2, M_1 and M_2 may be determined in the usual way.

A simultaneous solution for the whole system is now possible by combining equations (9.7.33), (9.7.35) and (9.7.36) to give:

$$V_{out} = C_3 + M_3(V_{in}) \qquad (9.7.37)$$

where

$$C_3 = \frac{C_1 + M_1 C_2}{1 + M_1((G2/T2) - M_2)}$$

and

$$M_3 = \frac{M_1}{1 + M_1((G2/T2) - M_2)}$$

After a solution of V_{out} is obtained it may be subjected to the limits of equation

(9.7.31). If it is necessary Vfb can now be evaluated from equations (9.7.33) and (9.7.36) using the limited value of V_{out}.

Where this simple controller model represents an AVR the input V_{in} may well be the (negated) deviation of terminal voltage V_t from its specified value (V_s). It is simpler to treat V_t as an extra nonintegrable variable rather incorporate

$$V_t = \sqrt{(V_r^2 + V_m^2)}$$

in the model.

The usual models of speed governors do not have feedback loops associated with them, but the input to the governor (machine speed) is related to the turbine output (mechanical power) by differential equations. It is therefore necessary to solve a set of simultaneous equations in a similar manner to the example above. The simultaneous solution should be first made at a point at which limits are applied (i.e. at the valve) and then, after ensuring the result conforms to the limits, all the other variables around the loop (including machine speed and rotor angle) can be evaluated.

SOLUTION FOR SATURATING AVR EXCITER

Another problem occurs when a nonlinear function is encountered. Equations (9.3.3) and (9.3.11) may be combined to form a single differential equation but this, then, involves a term in Ef^2 which complicates the evaluation of Ef. As the saturation function is approximate, it may be further simplified to give:

$$Ve = (k_1^* Ef^* - k_2^*)Ef \qquad (9.7.38)$$

where Ef^* is the value of Ef at the previous iteration and k_1^* and k_2^* are determined from Ef^*. The equation describing the saturating exciter is thus:

$$pEf = (Vam - (Ke - k_2^*)Ef - k_1^* Ef^* Ef)/Te \qquad (9.7.39)$$

and applying the trapezoidal rule the algebraic form of solution is:

$$Ef = [C_{ef} + M_{ef}(Vam)]/[1 + M_{ef}(k_1^* Ef^* - k_2^* - K_{ef})] \qquad (9.7.40)$$

where

$$C_{ef} = (1 - 2(Ke + K_{ef})M_{ef})Ef + M_{ef}Vam$$

$$M_{ef} = h/(2Te + h(Ke + K_{ef}))$$

$$K_{ef} = k_1 Ef - k_2$$

C_{ef}, M_{ef} and K_{ef} are evaluated once at the beginning of the step and, hence, only contain information obtained at the end of the previous step.

9.8. STRUCTURE OF A TRANSIENT STABILITY PROGRAM

Overall structure

An overview of the structure of a transient stability program is given in Fig. 9.19. Only the main parts of the program have been included, and as can be seen, the

241

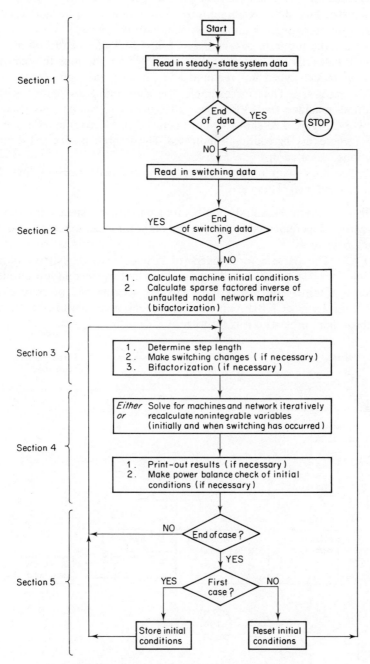

Fig. 9.19. Overview of a transient stability program structure

same system may have several case studies performed on it by repeatedly specifying switching data. When no further switching data is available, control returns to the start to see if another system is to be studied.

With care, the program can be divided into packages of subroutines each concerned with only one aspect of the system.[13] This permits the removal of component models when not required and the easy addition of new models whenever necessary. Thus for example, the subroutines associated with the synchronous machine, the AVRs, speed governors, etc., can be segregated from the network. Figure 9.20 shows a more detailed block diagram of the overall structure where this segregation is indicated. The diagram is subdivided into the five sections indicated in Fig. 9.19.

While the block diagrams are intended to be self-evident several logic codes need to be explained. These are:

KASE—This is the case study number for a particular system. It is initially set to zero and incremented by 1 at the end of the initialization and at the end of each case study.

KBIFA1—The sparse factored inverse of the nodal network matrix is obtained using three bifactorization subroutines. The first and second subroutines are integer routines which determine busbar ordering and nonzero element location. The code KBIFA1 is set to unity if it is necessary to enter these two subroutines, otherwise it is set to zero.

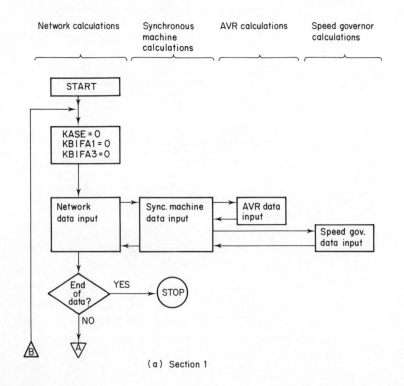

(a) Section 1

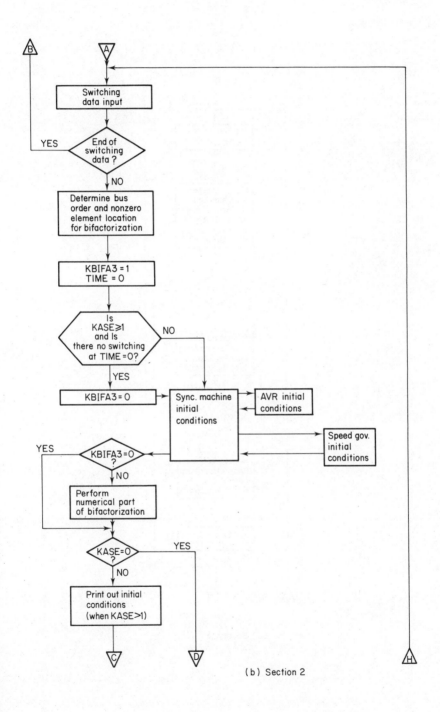

(b) Section 2

244

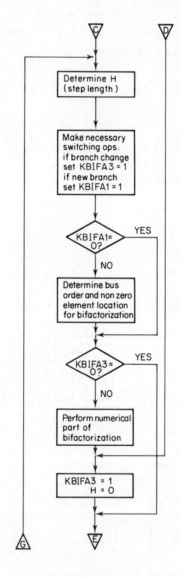

(c) Section 3

245

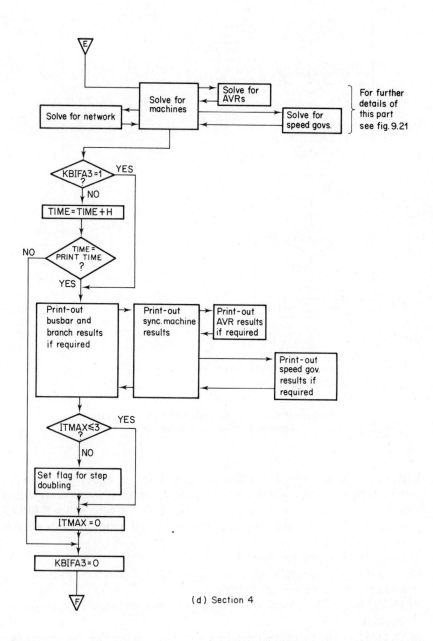

(d) Section 4

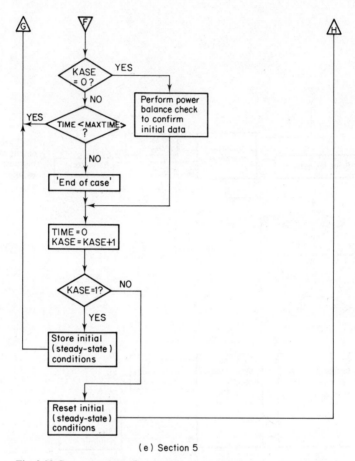

(e) Section 5

Fig. 9.20. Structure of transient stability program (a) Section 1 (b) Section 2
(c) Section 3 (d) Section 4 (e) Section 5

KBIFA3—The elements of the sparse-factored inverse are evaluated in the third bifactorization subroutine. The code KBIFA3 is set to unity if it is necessary to enter this subroutine, otherwise it is set to zero. When KBIFA3 is unity, it indicates that a network discontinuity has occurred and hence it is also used for this purpose.

TIME—The integration time.

H—The integration step length. Like KBIFA3, it is also used to indicate a discontinuity when it is set to zero.

PRINTTIME—The integration time at which the next print out of results is required.

MAXTIME—The predefined maximum integration time for the case study.

ITMAX—Maximum number of iterations per step since last print-out of results.

Note that many data error checks are required in a program of this type but they have been omitted from the block diagram for clarity.

Structure of machine and network iterative solution

The structure of this part of the program requires further description. Two forms of solution are possible depending on whether an integration step is being evaluated or the nonintegrable variables are being recalculated after a discontinuity. A block diagram is given in Fig. 9.21.

The additional logic codes used in this part of the program are:

ERROR—The maximum difference between any integrable variable from one iteration to another.

ITR—Number of iterations required for solution.

IHALF—Number of immediate step halving required for the solution.

TOLERANCE—Specified maximum value of ERROR for convergence.

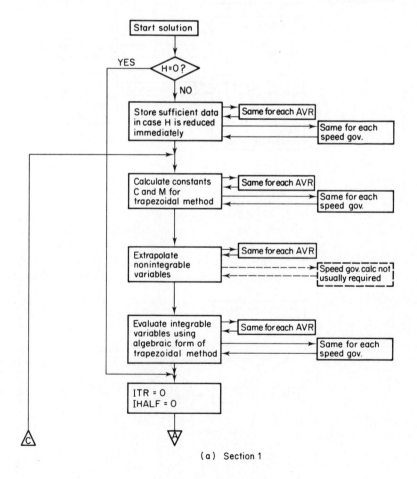

(a) Section 1

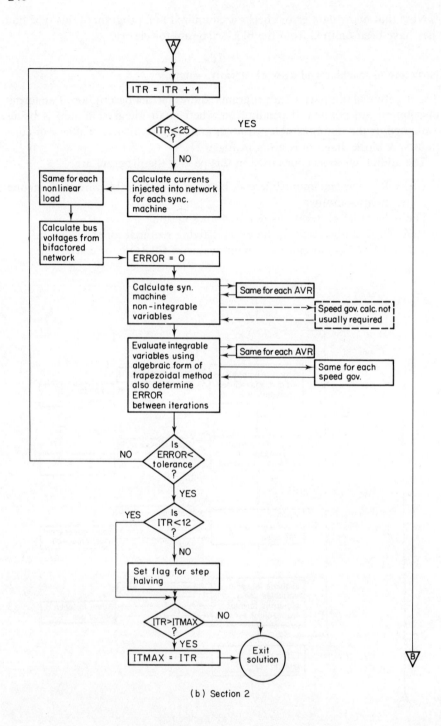

(b) Section 2

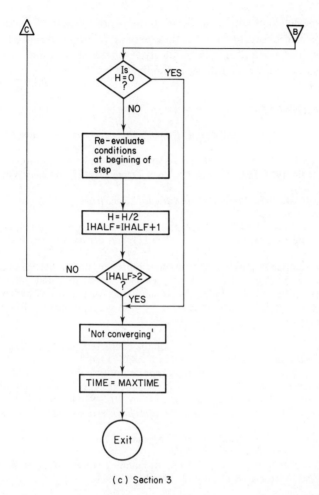

(c) Section 3

Fig. 9.21. Structure of machine and network iterative solution (a) Section 1
(b) Section 2 (c) Section 3

If convergence has not been achieved after a specified number of iterations, the case study is terminated. This is done by setting the integration time equal to the maximum integration time. The latest results are thus printed out and a new case study is attempted.

9.9. GENERAL CONCLUSIONS

The transient stability program described in this chapter is sufficient for many basic stability studies. It is more than adequate when first swing stability is being evaluated and the machine detail and controllers will allow second and subsequent swing stability to be examined also.

However, if synchronous machine saturation or compound thermal turbines have to be modelled, it will be necessary to incorporate parts of Chapter 10 into the program. The structure set out at the end of this chapter should allow changes of this sort to be made quite easily. Similarly, if other system components are to be included this can be done without difficulty.

9.10. REFERENCES

1. O. I. Elgerd, 1971. *Electrical Energy Systems Theory: An Introduction*. McGraw-Hill, New York.
2. B. M. Weedy, 1979. *Electric Power Systems*. Wiley and Sons, London.
3. C. Concordia, 1951. *Synchronous Machines. Theory and Performance*. Wiley and Sons, New York.
4. E. W. Kimbark, 1956. *Power System Stability : Synchronous Machines*. (Vol. 3). Wiley and Sons, New York.
5. P. L. Dandeno, *et al*. 1973. 'Effects of synchronous machine modeling in large-scale system studies', *IEEE Transactions on Power Apparatus and Systems*, **PAS-92** (2), 574–582.
6. IEEE Committee Report, 1968. 'Computer representation of exciter systems', *IEEE Transactions on Power Apparatus and Systems*, **PAS-87** (6), 1460–1464.
7. IEEE Committee Report, 1973. 'Dynamic models for steam and hydro turbines in power-system studies', *IEEE Transactions on Power Apparatus and Systems*, **PAS-92** (6), 1904–1915.
8. P. L. Dandeno and P. Kundur, 1973. 'A noniterative transient stability program including the effects of variable load-voltage characteristics', *IEEE Transactions on Power Apparatus and Systems*, **PAS-92** (5), 1478–1484.
9. G. L. Berg, 1973. 'Power system load representation', *Proc. IEE* **120** (3), 344–348.
10. H. W. Dommell and N. Sato, 1972. 'Fast transient stability solutions', *IEEE Transactions on Power Apparatus and Systems*, **PAS-91** (4), 1643–1650.
11. A. Brameller, *et al*. 1969. *Practical Diakoptics for Electrical Networks*. Chapman and Hall, London.
12. L. Lapidus and J. H. Seinfeld, 1971. *Numerical Solution of Ordinary Differential Equations*. Academic Press, New York.
13. C. P. Arnold, 1976. 'Solutions of the multi-machine power-system stability problem. Ph.D. thesis, Victoria University of Manchester, U.K.

10

Power System Stability—
Advanced Component
Modelling

10.1. INTRODUCTION

This chapter develops further some of the component models described in Chapter 9 and introduces new models needed to investigate the effects of other a.c. system plant components. Turbine generator models are extended by considering the effects of saturation in the synchronous machine and the response of compound thermal turbines. Detailed consideration is also given to the modelling of induction motors and static power converters. The chapter also deals with protective gear modelling and unbalanced faults.

The induction motor model allows for a good representation over the whole speed range so that motor starting can be investigated. The model can be created in three ways depending upon the induction motor data available.

Basic formulation of three-phase bridge rectification and inversion has already been considered in Chapter 3 and here it is extended so that the dynamic model can include abnormal operating conditions encountered during stability studies. It must be clarified however, that the controllability of h.v.d.c. links during large disturbances in either the a.c. or d.c. system cannot be determined by the type of transient stability program described in this chapter. These and other problems associated with transient stability analysis involving h.v.d.c. links are considered in Chapter 13.

The grouping of subroutines relevant to a particular component of the power system or aspect of the study, as developed in Chapter 9, should be retained for the models produced in this chapter. This ensures that additional models can be incorporated easily and models removed when not necessary.

10.2. SYNCHRONOUS MACHINE SATURATION

The relationship between mutual flux and the exciting MMF within a machine is not linear and some means of representing this nonlinearity is necessary if the results obtained from a stability study are to be accurate.

251

In most multimachine stability programs, each machine is represented by a voltage behind an impedance. As explained in Chapter 9 the impedance consists of armature resistance plus either transient or subtransient reactance. Also the voltage magnitude may be fixed or time-varying, depending on the complexity of the model.

Saturation may thus be taken into account by modifying the value of the reactance used in representing the machines. However, as explained for the model developed in Chapter 9, it is more convenient to fix the reactance and adjust the voltage accordingly.

Saturation is a part of synchronous machine modelling where there is still uncertainty as to the best method of simulation. The degree of saturation is not the same throughout the machine because the flux varies by the amount of leakage flux. Also, the saturation in the direct and quadrature axes are different, although this difference is small in the case of a cylindrical rotor.

Various methods have been adopted to account for saturation which differ not only in the model modification technique but also in the representation of the saturation characteristic of the machine.

Classical saturation model

Classical theory[1] for a cylindrical rotor machine assumes that the saturation is due to the total MMF produced in the iron and is the same in each axis.

It is necessary to make further assumptions in order to simplify the model as follows:

(1) The magnetic reluctance in both axes are equal. Thus, the synchronous reactances are equal, i.e. $X_d = X_q$.
(2) Saturation does not distort the sinusoidal variations assumed for rotor and stator inductances.
(3) Because load-test data is not usually available, saturation is determined using the open-circuit saturation curve.
(4) Potier reactance X_p may be used in calculating saturation.
(5) The total iron MMF (Fe) may be determined from

$$\text{Fe} = S \cdot If \qquad (10.2.1)$$

where S is the saturation factor, defined as:

$$S = 1 + \frac{\text{iron MMF}}{\text{air gap MMF}} \qquad (10.2.2)$$

(6) With reference to Fig. 10.1, the saturation factor S may be determined from:

$$S = \frac{AC}{AB} \qquad (10.2.3)$$

Figure 10.2 shows a typical voltage and MMF diagram for a round rotor synchronous machine. Potier voltage Ep, the voltage behind Potier reactance,

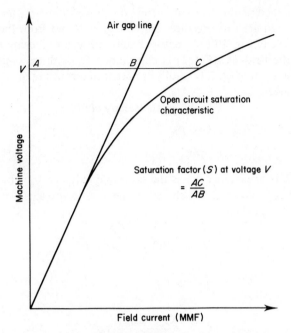

Fig. 10.1. Open circuit saturation characteristic of a synchronous machine

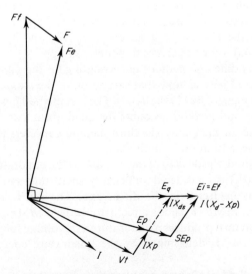

Fig. 10.2. Vector diagram of cylindrical rotor synchronous machine saturation

may be determined readily from the terminal voltage Vt and the terminal current I. The MMF required to produce this voltage is found from the open circuit saturation curve of Fig. 10.1. Armature reaction F is found using Assumption 4 from which the field MMF (Ff) is calculated. The voltage equivalent to Ff referred to the stator is Ef. It is readily apparent that rotating the MMF diagram through 90° gives:

$$Ff \propto Ei \text{ (in the steady state)}$$

$$Fe \propto S \cdot Ep \tag{10.2.4}$$

$$F \propto I(X_d - Xp)$$

In Fig. 10.2, the reactance X_{ds} is the saturated value of X_d. This produces an internal machine voltage E_q which lies on the quadrature axis. As $I(X_d - Xp)$ is parallel with $I \cdot X_{ds}$, then

$$\frac{Ep}{S \cdot Ep} = \frac{I(X_{ds} - Xp)}{I(X_d - Xp)} = \frac{E_q}{Ei} \tag{10.2.5}$$

and from this

$$Ei = S \cdot Eq \tag{10.2.6}$$

and

$$X_{ds} = \frac{(X_d - X_p)}{S} + Xp \tag{10.2.7}$$

All machine reactances subject to saturation are similarly modified.

Salient machine saturation

In the case of a salient synchronous machine, it may be assumed that the direct and quadrature axis armature reaction MMF's (F_d and F_q respectively) are proportional to the reactive voltage drops $I_d \cdot Xa_d$ and $I_q \cdot Xa_q$ respectively. Assumptions 3 and 4 for the classical model also apply.

There are many different methods of accounting for the saturation effect. The methods considered here assume that saturation in the d-axis is due at least in part to the component of flux in the d-axis. The first method ignores saturation in the q-axis, the second method accounts for quadrature axis saturation by the component of flux in the q-axis, the third method considers that the total flux contributes to the saturation in both axes.

In the first method, the density of the flux due to the quadrature axis armature reaction MMF (F_q) is considered sufficiently small that saturation effects on voltages are thus neglected in the direct axis. The other component of the armature reaction MMF (F_d) adds directly to the field MMF (Ff) to produce a main flux which in turn produces a quadrature axis voltage subject to saturation. The saturation level is determined by the quadrature component of Potier Voltage (Ep_q).

The second method,[2] allows for saturation in both the direct and quadrature axes components of the Potier voltage. It is assumed that the reluctances of the

d-axis and q-axis paths differ only because of the different air gaps in each axis. The d-axis component of Potier Voltage (Ep_d) is thus modified by the ratio X_q/X_d before the q-axis saturation factor is determined. Provided that it is assumed that the vector sum of the two saturated main flux components (Fe_q and Fe_d) is in phase with the MMF proportional to Potier Voltage, then the saturated d and q axis synchronous reactances (X_{ds} and X_{qs}) are:

$$X_{ds} = \frac{(X_d - Xp)}{S_d} + Xp \tag{10.2.8}$$

$$X_{qs} = \frac{(X_q - Xp)}{S_q} + Xp \tag{10.2.9}$$

A third method[1] distinguishes between the saturation in the rotor and stator, and saturation factors based on Ep and Ep_q are obtained. This method is difficult to implement because it is necessary to ensure that saturation is not applied twice to any part of the machine. That is, the saturation in the field poles must be isolated from that of the armature, giving two saturation curves. The two saturation factors for this case may be defined as:

$$S_{dq} = 1 + \frac{\text{iron MMF in the stator}}{\text{total air gap MMF}} \tag{10.2.10}$$

$$S_d = 1 + \frac{\text{iron MMF in the rotor}}{\text{direct axis air gap MMF}} \tag{10.2.11}$$

where S_{dq} acts equally on both d and q axes and S_d acts on the direct axis only.

Figure 10.3, 10.4 and 10.5 demonstrate the differences between the three methods of saturation representation.

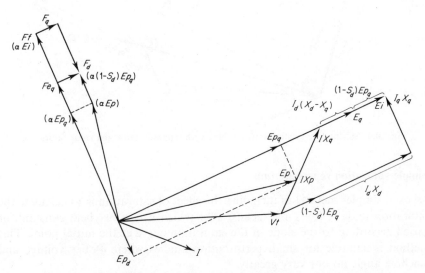

Fig. 10.3. Salient pole synchronous machine with direct axis saturation only

256

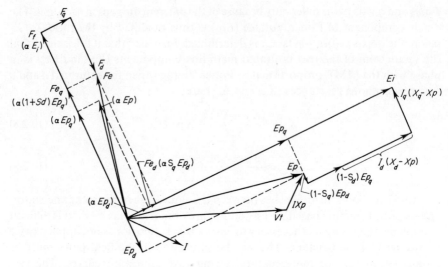

Fig. 10.4. Salient pole synchronous machine with direct and quadrature axis saturation

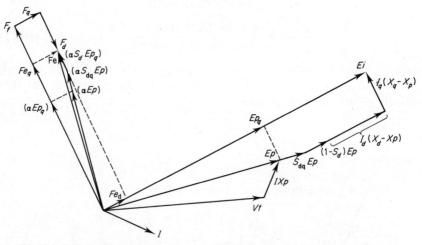

Fig. 10.5. Salient pole synchronous machine with separate stator and rotor saturation

Simple saturation representation

An even simpler method of including the effect of saturation is to calculate the saturation initially (by some means) after which it is either held constant, or varied according to the slope of the saturation curve at the initial point. This method is suitable for small perturbation studies where Potier voltage and machine angle do not vary greatly.

Saturation curve representation

The open-circuit saturation curve must be stored within the computer so that a new saturation factor can be determined at every stage of the study.

The most accurate method of storing this curve is to fit a polynomial of the form:

$$If = C_0 + C_1 V + C_2 V^2 + C_3 V^3 \ldots + C_n V^n \tag{10.2.12}$$

by taking $n + 1$ points on the curve. Normally n would be 5, 7 or 9. This is a clumsy method of both entering the data and storing it. In multimachine transient stability studies, where the machines are represented at best by subtransient parameters and an approximation to Potier reactance is made, nothing is achieved by such an elaborate method of representing the saturation curve.

The problem can be simplified by assuming that most of the coefficients of the polynomial are zero. A sufficiently good approximation is achieved with the equation[3]:

$$If = V + C_n V^n \tag{10.2.13}$$

where n is normally either 7 or 9. Only one point is needed to specify the curve and if If is always specified at a predetermined voltage, the data entries required per curve are reduced to one, from which C_n may be readily determined.

Potier reactance

The Potier reactance of a machine is rarely quoted, although the open-circuit saturation curve is normally available. In order to model the saturation effects, it is thus necessary to estimate this reactance.

From knowledge of the leakage reactance Xl, Beckwith[4] calculated that:

$$Xp = Xl + 0.63(X'_d - Xl) \tag{10.2.14}$$

and if Xl is not available, then:

$$Xp = 0.8X'_d \tag{10.2.15}$$

Equation (10.2.15) may be modified to account for the type of synchronous machine,[5] i.e.

$$Xp = 0.9X'_d \tag{10.2.16}$$

for a salient-pole machine, as most of the saturation occurs in the poles, and

$$Xp = 0.7X'_d \tag{10.2.17}$$

for a round rotor machine, as most of the saturation occurs in the rotor teeth and the Potier and leakage reactances have similar values.

The effect of saturation on the synchronous machine model

Saturation effectively modifies the ordinary differential equations describing the behaviour of the voltages used to model the synchronous machine. Equations (9.2.13), (9.2.14), (9.2.17) and (9.2.18) become respectively:

$$pE_q' = (Ef + (X_d - X_d')I_d - S_d E_q')/T_{d0}' \tag{10.2.18}$$

$$pE_d' = (-(X_q - X_q')I_q - S_q E_d')/T_{q0}' \tag{10.2.19}$$

$$pE_q'' = (S_d E_q' + (X_d - X_d'')I_d - S_d E_q'')/T_{d0}'' \tag{10.2.20}$$

$$pE_d'' = (S_q E_d' - (X_q' - X_q'')I_q - S_q E_d'')/T_{q0}'' \tag{10.2.21}$$

where S_d and S_q are the direct and quadrature axes saturation factors.

Where subtransients are considered then equations (9.2.15) and (9.2.16) are replaced by:

$$E_q'' - Vt_q = Ra \cdot I_q - X_{ds}'' \cdot I_d \tag{10.2.22}$$

$$E_d'' - Vt_d = Ra \cdot I_d + X_{qs}'' \cdot I_q \tag{10.2.23}$$

Representation of saturated synchronous machines in the network

Representation of a salient but unsaturated synchronous machine in the network has been discussed in Section 9.6. When saturation occurs, a double adjustment must be made at each step in the solution process.[6]

With the notation developed in Section 9.6, the fixed admittance (Y_0) which is included with the network is made up from unsaturated and nonsalient values of reactance, whereas the correct admittance (Y_{ts}) is made up from saturated and salient values. Using Fig. 9.14, as before the adjusting current to account for this change in admittance is

$$I_{adjs} = (\bar{Y}_{ts} - \bar{Y}_0)(\bar{E}'' - \bar{V}) \tag{10.2.24}$$

That is:

$$
\begin{bmatrix} I_{adjs_q} \\ I_{adjs_d} \end{bmatrix} = \begin{bmatrix} \dfrac{1}{(Ra^2 + X_{ds}''X_{qs}'')} \begin{bmatrix} Ra & X_{ds}'' \\ -X_{qs}'' & Ra \end{bmatrix} - \dfrac{1}{(Ra^2 + X_d''\cdot X_q'')} \begin{bmatrix} Ra & X_{dq} \\ -X_{dq} & Ra \end{bmatrix} \end{bmatrix} \begin{bmatrix} E_q'' - V_q \\ E_d'' - V_d \end{bmatrix}
$$

$$
\begin{bmatrix} I_{adjs_q} \\ I_{adjs_d} \end{bmatrix} = \begin{bmatrix} I_{adjs_q}^{(1)} \\ I_{adjs_d}^{(1)} \end{bmatrix} + \begin{bmatrix} \dfrac{1}{(Ra^2 + X_{ds}''X_{qs}'')} \begin{bmatrix} Ra & X_{dqs} \\ -X_{dqs} & Ra \end{bmatrix} - \dfrac{1}{(Ra^2 + X_d''X_q'')} \end{bmatrix}
$$

$$
\begin{bmatrix} Ra & X_{dq} \\ -X_{dq} & Ra \end{bmatrix} \begin{bmatrix} E_q'' - V_q \\ E_d'' - V_d \end{bmatrix} \tag{10.2.25}
$$

The current $\bar{I}^{(1)}_{adjs}$ is similar to $\bar{I}_{adj}$ developed in Section 9.6:

$$
\begin{array}{|c|}
\hline
I^{(1)}_{adjs_q} \\
\hline
I^{(1)}_{adjs_d} \\
\hline
\end{array}
=
\frac{\frac{1}{2}(X''_{ds} - X''_{qs})}{(Ra^2 + X''_{ds}X''_{qs})}
\begin{array}{|c|c|}
\hline
0 & 1 \\
\hline
1 & 0 \\
\hline
\end{array}
\begin{array}{|c|}
\hline
E''_q - V_q \\
\hline
E''_d - V_d \\
\hline
\end{array}
\tag{10.2.26}
$$

The current injected into the network is given by equation (9.6.4) and as the terms in the brackets contain no saliency then in the real and imaginary axis of the network:

$$
\begin{array}{|c|}
\hline
I_{inj_r} \\
\hline
I_{inj_m} \\
\hline
\end{array}
=
\begin{array}{|c|}
\hline
I^{(1)}_{adjs_r} \\
\hline
I^{(1)}_{adjs_m} \\
\hline
\end{array}
+
\begin{array}{|c|}
\hline
I^{(2)}_{adjs_r} \\
\hline
I_{adjs_m} \\
\hline
\end{array}
+
\frac{1}{(Ra^2 + X''_d X''_q)}
\begin{array}{|c|c|}
\hline
Ra & X_{dq} \\
\hline
-X_{dq} & Ra \\
\hline
\end{array}
\cdot
\begin{array}{|c|}
\hline
V_r \\
\hline
V_m \\
\hline
\end{array}
\tag{10.2.27}
$$

where $\bar{I}^{(2)}_{adjs}$ contains saturated but nonsalient reactance terms and $I^{(1)}_{adjs}$ contains salient and saturated reactance terms.

$$
\begin{array}{|c|}
\hline
I^{(1)}_{adjs_r} \\
\hline
I^{(1)}_{adjs_m} \\
\hline
\end{array}
=
\frac{\frac{1}{2}(X''_{ds} - X''_{qs})}{(Ra^2 + X''_{ds}X''_{qs})}
\begin{array}{|c|c|}
\hline
-\sin 2\delta & \cos 2\delta \\
\hline
\cos 2\delta & \sin 2\delta \\
\hline
\end{array}
\cdot
\begin{array}{|c|}
\hline
E''_r - V_r \\
\hline
E''_m - V_m \\
\hline
\end{array}
\tag{10.2.28}
$$

$$
\begin{array}{|c|}
\hline
I^{(2)}_{adjs_r} \\
\hline
I^{(2)}_{adjs_m} \\
\hline
\end{array}
=
\frac{1}{(Ra^2 + X''_{ds}X''_{qs})}
\begin{array}{|c|c|}
\hline
Ra & X_{dqs} \\
\hline
-X_{dqs} & Ra \\
\hline
\end{array}
\cdot
\begin{array}{|c|}
\hline
E''_r - V_r \\
\hline
E''_m - V_m \\
\hline
\end{array}
\tag{10.2.29}
$$

Note that the third part of equation (10.2.27) is $\bar{Y}_0 \bar{V}$ and not $\bar{Y}_0 \bar{E}''$. This part of the injected current is merely the current flowing through $\bar{Y}_0$ and could be eliminated if $\bar{Y}_0$ was not included in the network. The conditioning of the network would be reduced however, and in certain systems this could lead to numerical problems.

Inclusion of synchronous machine saturation in the transient stability program

Only two subroutines need modification to allow saturation effects in synchronous machines to be modelled. In both cases, an iterative solution is necessary for each saturating machine, although in most instances the number of iterations is small.

Saturation is a function of the voltage behind armature resistance and Potier reactance. Assuming the second method of salient machine saturation is being used, then from equation (10.2.13):

$$S_d = 1 + C_{n_d}(Ep_q)^{n-1} \tag{10.2.30}$$

$$S_q = 1 + C_{n_q}(Ep_d)^{n-1}$$

where

$$C_{n_q} = \frac{X_q}{X_d} \cdot C_{n_d} \tag{10.2.31}$$

and

$$Ep_q = V_q + Ra \cdot I_q - Xp \cdot I_d \tag{10.2.32}$$

$$Ep_d = V_d + Ra \cdot I_d + Xp \cdot I_q$$

and I_d and I_q are given by equations (10.2.22) and (10.2.23).

A Jacobi iterative technique is quite adequate to establish the initial conditions of the synchronous machine and this can be incorporated in the relevant subroutine shown in Section 2 of Fig. 9.20.

During the time solution, however, saturation can vary over a large range of values and a Newton form of iteration is an advantage especially if large integration steps are used.

Redefining equation (10.2.30) as:

$$f1 = 1 - S_d + C_{n_d}(Ep_q)^{n-1} \tag{10.2.33}$$

$$f2 = 1 - S_q + C_{n_q}(Ep_d)^{n-1}$$

the elements of a 2×2 Jacobian matrix can be found. However, elements $\partial f1/\partial S_q$ and $\partial f2/\partial S_d$ are small with respect to the other two elements and if Ra is considered to be zero then the four elements reduce to:

$$\frac{\partial f1}{\partial S_d} = \frac{[(n-1) \cdot C_{n_d}(Ep_q)^{n-2} \cdot Xp(X_d'' - Xp)(E_q'' - Vt_q)]}{[(S_d - 1)Xp + X_d'']} - 1$$

$$\frac{\partial f2}{\partial S_q} = \frac{[(n-1) \cdot C_{n_q}(Ep_d)^{n-2} \cdot Xp(X_q'' - Xp)(E_d'' - Vt_d)]}{[(S_q - 1)Xp + X_q'']} - 1 \tag{10.2.34}$$

$$\frac{\partial f1}{\partial S_q} = \frac{\partial f2}{\partial S_d} = 0$$

This decouples the Newton method and each saturation factor may be solved independently[7]

$$S_d^{(p+1)} = S_d^{(p)} - f1^{(p)}/(\partial f1/\partial S_d)^{(p)} \tag{10.2.35}$$

$$S_q^{(r+1)} = S_q^{(r)} - f2^{(r)}/(\partial f2/\partial S_q)^{(r)}$$

Despite the advantages of a Newton form of solution, it can be found to be divergent if too great an integration step length is used. Analysis of the functions

$f1$ and $f2$ show that they have discontinuities, when $X_d'' = (S_d - 1)Xp$ and $X_q'' = (S_q - 1)Xp$ respectively, although otherwise are almost linear. It is therefore necessary to monitor this iterative procedure and modify the step length if necessary.

The evaluation of S_d and S_q should be performed twice during each iteration. Considering Fig. 9.21, this is during the calculation of the injected currents into the network and the calculation of the nonintegrable variables. Provided the discontinuity is not encountered, convergence is achieved in one or two iterations at each re-evaluation especially if the saturation factors are extrapolated at the beginning of each step.

10.3. DETAILED TURBINE MODEL

More detailed turbine models than the one described in the previous chapter are often required for the following reasons:

(1) A longer-term transient stability study or a dynamic stability study is to be made.
(2) The turbine is a two-shafted cross-compound machine which has a separate generator on each shaft.
(3) Generator overspeed is such that an interceptor valve may operate during the study.

A generalized model to accommodate the different types of compound turbine has been developed by the IEEE.[8] As with the generalized AVR model, by setting certain gains to either zero or unity and time constants to either infinity (very large) or zero, the model can be reduced to any desired form. An interceptor valve can easily be incorporated as shown in Fig. 10.6.

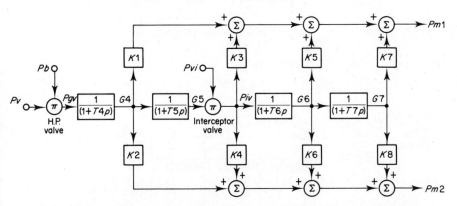

Fig. 10.6. Generalized detailed turbine model including H.P. and interceptor valves

Some normal compound turbine configurations are shown in Fig. 10.7 and Table 10.1 gives typical values for these configurations using the generalized model. A hydroturbine can also be represented and the values given in Table 10.1

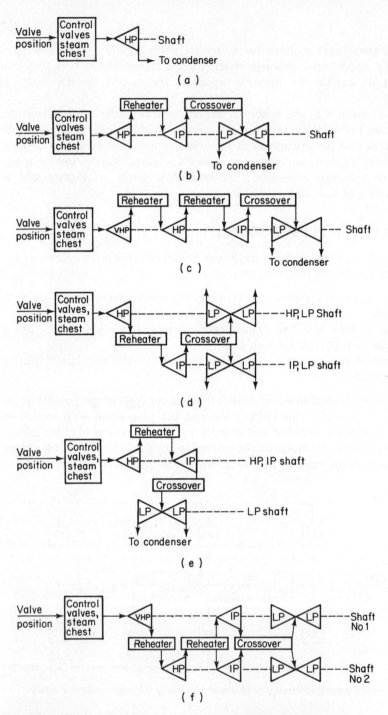

Fig. 10.7. Common steam turbine configurations[8] (© 1982 IEEE)

(a) Nonreheat;

(b) Tandem compound, single reheat;

(c) Tandem compound, double reheat;

(d) Cross compound, single reheat;

(e) Cross compound, single reheat;

(f) Cross compound, double reheat

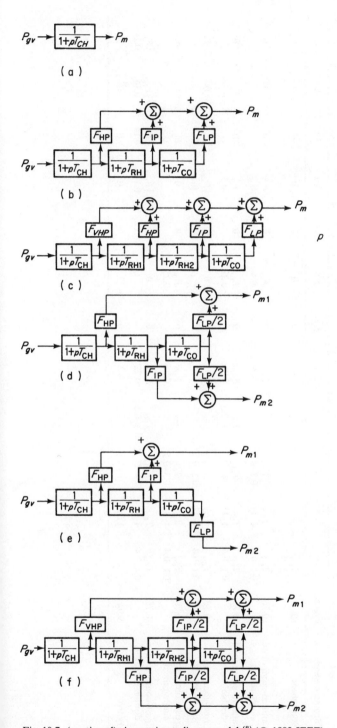

Fig. 10.7. (continued). Approximate linear models[8] (© 1982 IEEE)

Table 10.1. Parameters used in generalized detailed turbine model.[8] (© 1982 IEEE)

Turbine system	Figure	Time constants with typical values (seconds)				Fractions with typical values (p.u.)							
		T_4	T_5	T_6	T_7	K_1	K_2	K_3	K_4	K_5	K_6	K_7	K_8
Non-reheat	10.7a	T_{CH} 0.2–0.5	—	—	—	1	0	0	0	0	0	0	0
Tandem-compound single-reheat	10.7b	T_{CH} 0.1–0.4	T_{RH} 4–11	T_{CO} 0.3–0.5	—	F_{HP} 0.3	0	F_{IP} 0.4	0	F_{LP} 0.3	0	0	0
Tandem-compound double-reheat	10.7c	T_{CH} 0.1–0.4	T_{RH1} 4–11	T_{RH2} 4–11	T_{CO} 0.3–0.5	F_{VHP} 0.22	0	F_{HP} 0.22	0	F_{IP} 0.3	0	F_{LP} 0.26	0
Cross-compound single-reheat	10.7d	T_{CH} 0.1–0.4	T_{RH} 4–11	T_{CO} 0.3–0.5	—	F_{HP} 0.3	0	0	F_{IP} 0.3	$\frac{1}{2}F_{LP}$ 0.2	$\frac{1}{2}F_{LP}$ 0.2	0	0
Cross-compound single-reheat	10.7e	T_{CH} 0.1–0.4	T_{RH} 4–11	T_{CO} 0.3–0.5	—	F_{HP} 0.25	0	F_{IP} 0.25	0	0	F_{LP} 0.5	0	0
Cross-compound double-reheat	10.7f	T_{CH} 0.1–0.4	T_{RH1} 4–11	T_{RH2} 4–11	T_{CO} 0.3–0.5	F_{VHP} 0.22	0	0	F_{HP} 0.22	$\frac{1}{2}F_{IP}$ 0.14	$\frac{1}{2}F_{IP}$ 0.14	$\frac{1}{2}F_{LP}$ 0.14	$\frac{1}{2}F_{LP}$ 0.14
Hydro	9.9a	0	$\frac{1}{2}T_W$	—	—	-2	0	3	0	0	0	0	0

are justified by the method of representing a lead-lag circuit described in Chapter 9, with the time constant T_5 set at $\frac{1}{2} Tw$ in the case of the simplest model.

The full set of equations for the detailed turbine model is:

$$pG4 = (P_{gv} - G4)/T4 \tag{10.3.1}$$

$$pG5 = (G4 - G5)/T5 \tag{10.3.2}$$

$$Piv = G5 \cdot Pvi \tag{10.3.3}$$

$$pG6 = (Piv - G6)/T6 \tag{10.3.4}$$

$$pG7 = (G6 - G7)/T7 \tag{10.3.5}$$

$$P_{m1} = K1 \cdot G4 + K3 \cdot Piv + K5 \cdot G6 + K7 \cdot G7 \tag{10.3.6}$$

$$P_{m2} = K2 \cdot G4 + K4 \cdot Piv + K6 \cdot G6 + K8 \cdot G7 \tag{10.3.7}$$

Also note that:

$$\sum_{n=1}^{8} Kn = 1 \tag{10.3.8}$$

and that, in the initial steady state, the interceptor valve, if present, will be fully open ($Pvi = 1$) in which case:

$$G4 = G5 = Piv = G6 = G7 = P_{m1} + P_{m2} \tag{10.3.9}$$

The speed governor controlling the interceptor valve is similar to that controlling the HP turbine except that it is set to operate at some overspeed value of slip (k_ω) and not about synchronous speed. Equation (9.3.12) can be modified in this case to:

$$pG1 = [R(1 + T2p) \cdot (2\pi f\, 0(1 + k_\omega) - \omega) - G1]/T1 \tag{10.3.10}$$

10.4. INDUCTION MACHINES

An approach similar to that used to construct the synchronous machine models is required if induction machines are to be explicitly modelled.[2],[9] However, speed cannot be assumed to vary only slightly and this basic difference requires that the equations describing the behaviour of induction machines be somewhat different from those developed for a synchronous machine.

Mechanical equations

It is necessary to express the equation of motion of an induction machine in terms of torque and not power. Also symmetry of the rotor makes its angular position unimportant, and slip (S) usually replaces angular velocity (ω) as the variable, where:

$$S = (\omega_0 - \omega)/\omega_0 \tag{10.4.1}$$

Assuming negligible windage and friction losses and smooth mechanical shaft power, the equation of motion is:

$$pS = (Tm - Te)/(2Hm) \qquad (10.4.2)$$

where Hm is the inertia constant measured in kWs/kVA established at synchronous speed. The mechanical torque (Tm) and electrical torque (Te) are assumed to be positive when the machine is motoring.

The mechanical torque Tm will normally vary with speed, the relationship depending on the type of load.

A commonly used characteristic is:

$$Tm \alpha \{(\text{speed})^k\}$$

where $k = 1$ for fan-type loads and $k = 2$ for centrifugal pumps. A more elaborate torque/speed characteristic can be used for a composite load, i.e.:

$$Tm \alpha \{a + b(\text{speed}) + c(\text{speed})^2\} \qquad (10.4.3)$$

which can include the effect of striction when start-up is being considered.

In terms of slip the torque is thus:

$$Tm = A + B \cdot S + C \cdot S^2 \qquad (10.4.4)$$

where

$$A \; \alpha\{a + b + c\}$$

$$B \; \alpha\{b + 2c\}$$

$$C \; \alpha \; c$$

The values of A, B and C are determined from the initial (steady-state) loading of the motor and hence its initial value of slip.

The electrical torque Te is related to the air gap electrical power by the electrical frequency which is assumed constant and hence:

$$Te = Re(\bar{E} \cdot \bar{I}_1{}^*)/2\pi f0 \qquad (10.4.5)$$

Electrical equations

A simplified equivalent circuit for a single-cage induction motor is shown in Fig. 10.8, with $R1$ and $X1$ referring to the stator and $R2$ and $X2$ referring to the

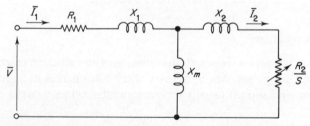

Fig. 10.8. Steady-state equivalent circuit of a single cage induction motor

rotor resistance and reactance respectively. In a similar manner to the transient model of a synchronous machine, an induction motor may be modelled by a Thevenin equivalent circuit of a voltage E' behind the stator resistance $R1$ and a transient reactance X'. The transient reactance is the apparent reactance when the rotor is locked stationary and the slip (S) is unity and is given by:

$$X' = X1 + \frac{X2 \cdot Xm}{(X2 + Xm)} \tag{10.4.6}$$

where Xm is the magnetizing reactance of the machine. The rate of change of transient voltage is given by:

$$p\bar{E}' = -j2\pi f \cdot S\bar{E}' - (\bar{E}' - j(X0 - X')\bar{I}1)/T_0' \tag{10.4.7}$$

where the rotor open-circuit time constant T_0' is:

$$T_0' = \frac{(X2 + Xm)}{2\pi f 0 R2} \tag{10.4.8}$$

and the open-circuit reactance Xo is:

$$Xo = X1 + Xm \tag{10.4.9}$$

The reactances are unaffected by rotor position and the model is described in the real and imaginary components used for the network, that is, in the synchronously rotating frame of reference. Thus, for a full description of the model, the following equations are used:

$$V_r - E_r' = R1 \cdot I1_r - X' \cdot I1_m \tag{10.4.10}$$

$$V_m - E_m' = R1 \cdot I1_m + X' \cdot I1_r \tag{10.4.11}$$

$$pE_r' = 2\pi f 0 S E_m' - (E_r' + (X0 - X') I1_m)/T'0 \tag{10 4.12}$$

$$pE_m' = -2\pi f 0 S E_r' - (E_m' - (X0 - X') I1_r)/T'0 \tag{10.4.13}$$

A transient-stability program incorporating an induction motor model uses the transient and open-circuit parameters, but it is often convenient to allow the stator, rotor and magnetizing parameters to be specified and let the program derive the former parameters.

For completeness, the electrical torque may now be written as:

$$Te = (E_r' \cdot I1_r + E_m' \cdot I1_m)/\omega_0 \tag{10.4.14}$$

Electrical equations when the slip is large

Single-cage induction motors have low-starting torques and it is often difficult to bring them to speed without either reducing the load or inserting external resistance in the rotor circuit. As a result of the low-starting torque, when the slip exceeds the point of maximum torque, the single-case model is often insufficiently accurate. These problems are overcome by the use of a double-cage or deep-bar rotor model.

268

When a torque slip characteristic of the motor is available, then a simple solution is to modify the torque-slip characteristic of the single-cage motor model. Double-cage or deep-bar rotors have a resistance and reactance which varies with slip. A cage factor Kg can be included which allows for the variations of rotor resistance.

$$R2 = R2(0) \cdot (1 + Kg \cdot S) \tag{10.4.15}$$

where $R2(0)$ is the rotor resistance at zero slip.

It is usually convenient to make the cage factor larger than that necessary to describe the change in rotor resistance. In this way, the torque-slip characteristics of the model can be made similar to that of the motor without the need to vary the rotor reactance with slip. The result of varying the rotor resistance is to modify the open-circuit transient time constant only, and this can be done quite simply at each integration step of the simulation.

Rotor reactance does not vary with slip as greatly as rotor resistance, provided saturation effects are ignored, and its effect on the open-circuit transient time constant is thus small. Transient reactance (X') varies with rotor reactance. However, this variation on the term $(X_0 - X')$ in equations (10.4.12) and (10.4.13) is insignificant. Thus, the only major effect of varying rotor reactance is in equations (10.4.10) and (10.4.11) which requires a technique similar to that adopted in the synchronous machine model to account for saturation and saliency. However, the gains obtained in using two-cage factors are insignificant and a single cage factor varying rotor resistance is usually adopted.

DOUBLE-CAGE ROTOR MODEL

An alternative to the cage factor is the use of a better rotor model, though this is often restricted by the unavailability of suitable data.

Induction motor loads having double-cage or deep-bar rotors can be represented in a similar manner to a single cage motor.[10],[11] It is assumed that the end-ring resistance and that part of the leakage flux which links the two secondary windings, but not the primary, are neglected. The steady-state

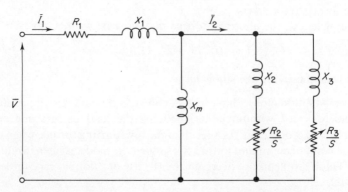

Fig. 10.9. Steady-state equivalent circuit of a double cage induction motor

equivalent circuit shown in Fig. 10.9 can thus be obtained where $R3$ and $X3$ are the resistance and reactance of the additional rotor winding. A circuit similar to that of Fig. 10.8 can be obtained by substituting the two parallel rotor circuit branches by a single series circuit, where:

$$R2(S) = \frac{R2 \cdot R3(R2 + R3) + S^2(R2 \cdot X3^2 + R3 \cdot X2^2)}{(R2 + R3)^2 + S^2(X2 + X3)} \qquad (10.4.16)$$

$$X2(S) = \frac{R2^2 X3 + R3^2 X2 + S^2(X2 + X3)X2 \cdot X3}{(R2 + R3)^2 + S^2(X2 + X3)^2} \qquad (10.4.17)$$

At any instant during a transient stability study, the rotor impedance may be assumed to be the steady-state value given above.

Analysis similar to that used in developing equations (10.4.10) to (10.4.13) gives:

$$V_r - E_r'' = R1 \cdot I1_r - X'' \cdot I1_m \qquad (10.4.18)$$

$$V_m - E_m'' = R1 \cdot I1_m + X'' \cdot I1_r \qquad (10.4.19)$$

$$pE_r'' = -2\pi f0 \cdot S(E_m' - E_m'') + pE_r' + (E_r' - E_r'' - (X' - X'')I1_m)/T_0'' \qquad (10.4.20)$$

$$pE_m'' = 2\pi f0 \cdot S(E_r' - E_r'') + pE_m' + (E_m' - E_m'' + (X' - X'')I1_r)/T_0'' \qquad (10.4.21)$$

$$Te = E_r'' \cdot I1_r + E_m'' \cdot I1_m \qquad (10.4.22)$$

with equations (10.4.12) and (10.4.13) applying also.

The parameters for the model, when the motor has a double-cage rotor are given by equations (10.4.6), (10.4.8), (10.4.9) and:

$$X'' = X1 + \frac{X2 \cdot X3 \cdot Xm}{(X2 \cdot X3 + X2 \cdot Xm + X3 \cdot Xm)} \qquad (10.4.23)$$

$$T_0'' = \frac{X3 + (X2 \cdot Xm)/(X2 + Xm)}{2\pi f0R3} \qquad (10.4.24)$$

If the rotor is of the deep-bar type, then the parameters of the equivalent double-cage type may be determined using equation (10.4.16) and (10.4.17). The rotor parameters at zero slip are:

$$R2(0) = \frac{R2 \cdot R3}{(R2 + R3)} \qquad (10.4.25)$$

$$X2(0) = \frac{(R2^2 \cdot X3 + R3^2 \cdot X2)}{(R2 + R3)} \qquad (10.4.26)$$

and at standstill are:

$$R2(1) = \frac{X2^2 \cdot R3 + X3^2 \cdot R2 + R2 \cdot R3(R2 + R3)}{(X2 + X3)^2 + (R2 + R3)^2} \qquad (10.4.27)$$

$$X2(1) = \frac{X2 \cdot X3(X2 + X3) + R2^2 \cdot X3 + R3^2 \cdot X2}{(X2 + X3)^2 + (R2 + R3)^2} \qquad (10.4.28)$$

This set of nonlinear equations may be solved using Newtonian techniques but by substituting:

$$R3 = \frac{R2 \cdot R2(0)}{(R2 - R2(0))} \qquad (10.4.29)$$

and

$$X3 = \frac{X2 \cdot Xx}{(X2 - Xx)} \qquad (10.4.30)$$

where

$$Xx = \frac{X2(1) - (R2(1) - R2(0))^2}{(X2(0) - X2(1))} \qquad (10.4.31)$$

the number of variables reduces to two and a simple iterative procedure yields a result in only a few iterations.[12] A reasonable starting value is $X2 \simeq \frac{3}{2} X_0$ derived from assuming $R2 \simeq \frac{1}{5} R3$ and $X2 \simeq \frac{5}{2} X3$.

Representation of induction machines in the network

This is quite simple compared to the representation of a synchronous machine as neither saliency nor saturation are normally considered in the induction machine models. They may, therefore, be considered as injected currents in parallel with fixed admittances.

Modifying equations (10.4.18) and (10.4.19) gives a machine current of:

$$\bar{I}1 = \bar{Y}(\bar{V} - \bar{E}'') \qquad (10.4.32)$$

or

$$
\begin{vmatrix} I1_r \\ I1_m \end{vmatrix} = \frac{1}{(R1^2 + X''^2)} \begin{vmatrix} R1 & X'' \\ -X'' & R1 \end{vmatrix} \begin{vmatrix} V_r - E_r'' \\ V_m - E_m'' \end{vmatrix} \qquad (10.4.33)
$$

The injected current into the network which includes $\bar{Y}$ is thus:

$$
\begin{vmatrix} I_{\mathrm{inj}_r} \\ I_{\mathrm{inj}_m} \end{vmatrix} = \frac{-1}{(R1^2 + X''^2)} \begin{vmatrix} R1 & X'' \\ -X'' & R1 \end{vmatrix} \begin{vmatrix} E_r'' \\ E_m'' \end{vmatrix} \qquad (10.4.34)
$$

where the minus sign confirms the induction machine as being assumed to be motoring.

Inclusion of induction machine in the transient stability program

This is relatively straightforward using the same format as developed for synchronous machines. Most induction machines are equipped with contactors which respond to terminal conditions such as undervoltage and it is sometimes necessary to model this equipment. The characteristics and logic associated with contactors are included in Section 10.7 (Relays).

10.5. A.C.–D.C. CONVERSION

The use of high-voltage and/or high-current direct current systems is now sufficiently widespread to require the inclusion of d.c. converter models as a standard part of a comprehensive transient stability program. Further, rectification equipment is also required in many industrial processes, notably smelters and chlorine producers, and these are sufficiently large-load items to warrant good modelling.

The dynamic behaviour of h.v.d.c. links immediately after a large disturbance either on the d.c. side or close to the converter a.c. terminals requires much more elaborate models, as discussed in Chapters 11 and 12.

When analysing small perturbations and dynamic stability, it is often assumed that the converter equipment operates in a controlled manner almost instantaneously when compared with the relatively slow a.c. system dynamics. In these cases, it is quite acceptable to use a modified steady-state (or quasi-steady-state) model, the modifications being due to the different constraints imposed by the load-flow and stability studies. Such a model is also suitable for representing large rectifier loads during a.c. system disturbances, with further modifications necessary to represent abnormal rectifier operating modes.

Further to the basic assumptions listed in Chapter 3, the following need to be made here:

— The implementation of delay angle control is instantaneous.
— The transformer tap position remains unchanged throughout the stability study unless otherwise specified.
— The direct current is smooth, though its actual value may change during the study.

Rectifier loads

Large rectifier loads generally consist of a number of series and/or parallel connected bridges, each bridge being phase shifted relative to the others. With these configurations, high-pulse numbers can be achieved resulting in minimal distortion of the supply voltage without filtering. Rectifier loads can therefore be modelled as a single equivalent bridge with a sinusoidal supply voltage at the terminals but without representation of passive filters. This model is shown in Fig. 10.10.

Rectifier loads can utilize a number of control methods. They can use diode and thyristor elements in full or half-bridge configurations. In some cases, diode bridges are used with tap changer and saturable reactor control. The effect of the saturable reactors on diode conduction is identical to delay angle control of a thyristor over a limited range of delay angles. All these different control methods

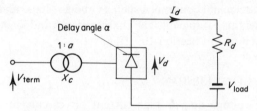

Fig. 10.10. Static rectifier load equivalent circuit

can be modelled using a controlled rectifier with suitable limits imposed on the delay angle (α).[13]

STATIC LOADS

Operating under constant current control, the d.c. equations are

$$V_d = I_d R_d + V_{\text{load}} \tag{10.5.1}$$

$$I_d = \frac{(A \cdot I_{d_s} - V_{\text{load}})}{(A + R_d)} \tag{10.5.2}$$

where A is the constant current controller gain and I_{d_s} is the nominal d.c. current setting as shown in Fig. 10.11.

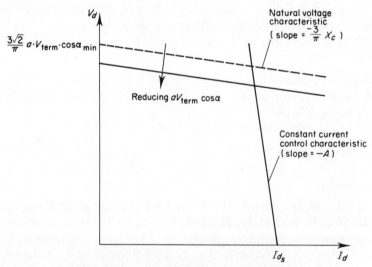

Fig. 10.11. Simple rectifier control characteristic

Constant current cannot be maintained during a large disturbance as a limit of delay angle will be reached. In this event, the rectifier control specification will become one of constant delay angle and equation (10.5.2) becomes:

$$I_d = \frac{[(3\sqrt{2}/\pi)aV_{\text{term}}\cos\alpha_{\text{min}} - V_{\text{load}}]}{[R_d + (3X_c/\pi)]} \tag{10.5.3}$$

Protection limits and disturbance severity determines the rectifier operating characteristics during the disturbance. Shutdown occurs if I_d reaches a set minimum or zero and the voltage V_{load} will cause shutdown before the a.c. terminal voltage reaches zero. The action of the rectifier load system is thus described by equations (3.2.5), (3.2.7), (3.2.10), (3.2.12), (3.2.13), (10.5.1) and either (10.5.2) or (10.5.3).

DYNAMIC LOADS

The basic rectifer load model assumes that current on the d.c. side of the bridge can change instantaneously. For some types of rectifer loads, this may be a valid assumption, but the d.c. load may well have an overall time constant which is significant with respect to the fault clearing time. In order to realistically examine the effects which rectifiers have on the transient-stability of the system, this time constant must be taken into account. This requires a more complex model to account for extended overlap angles, when low commutating voltages are associated with large d.c. currents.

When the delay angle (α) reaches a limiting value, the dynamic response of the d.c. current (I_d) is given by:

$$V_d = I_d \cdot R_d + V_{\text{load}} + L_d \cdot pI_d \tag{10.5.4}$$

where L_d represents the equivalent inductance in the load circuit. Substituting for V_d using equation (3.2.5) gives:

$$pI_d = \frac{1}{T_{dc}} \left\{ \frac{3\sqrt{2}}{\pi R_d} \cdot a \cdot V_{\text{term}} \cdot \cos\alpha - \left(\frac{3X_c}{R_d \pi} + 1 \right) I_d - \frac{V_{\text{load}}}{R_d} \right\} \tag{10.5.5}$$

where $T_{dc} = L_d / R_d$.

The controller time-constant may also be large enough to be considered. However, in transient stability studies where large disturbances are usually being investigated, faults close to the rectifier load force the delay angle (α) to minimum very quickly. Provided the rectifier load continues to operate, the delay angle will remain at its minimum setting throughout the fault period and well into the post-fault period until the terminal voltage recovers. The controller will, therefore, not exert any significant control over the d.c. load current. Ignoring the controller time constant can therefore be justified in most studies.

ABNORMAL MODES OF OPERATION

The slow response of the d.c. current when a large disturbance has been applied to the a.c. system can cause the rectifier to operate in an abnormal mode.

After a fault application near the rectifier, the near normal value of d.c. current (I_d) needs to be commutated by a reduced a.c. voltage. This causes the commutation angle (μ) to increase and it is possible for it to exceed $60°$. This mode of operation is beyond the validity of the equations and to model the dynamic load effects accurately it is necessary to extend the model.

The full range of rectifier operation can be classified into four modes.[14]

Mode 1—Normal operation. Only two valves in the bridge are involved in simultaneous commutation at any one time. This mode extends up to a commutation angle of 60°.

Mode 2—Enforced delay. Although a commutation angle greater than 60° is desired, the forward voltage across the incoming thyristor is negative until either the previous commutation is complete or until the firing angle exceeds 30°. In this mode, μ remains at 60° and α ranges up to 30°.

Mode 3—Abnormal operation. In this mode, periods of three-phase short circuit and d.c. short circuit exist when two commutations overlap. During this period there is a controlled safe short circuit which is cleared when one of the commutations is complete. During the short-circuit periods, four valves are conducting. Commutation cannot commence until 30° after the voltage crossover.

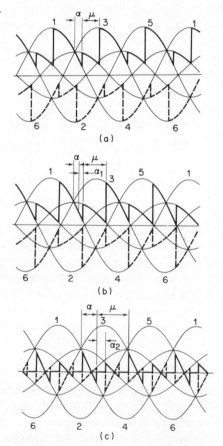

Fig. 10.12. Rectifier voltage waveforms showing different modes of operation. (a) Mode 1, $\mu < 60°$; (b) Mode 2, $\mu = 60°$ with enforced delay α_1; (c) Mode 3, $\mu > 60°$ with short circuit period α_2

Mode 4—Continuous three-phase and d.c. short circuit caused by two commutations taking place continuously. In this mode, the commutation angle is
120° and the a.c. and d.c. current paths are independent.

The waveforms for these modes are shown in Fig. 10.12 and Table 10.2
summarizes the conditions

Table 10.2. Rectifier modes of operation

Mode	Firing angle	Overlap angle
1	$0° \leq \alpha \leq 90°$	$0° \leq \mu \leq 60°$
2	$0° \leq \alpha \leq 30°$	$60°$
3	$30° \leq \alpha \leq 90°$	$60° \leq \mu < 120°$
4	$30° \leq \alpha \leq 90°$	$120°$

for the different modes of operation. Equations (3.2.5) and (3.2.7) do not apply
for a rectifier operating in Mode 3 and they must be replaced by:

$$V_d = \frac{3\sqrt{6}}{\pi} a \cdot V_{\text{term}} \cdot \cos \alpha' - \frac{9}{\pi} X_c \cdot I_d \qquad (10.5.6)$$

and

$$I_d = \frac{a}{\sqrt{6}X_c} V_{\text{term}}(\cos \alpha' - \cos \gamma') \qquad (10.5.7)$$

Fourier analysis of the waveform leads to the relationship between a.c. and d.c.
current given by equation (3.2.9) where the factor k is now:

$$k = \frac{3(\underline{/-2\alpha'} - \underline{/-2\gamma'} - j2\mu)}{4(\cos \alpha' - \cos \gamma')} \qquad (10.5.8)$$

where

$$\alpha' = \alpha - 30°$$

and

$$\gamma' = \gamma + 30° \qquad (10.5.9)$$

A graph showing the value of k for various delay angles (α) and commutation
angles (μ) is shown in Fig. 10.13.

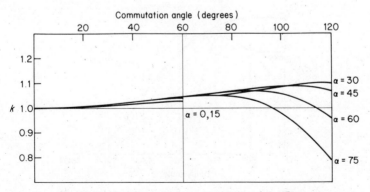

Fig. 10.13. Variation of k in expression $I_p = k(3\sqrt{2}/\pi)I_d$

IDENTIFICATION OF OPERATING MODE

The mode in which the rectifier is operating can be determined simply by use of a current factor K_I. The current factor is defined as:

$$K_I = \frac{\sqrt{2} \cdot X_c \cdot I_d}{a \cdot V_{\text{term}}} \tag{10.5.10}$$

Substitution in this, using the relevant equations, yields limits for the modes.

Mode 1:

$$K_I \leq \cos(60° - \alpha) \tag{10.5.11}$$

and

$$K_I \leq 2\cos(\alpha) \quad \text{for rectifier operation}$$

Mode 2:

$$K_I \leq \frac{\sqrt{3}}{2} \tag{10.5.12}$$

Mode 3:

$$K_I < \frac{2}{\sqrt{3}} \qquad \text{when } \alpha \leq 30°$$

$$K_I < \frac{2}{\sqrt{3}}\cos(\alpha - 30°) \quad \text{when } \alpha > 30° \tag{10.5.13}$$

Mode 4:

$$K_I = \frac{2}{\sqrt{3}} \qquad \text{when } \alpha \leq 30°$$

$$K_I = \frac{2}{\sqrt{3}}\cos(\alpha - 30°) \quad \text{when } \alpha > 30° \tag{10.5.14}$$

This can be demonstrated in the curve of converter operation shown in Fig. 10.14.

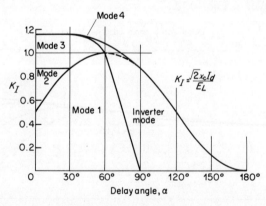

Fig. 10.14. Converter operation [14]

It can thus be seen that the mode of operation can be established prior to solving for the rectifier load equations at every step in the solution.

D.C. link

Provided that it can be safely assumed that a d.c. link is operating in the Quasi Steady State (QSS) Mode 1, the equations developed for converters in Chapter 3 can be used. That is, the converters are considered to be controllable and fast acting so that the normal steady-state type of model can be used at each step in the transient stability study.

The initial steady-state operating conditions of the d.c. link will have been determined by a load flow and in this, the control type, setting and margin will have been established.

CONSTANT CURRENT CONTROL

During the solution process at each iteration the control Mode must be established. This can be done by assuming mode 1 (i.e. with the rectifier on c.c. control) and by combining equations (3.3.2), (3.5.1) and (3.5.2) a d.c. current can be determined as:

$$I_{d_{\text{mode 1}}} = \frac{I_{d_{s_r}} - [(3\sqrt{2}/\pi)a_i \cdot V_{\text{term}_i} \cdot \cos\gamma_{i_c}]/A_r}{[1 + (R_d - (3/\pi)X_{c_i})/A_r]} \qquad (10.5.15)$$

Assuming this current to be valid, then d.c. voltages at each end of the link can be calculated using equations (3.2.5) and (3.3.2). The d.c. link is operating in mode 2 (i.e. with the inverter on C.C. control) if:

$$V_{d_{r\text{mode 1}}} - V_{d_{i\text{mode 1}}} < 0 \qquad (10.5.16)$$

The d.c. current for mode 2 operation is given by:

$$I_{d_{\text{mode 2}}} = \frac{I_{d_{s_i}} + \left(\dfrac{3\sqrt{2}}{\pi}a_r \cdot V_{\text{term}_r} \cdot \cos\alpha_{r\min}\right)\Big/ A_i}{[(1 + (3/\pi)X_{c_i})/A_i]} \qquad (10.5.17)$$

CONSTANT POWER CONTROL

For constant power control, under control mode 1, the d.c. current may be determined from the quadratic equation:

$$\left(\frac{R_d}{2} - \frac{3}{\pi}X_{c_i}\right)I_{d\text{ mode 1}}^2 + \left(\frac{3\sqrt{2}}{\pi}\alpha_i \cdot V_{\text{term}_i} \cdot \cos\gamma_{i_c}\right)I_{d\text{ mode 1}} - P_{d_s} = 0 \qquad (10.5.18)$$

where P_{d_s} is the setting at the electrical mid-point of the d.c. system, that is:

$$P_{d_s} = (P_{d_{s_r}} + P_{d_{s_i}})/2 \qquad (10.5.19)$$

The correct value for $I_{d\text{mode 1}}$ can then be found from Table 10.3. Control mode 2 is determined using equation (10.5.16) and in this case the following quadratic

Table 10.3. Current setting for constant power control from quadratic equation

I_{d_1}	I_{d_2}	I_d
Within	Outside	I_{d_1}
Outside	Within	I_{d_2}
Within	Within	Greater of I_{d_1} and I_{d_2}
Greater	Greater	$I_{d_{max}}$
Greater	Less	$I_{d_{max}}$
Less	Greater	$I_{d_{max}}$
Less	Less	0

Within $\equiv$ within the range $I_{d_{min}}$ to $I_{d_{max}}$;

Outside $\equiv$ outside the range $I_{d_{min}}$ to $I_{d_{max}}$;

Greater $\equiv$ greater than $I_{d_{max}}$

Less $\equiv$ less than $I_{d_{min}}$

equation is required to be solved.

$$k_r \cdot I^2_{d_{\text{mode }2}} - k_v \cdot I_{d_{\text{mode }2}} - P_{d_{\text{marg}}} - P_{d_s} = 0 \tag{10.5.20}$$

where

$$k_r = \frac{R_d}{2} + \frac{3}{\pi} X_{c_r} \tag{10.5.21}$$

$$k_v = \frac{3\sqrt{2}}{\pi} a_r \cdot V_{\text{term}_r} \cdot \cos \alpha_{r_{\min}} \tag{10.5.22}$$

If the link is operating under constant power control but with a current margin then for control mode 2

$$-k_r I^2_{d_{\text{mode }2}} + (k_v - k_r I_{d_{\text{marg}}}) I^2_{d_{\text{mode }2}} + k_v I_{d_{\text{marg}}} - P_{d_{s_r}} = 0 \tag{10.5.23}$$

It is possible for the d.c. link to be operating in control mode 2 despite satisfying the inequality of equation (10.5.16). This occurs when the solution indicates that the rectifier firing angle (α_r) is less than the minimum value ($\alpha_{r_{\min}}$). In this case the delay angle should be set to its minimum and a solution in mode 2 is obtained.

It is also possible, that when the link is operating close to the changeover between modes, convergence problems will occur in which the control mode changes at each iteration. This can easily be overcome by retaining mode 2 operation whenever detected for the remaining iterations in that particular time step.

D. C. POWER MODULATION

It has been shown in the previous Section that under the constant power control mode, the d.c. link is not responsive to a.c. system terminal conditions, i.e. the d.c. power transfer can be controlled disregarding the actual a.c. voltage angles. Since, generally, the stability limit of an a.c. line is lower than its thermal limit, the former can be increased in systems involving d.c. links, by proper utilization of the fast converter controllability.

The d.c. power can be modulated in response to a.c. system variables to increase system damping. Optimum performance can be achieved by controlling the d.c. system so as to maximize the responses of the a.c. system and d.c. line simultaneously following the variation of terminal conditions.

The dynamic performance under d.c. power modulation is best modelled in three separate levels[15]. These levels, illustrated in Fig. 10.15, are the a.c. system controller (i), the d.c. system controller (ii) and the a.c.–d.c. network (iii).

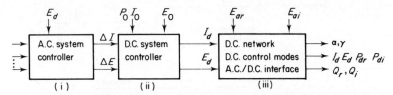

Fig. 10.15. A.c.–d.c. dynamic control structure. (i) A.c. system controller; (ii) D.c. system controller; (iii) A.c.–d.c. network

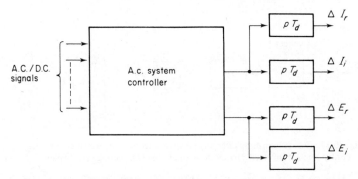

Fig. 10.16. A.c. system controller

(i) The a.c. system controller uses a.c. and/or d.c. system information to derive the current and voltage modulation signals. A block diagram of the controller and a.c.–d.c. signal conditioner is shown in Fig. 10.16.

(ii) The d.c. system controller receives the modulation signals ΔI and ΔE and the steady-state specifications for power P_0 current I_0 and voltage E_0. Figure 10.17 (a) illustrates the power controller model, which develops the scheduled current setting; it is also shown that the current order undergoes a gradual increase during restart, after a temporary blocking of the d.c. link.

The rectifier current controller, Figure 10.17(b), includes signal limits and rate limits, transducer time constant, band-pass filtering and a voltage dependent current order limit (VDCOL).

The inverter current controller, Fig. 10.17(c), includes similar components plus a communications delay and the system margin current (Im).

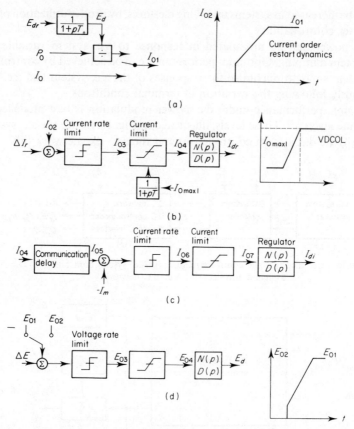

Fig. 10.17. D.c. system controller. (a) Power controller; (b) Rectifier current controller; (c) Inverter current controller; (d) D.c. voltage controller

Finally the d.c. voltage controller, including voltage restart dynamics, is illustrated in Fig. 10.17(d).

(iii) The d.c. current I_d and voltage E_d derived in the d.c. system controller constitute the input signals for the a.c.–d.c. network model which involves the steady-state solution of the d.c. system (neglecting the d.c. line dynamics which are included in the d.c. system controller). Here the actual a.c. and d.c. system quantities are calculated, i.e. control angles, d.c. current, voltage, active and reactive power. The converter a.c. system constraints are the open circuit secondary voltages E_{ar} and E_{ai}.

Representation of converters in the network

RECTIFIERS

The static-load rectifier model can be included in the overall solution of the transient stability program in a similar manner to the basic loads described in Chapter 9.

From the initial load flow, nominal bus shunt admittance $(\bar{y}_0)$ can be calculated for the rectifier. This is included directly into the network admittance matrix $[Y]$. The injected current into the network in the initial steady state is therefore zero. In general:

$$I_{inj} = (\bar{y}_0 - \bar{y})\bar{V}_{term} \qquad (10.5.24)$$

where

$$\bar{y}_0 = \frac{\bar{S}_0^*}{|V_{term_0}|^2} \qquad (10.5.25)$$

and

$$\bar{y} = \frac{\bar{S}^*}{|V_{term}|^2} \qquad (10.5.26)$$

The static-load rectifier model does not depart greatly from an impedance characteristic and is well behaved for low-terminal voltages, the injected current tending to zero as the voltage approaches zero. Figure 10.18 compares the current

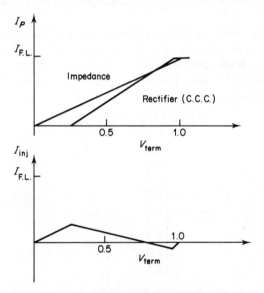

Fig. 10.18. Difference between impedance and static load rectifier characteristic (for $V_{load} \neq 0$)

due to a rectifier with that due to a constant impedance load. As the injected current is never large, the iterative solution for all a.c. conditions is stable.

When the rectifier model is modified to account for the dynamic behaviour of the d.c. load its characteristic departs widely from that of an impedance. Immediately after a fault application, the voltage drops to a low value but the injected current magnitude does not change significantly. Similarly on fault clearing, the voltage recovers instantaneously to some higher value while the current remains low.

When the load characteristic differs greatly from that of an impedance, the sequential solution technique can exhibit convergence problems,[6] especially when the voltage is low. With small terminal voltages, the a.c. current magnitude of the rectifier load is related to the d.c. current but the current phase is greatly affected by the terminal voltage. Small voltage changes in the complex plane can result in large variations of the voltage and current phase angles.

To avoid the convergence problems of the sequential solution, an alternative algorithm has been developed.[13] This combines the rectifier and network solutions into a unified process. It, however, does not affect the sequential solution of the other components of the power system with the network.

The basis of this approach is to reduce the a.c. network, excluding the rectifier, to an equivalent Thevenin source voltage and impedance as viewed from the primary side of the rectifier transformer terminals. This equivalent of the system, along with the rectifier, can be described by a set of nonlinear simultaneous equations which can be solved by a standard Newton–Raphson algorithm. The solution of the reduced-system yields the fundamental a.c. current at the rectifier terminals.

To obtain the network equivalent impedance, it is only necessary to inject 1 p.u. current into the network at the rectifier terminals while all other nodal injected currents are zero. With an injected current vector of this form, a solution of the network equation (9.5.1) gives the driving point and transfer impedances in the resulting voltage vector.

$$[\bar{Z}] \equiv [\bar{V}'] = [\bar{Y}]^{-1}[\bar{I}'_{\text{inj}}] \tag{10.5.27}$$

where

$$[\bar{I}'_{\text{inj}}] = \begin{bmatrix} 0 \\ \vdots \\ 0 \\ \bar{I}'_r \\ 0 \\ \vdots \\ 0 \end{bmatrix} \quad \text{and} \quad \bar{I}'_r = 1 + j0 \tag{10.5.28}$$

The equivalent circuit shown in Fig. 10.19 can now be applied to find the rectifier current ($\bar{I}_p$) by using the Newton–Raphson technique.

The effect of the rectifier on the rest of the system can be determined by superposition:

$$[\bar{V}] = [\bar{V}^\sigma] + [\bar{Z}]I_p \tag{10.5.29}$$

where

$$[\bar{V}^\sigma] = [\bar{Y}]^{-1}[\bar{I}^\sigma_{\text{inj}}] \tag{10.5.30}$$

and $[\bar{I}^\sigma_{\text{inj}}]$ are the injected currents due to all other generation and loads in the system.

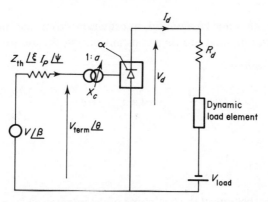

Fig. 10.19. Equivalent system for Newton–Raphson solution

If the network remains constant, vector $[\bar{Z}]$ is also constant and thus only needs re-evaluation on the occurrence of a discontinuity.

Thus the advantages of the unified and sequential methods are combined. That is, good convergence for a difficult element in the system is achieved while the programming for the rest of the system remains simple and storage requirements are kept low.

The equivalent system of Fig. 10.19 contains seven variables ($V_{\text{term}}, I_p, \theta, \psi, \alpha, V_d$ and I_d). With these variables four independent equations can be formed. They are equation (3.2.5) and:

$$V\underline{/\beta} - V_{\text{term}}\underline{/\theta} - Z_{\text{th}}\ \underline{/\xi} \cdot I_p\underline{/\psi} = 0 \qquad (10.5.31)$$

$$V_d \cdot I_d - \sqrt{3}\,\alpha \cdot V_{\text{term}} \cdot I_p \cdot \cos{(\theta - \psi)} = 0 \qquad (10.5.32)$$

Equation (10.5.31) is complex and represents two equations. Substituting for V_d and I_p using equations (3.2.11) and (10.5.1) reduces the number of variables to five. A fifth equation is necessary and with constant current control, that is with the delay angle (α) within its limits, this can be written as:

$$I_d - I_{d_{\text{sp}}} = 0 \qquad (10.5.33)$$

Equation (3.2.5) suitably reorganized and equations (10.5.31) to (10.5.33) represent $[F(X)] = 0$ of the Newton–Raphson process and:

$$[X]^T = [E_r, \theta, \psi, \alpha, I_d] \qquad (10.5.34)$$

When the delay angle reaches a specified lower limit (α_{min}), the control specification, given by equation (10.5.33), changes to:

$$\alpha - \alpha_{\text{min}} = 0 \qquad (10.5.35)$$

and equation (3.2.5) is no longer valid. The d.c. current (I_d) is now governed by the differential equation (10.5.5). If the trapezoidal method is being used, this equation can be transformed into an algebraic form similar to that described in Chapter 9. Equation (3.2.5) is replaced by:

$$I_d = ka \cdot E_r \cdot \cos{\alpha} - kb = 0 \qquad (10.5.36)$$

The variables ka and kb contain information from the beginning of the integration step only and are thus constant during the iterative procedure.

$$ka = h/(2 + kc \cdot h) \tag{10.5.37}$$

$$kb = (1 - 2\,kc \cdot ka)I_d(t) + \frac{3\sqrt{2}}{\pi T_{dc} \cdot R_d} a \cdot ka \cdot V_{\text{term}}(t) \cos \alpha(t) + \frac{2 \cdot ka \cdot V_{\text{load}}}{T_{dc} \cdot R_d} \tag{10.5.38}$$

where

$$kc = \left(\frac{3X_c}{\pi R_d}\right) + 1/T_{dc} \tag{10.5.39}$$

and t represents the time at the beginning of the integration step and h is the step length.

Commutation angle μ is not explicitly included in the formulation, and since these equations are for normal operation, the value of k in equation (3.2.11) is close to unity and may be considered constant at each step without loss of accuracy. On convergence, μ may be calculated and a new k evaluated suitable for the next step.

In Mode 3 operation, the value of k becomes more significant and for this reason the number of variables is increased to six to include the commutation angle μ. The equations $[F(X)] = 0$ for the Newton–Raphson method in this case are:

$$V\underline{/\beta} - V_{\text{term}}\underline{/\theta} - Z_{\text{th}}\underline{/\xi} \cdot \frac{\sqrt{6}}{\pi} f(\mu) \cdot I_d \underline{/\psi} = 0 \tag{10.5.40}$$

$$\frac{\sqrt{2}}{\pi} a \cdot V_{\text{term}} \cdot \cos(\theta - \psi) \cdot f(\mu) - \frac{\sqrt{6}}{\pi} a \cdot V_{\text{term}} \cos \alpha' + \frac{3X_c}{\pi} I_d = 0 \tag{10.5.41}$$

$$I_d - ka \cdot a \cdot V_{\text{term}} \cos \alpha' - kb = 0 \tag{10.5.42}$$

$$\cos(\alpha + \mu + 30) - \cos \alpha' + \frac{\sqrt{6}X_c}{a \cdot V_{\text{term}}} \cdot I_d = 0 \tag{10.5.43}$$

$$\alpha - \alpha_{\min} = 0 \tag{10.5.44}$$

Although k can be calculated explicitly, a linearized form of equation (10.5.8) obtained for $\alpha = 30°$ can be used to simplify the expression. In the range $60° < \mu < 120°$, the value of k can be obtained from:

$$f(\mu) = 1.01 - 0.0573\mu \tag{10.5.45}$$

where μ is measured in radians.

In Mode 4, the a.c. and d.c. systems are both short circuited at the rectifier and operate independently. In this case the system equivalent of Fig. 10.19 reduces to that shown in Fig. 10.20. The network equivalent can be solved directly and the d.c. current is obtained from the algebraic form of the differential equation (10.5.5).

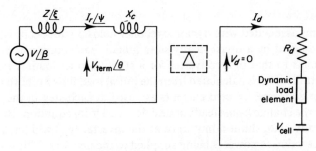

Fig. 10.20. Rectifier load equivalent in Mode 4 operation

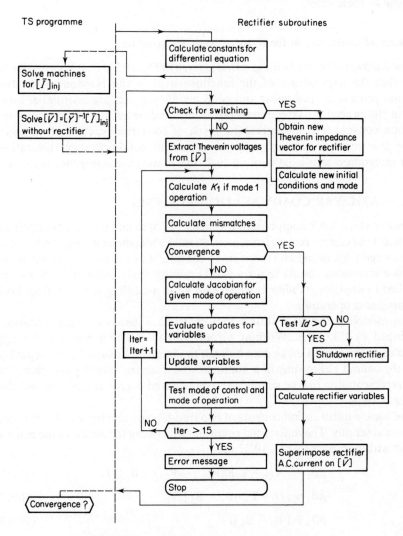

Fig. 10.21. Unified algorithm flow diagram

D.C. LINKS

The problems associated with dynamic rectifier loads do not occur when the d.c. link is represented by a quasi steady-state model. Each converter behaves in a manner similar to that of a converter for a static rectifier load. A nominal bus shunt admittance $(\bar{y}_0)$ is calculated from the initial load flow for both the rectifier and inverter ends and injected currents are used at each step in the solution to account for the change from steady state calculated from equation (10.5.24). Note that the steady-state shunt admittance at the inverter $(\bar{y}_{0_i})$ will have a negative conductance value as power is being supplied to the network. This is not so for a synchronous or induction generator as the shunt admittance serves a different purpose in these cases.

Inclusion of converters in the transient stability program

A flow diagram of the unified algorithm is given in Fig. 10.21.[16] It is important to note that the hyperplanes of the functions used in the Newtonian iterative solution process are not linear and good initial estimates are essential at every step in the procedure. A common problem in converter modelling is that the solution converges to the unrealistic result of converter reactive power generation. It is therefore necessary to check against this condition at every iteration. With integration step lengths of up to 25 mS however, convergence is rapid.

10.6. STATIC VAR COMPENSATION SYSTEMS

The use of static VAR compensation systems (SVS) to maintain an even voltage profile at load centres remote from generation has become common. An SVS can have a large VAR rating and therefore to consider it as a fixed shunt element can produce erroneous results in a transient stability study. Also an SVS may be installed to improve stability in which case good modelling is essential for both planning and operation.

The model of the SVS shown in Fig. 10.22 is based on representations developed by C.I.G.R.E. Working Group 31–01.[17] The model is not overly complex as this would make data difficult to obtain and would be incompatible with the overall philosophy of a multimachine transient stability program. The SVS representation can be simplified to any desired degree however by suitable choice of data.

The basic control circuit consists of two lead-lag and one lag transfer function connected serially. The differential equations describing the action of the control circuit with reference to Fig. 10.22 are:

$$pB_1 = [K(1 + T_2 p)(V_{SV_{set}} - V'_{SV}) - B_1]/T_1 \qquad (10.6.1)$$

$$pB_2 = [(1 + T_4 p)B_1 - B_2]/T_3 \qquad (10.6.2)$$

$$pB_3 = [B_2 - B_3]/T_5 \qquad (10.6.3)$$

Although electronically produced, the dead band may be considered as a

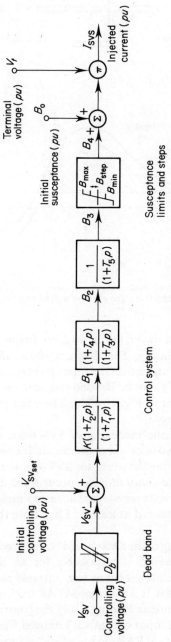

Fig. 10.22. Composite static var compensation system (SVS) model

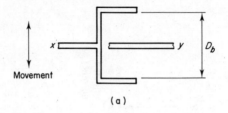

(a)

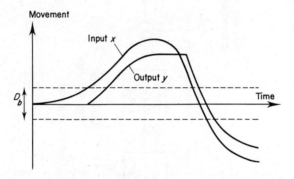

Fig. 10.23. Dead band analogy and effect. (a) Physical analogy of dead band; (b) The effect of a dead band on output

physical linkage problem as shown in Fig. 10.23(a). In this example, the input (x)
and output (y) move vertically. The diagram shows the initial steady-state
condition in which x and y are equal. The input (x) may move in either direction
by an amount $D_b/2$ before y moves. Beyond this amount of travel, y follows x,
lagging by $D_b/2$ as depicted in Fig. 10.23(b). The effect of a dead band can be
ignored by setting D_b to zero.

Stepped output permits the modelling of SVS when discrete capacitor (or
inductor) blocks are switched in or out of the circuit. It is usual to assume that all
blocks are of equal size. During the study, the SVS operates on the step nearest to
the control setting. Iterative chattering can occur if the control-system output
(B_3) is on the boundary between two steps. The most simple remedy is to prevent
a step change until B_3 has moved at least $0.55 B_{step}$ from the mean setting of the
step.

The initial MVAR loading of the SVS should be included in the busbar loading
schedule data input. However, it is possible for an SVS to contain both
controllable and uncontrollable sections (e.g. variable reactor in parallel with
fixed capacitors or vice versa). It is the total MVAR loading of the SVS which is,
therefore, included in the busbar loading. Only the controllable part should be
specified in the SVS model input and this is removed from the busbar loading
leaving an uncontrollable MVAR load which is converted into a fixed
susceptance associated with the network.

Note that busbar load is assumed positive when flowing out of the network.
The sign is therefore opposite to that for the SVS loading.

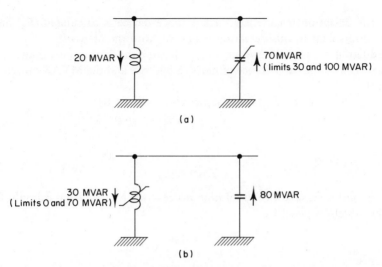

Fig. 10.24. Example of an overall SVS controllable and uncontrollable sections. (a) Example of overall SVS using controllable capacitors; (b) Alternative to overall SVS in (a) using a controllable reactor

In order to clarify this, consider an overall SVS operating in the steady state as shown in Fig. 10.24(a). The busbar loading in this case must be specified as -50 MVAR and it may be varied between -10 MVAR and -80 MVAR provided the voltage remains constant.

The SVS may be specified in a variety of ways, some more obvious than others, the response of the system being identical. Three possible specifications are given in Table 10.4. In the first example, the network static load will be $+20$ MVAR while in the second case the static load will be zero. The third example may be represented by an overall SVS as shown in Fig. 10.24(b).

Table 10.4. Examples of MVAR loading specification for SVS shown in Fig. 10.24(a)

Example	Initial loading	Maximum limit	Minimum limit
1	70	100	30
2	50	80	10
3	-30	0	-70

It is convenient, when specifying the initial steady-state operation, to use MVAR. However, this is a function of the voltage and hence all MVAR settings must be converted to their equivalent per unit susceptance values prior to the start of the stability study.

Representation of SVS in the overall system

The initial MVAR loading of the SVS is converted into a shunt susceptance (B_0) and added to the total susceptance at the SVS terminal busbar. During the system study, the deviation from a fixed susceptance device is calculated (B_4) and a current equivalent to this deviation is injected into the network.

A reduction in controlling voltage V_{sv} will cause the desired susceptance B_4 to increase. That is the capacitance of the SVS will rise and the MVAR output will increase.

The injected current $(\bar{I}_{inj})$ into the network is given by:

$$\bar{I}_{inj} = -\bar{V} \cdot \bar{Y} \tag{10.6.4}$$

where

$$Y = 0 + jB_4$$

Although not necessary for the solution process, the MVA output from the SVS into the system is given by:

$$\bar{S} = \bar{V}\bar{I}_{SVS}^*$$

and hence

$$Q = |V|^2(B_4 + B_0) \tag{10.6.5}$$

10.7. RELAYS

Relay characteristics may be applied to a transient stability program and the effect of relay operation automatically included in system studies. This permits checking of relay settings and gives more realistic information as to system behaviour after a disturbance, assuming 100 percent reliability of protective equipment. Reconstruction of the events after fault occurrences may also be carried out.

Unit protection only responds to faults within a well-defined section of a power system and as the faults are prespecified, the operation of unit protection schemes can equally be specified in the switching data input. Thus, only nonunit protection needs to be modelled and of these overcurrent, undervoltage and distance schemes are the most common.

Instantaneous overcurrent relays

Instantaneous or fixed time-delay overcurrent relays are readily modelled. The operating point of the relay should be specified in terms of p.u. primary current thus avoiding the need to specifically model the current transformer. However, the location of the current transformer must be specified, e.g. at busbar A on branch to B, so that the correct signals are used by the relay model. The only other piece of information required is delay-time (t_{del}) between the relay operation time and the circuit-breaker arc extinction time (t_{cb}).

Initially, the circuit-breaker operating time (t_{cb}) is set to some large value as it must be assumed that the steady-state current is less than the relay setting. At the end of each time step (e.g. at time t) the current at the current transformer location is evaluated and if it exceeds the relay setting, the effective circuit-breaker time is set to:

$$t_{cb} = t + t_{del} \tag{10.7.1}$$

The integration then proceeds until the time-step nearest to t_{cb} when circuit-breaker opening is simulated by reducing the relevant branch admittance to a very small value. Alternatively, the integration step length can be adjusted to open the circuit at time t_{cb}.

During the period between relay operation and t_{cb} the simulation of relay drop off may be desired. In this case, if current falls below a prespecified percentage of relay setting current then t_{cb} is reset to a large value.

Inverse definite minimum time lag overcurrent relays

The inverse time characteristics of induction disc and similar relays may easily be included in an overcurrent relay. This may be accomplished by defining several points on the characteristic and interpolating, but curve fitting is better if a simple function can be found.

For example, an overcurrent relay conforming to British Standard B.S. 142 would appear to be accurately modelled by defining seven points on the curve as shown in Fig. 10.25(a). However, when plotted on a log-log graph as in Fig. 10.25(b), the errors are more obvious and can exceed the accuracy limits laid down in the Standard if care is not taken. However, acceptable accuracy can be obtained by using the approximation:

$$t_{op} = 3.0/[\log(I)] \quad \text{for } 1 \cdot 1 \le I \le 20 \tag{10.7.2}$$

and

$$t_{op} = \infty \quad \text{for } I < 1 \cdot 1$$

where t_{op} is the operating time of the relay for a current of I.

Plug bridge setting (S_{pb}) and time multiplier setting (S_{tm}), both measured in per unit, can be incorporated into the relay characteristic and the relay induction disc travel (D_t) at time t due to a current I_t may be determined from the previous travel at time $t - h$ by:

$$D_t = D_{t-h} + \frac{h \cdot \log\{\frac{1}{2}(I_t + I_{t-h})/S_{pb}]}{3 \cdot S_{tm}} \tag{10.7.3}$$

provided that $I_t \ge 1 \cdot 1 S_{pb}$.

For currents less than the definite minimum value the relay may be assumed to reset by spring action. Assuming that resetting from full travel takes $2s$ then:

$$D_t = D_{t-h} - \frac{h}{2 \cdot S_{tm}} \tag{10.7.4}$$

when

$$I_t < 1 \cdot 1 S_{pb}$$

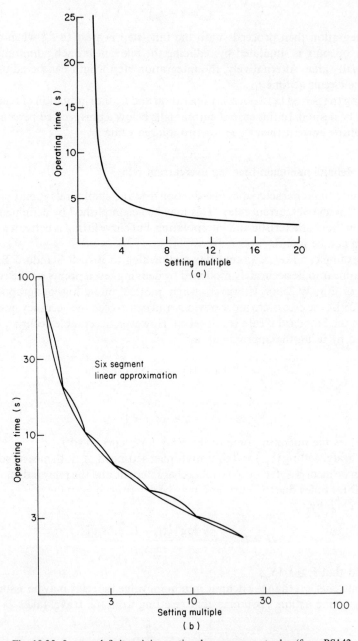

Fig. 10.25. Inverse definite minimum time lag overcurrent relay (from BS142 (1966)). (a) Linear scale; (b) Logarithmic scale

Initially, travel is set to zero and relay operation is assumed when D equals or exceeds 1 p.u. If necessary, the relay operating time may be determined by linear interpolation backwards over the last time interval.

$$t_{op} = t - \frac{(D_t - 1 \cdot 0)}{(D_t - D_{t-h})} \cdot h \qquad (10.7.5)$$

when

$$D_t \geq 1 \cdot 0$$

and

$$D_{t-h} < 1 \cdot 0$$

and from this the circuit-breaker operating time is given by:

$$t_{cb} = t_{op} + t_{del} \qquad (10.7.6)$$

Many static relays have been designed which conform to mechanical characteristics but they have also permitted different and more suitable characteristics to be developed. These may be modelled in a similar manner.

Undervoltage relays

Apart from the fact that the relay-operating current is proportional to primary voltage and not primary current, these relays should be modelled in the same manner as instantaneous or fixed time-delay overcurrent relays.

Induction machine contactors

The transient analysis of industrial power systems usually require that many induction machines are modelled. During a disturbance the voltage levels throughout the system will fluctuate and may result in the machine being disconnected from the system, albeit temporarily. Other machines may be on automatic stand-by to maintain essential services.

It is, therefore, necessary to include models of induction machine contactors in a transient stability program. Undervoltage protection is usually associated with the contactors and can be modelled in the normal manner.

Directional overcurrent relay

A directional overcurrent relay requires a voltage signal as well as current. The relay may operate only when the phase difference of the two signals is within prescribed limits and all other constraints are satisfied.

Distance relays

As in practice, both busbar voltage and branch current signals are required, from which an apparent impedance $Z_s \underline{/\theta_s}$ of the system at the relaying point can be calculated. This is then compared with the relay characteristic to determine operation.

A typical three-zone distance protection relay is shown in Fig. 10.26. Assuming circular characteristics, then the settings of the relay may be identified by forward reach $Z_{rf}/\underline{\theta_{rf}}$ measured in impedance (complex) coupled with backward reach R_b expressed as a per unit of forward reach. From this information, the centre $(p + jq)$ and radius (a) of each of the three circles in the impedance plane can be established:

$$a = \tfrac{1}{2}Z_{rf}(1 + R_b) \qquad (10.7.7)$$

$$p = \tfrac{1}{2}Z_{rf}(1 - R_b)\cos\theta_{rf}$$

$$q = \tfrac{1}{2}Z_{rf}(1 - R_b)\sin\theta_{rf}$$

In the example in Fig. 10.26, R_b for zones 1 and 2 is zero.

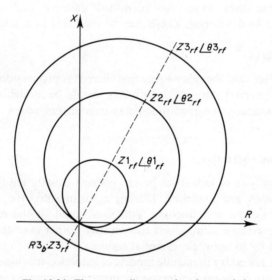

Fig. 10.26. Three-zone distance relay characteristic

The equation of the boundary of an operating zone is:

$$(Z\cdot\cos\theta - p)^2 + (Z\cdot\sin\theta - q)^2 - a^2 = 0 \qquad (10.7.8)$$

and hence operation is defined when:

$$Z_s^2 - 2Z_s(p\cdot\cos\theta_s + q\cdot\sin\theta_s) + (p^2 + q^2 - a^2) \le 0 \qquad (10.7.9)$$

Tomato, lens, quadrilateral or other complex characteristics may be constructed by combining several simple characteristics of this type.

Each zone has a fixed time-delay associated with it so that the timing for circuit-breaker action is the same as that described previously.

Incorporating relays in the transient stability program

Nonunit protection equipment usually only trips the local circuit breaker. Therefore, it is necessary to create dummy busbars so that a faulted branch can be switched out correctly. This can be done automatically during the data input stage in the same manner as described in Chapter 9 for faults located on branches. Thus, a faulted branch may have several dummy busbars associated with it and care should be taken to ensure all are adequately identified. Protected branches which are not directly faulted need not be modelled as accurately and the whole branch may be removed if the circuit breaker at either end is opened.

Relay characteristics should be checked at the end of each time solution and reconvergence after a discontinuity. It is not necessary to perform the check at each iteration however. This reduces the computational effort associated with relays and permits more complex relay characteristics to be modelled at critical points in the system.

Induction machine switching should not be simulated by creating dummy branches which can be removed from the network whenever necessary. While this is a feasible solution, it is extremely wasteful of computational storage and effort. A more satisfactory method is to identify the state of the machine, i.e. either switched in or out, by a simple flag and when switched out to solve for the machine with zero stator current and likewise remove its injected current from the network.

The network, however, usually includes a shunt admittance representing the machine in the initial steady state. This problem may be overcome by injecting another current to compensate for this admittance whenever a machine is switched out. Alternatively, a machine liable to switching need not have its equivalent shunt admittance included in the network at any time during the study. This simplifies periods when the machine is switched out, but requires a different injected current to the usual when switched in.

A minor problem occurs when induction machines, which are initially switched out of service, are included in the input data. An estimate of the full-load active power of the machine must be specified so that the load characteristics of the machine can be adequately defined. Also induction motors on stand by for automatic start-up must be modelled accurately if sensible run-up simulation is to be achieved.

10.8. UNBALANCED FAULTS

The models developed so far for transient stability analysis have assumed balanced three-phase operation even during the fault period. Although three-phase faults are the most onerous, there are occasions when unsymmetrical fault conditions need to be analysed. It is possible to develop three-phase models of all power-system equipment but the development effort plus the extra computational costs restrict this type of program to very simple systems. Unbalanced fault studies are relatively rare and the unbalance only occurs for a short period of the study thus the need for a three-phase model is limited, and makes full scale development unattractive.

A more practical approach is the use of symmetrical components. The negative- and zero-sequence component system models can be added to the existing single-phase (positive-sequence) model without major disruption and can be easily removed when not required.

Negative sequence system

Of the two additional symmetrical component systems the negative sequence is the easier. It is very similar to the positive-sequence system.

The negative-sequence impedances of the components of the transmission network and static loads are usually the same as for the positive-sequence impedances and hence no additional storage is required. Phase displacement in transformer banks is of the opposite sign to that for the positive sequence. While phase displacement can be ignored during balanced operation, it must be established if phase quantities are to be calculated during unbalanced operation. A simple clock notation with each hour representing 30 degree shift is suitable for this purpose.

The negative-sequence impedance of synchronous machines is different from the positive-sequence impedance. The flux produced by negative-sequence armature current rotates in the opposite direction to the rotor, unlike that produced by positive-sequence current, which is stationary with respect to the rotor. Rotor currents induced by this flux prevent it from penetrating deeply into the rotor. The flux path oscillates rapidly between the positive-sequence d and q axis subtransient flux paths and the negative-sequence reactance X_2 may conveniently be defined as:

$$X_2 = (X_d'' + X_q'')/2 \qquad (10.8.1)$$

This reactance is the same as the reactance which represents the machine in the positive-sequence network. The negative-sequence resistance is given by[18]:

$$R_2 \simeq Ra + \tfrac{1}{2}Rr \qquad (10.8.2)$$

where Rr is the rotor resistance. While R_2 and Ra will differ, the overall difference between the negative-sequence impedance (Z_2) and positive-sequence impedance representing the machine is so small as to be neglected in most cases. Further, rotating machinery do not generate negative-sequence e.m.f.s and hence there is no negative-sequence Norton injected current.

Thus, ignoring d.c. equipment, the overall negative-sequence network is identical to the positive-sequence network with all injected currents set to zero.

Negative-sequence currents have a braking effect on the dynamic behaviour of rotating machinery. For a synchronous machine, where torque and power may be assumed to be equivalent, the mechanical breaking power (Pb) is[18]:

$$Pb = I_2^2(R_2 - Ra) \qquad (10.8.3)$$

which may be added directly into the mechanical equation of motion given by equation (9.2.4).

A similar expression can be found for negative-sequence-breaking torque in an induction machine where the speed of the rotor and the negative-sequence currents are taken into account.

Zero sequence system

The zero-sequence system differs greatly from the other two sequence systems.

The zero-sequence impedance of transmission lines is higher and for a transformer its value and location depends on the phase connection and neutral arrangements. Fig. 10.27 shows the zero-sequence models for various typical transformer connections. By replacing the open circuit of transformer types 2, 3 and 4 with a very low admittance, the topology of the zero-sequence network can be made the same as for the other sequence networks.

The zero-sequence impedance of rotating machinery must be specified in the data input so that its inverse can be included in the zero-sequence system admittance matrix. As with the negative-sequence system model, there is no zero-sequence e.m.f.,s generated and hence there is no Norton injected currents into this system.

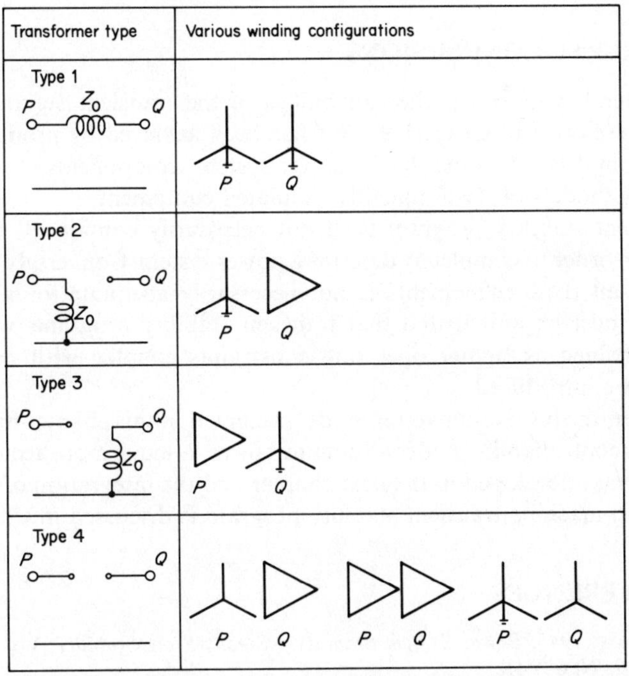

Transformer type	Various winding configurations
Type 1	
Type 2	
Type 3	
Type 4	

Fig. 10.27. Modelling of zero-sequence equivalent networks of transformers

Inclusion of negative and zero sequence systems for unsymmetrical faults

The major effect of unsymmetrical faults is to increase the apparent fault impedance. On fault application, the negative- and zero-sequence impedances of the system at the point of fault are calculated. These are simply the inverse of the self-admittances at the point of fault and are determined in the same manner as described by equations (10.5.27) and (10.5.28). Depending on the type of fault, the fault impedance is modified to include the negative- and zero-sequence imped-ance. The fault impedance then remains constant until changed by either branch switching or fault removal.

If negative-sequence-braking effects are to be included, it is necessary to evaluate the negative-sequence current in the relevant machines at each iteration. This is done by injecting the negative-sequence current, determined at the point of fault, into the negative sequence system admittance matrix $[\bar{Y}_2]$.

$$[\bar{Y}_2][\bar{V}_2] = [\bar{I}f_2] \qquad (10.8.4)$$

where $[\bar{I}f_2]$ is a zero vector except at the point of fault. The vector $[\bar{V}_2]$ contains the negative-sequence voltages at all busbars from which the machine negative-sequence currents are readily obtained.

If phase information is required, then the zero-sequence voltages at all busbars also need to be determined, depending on the type of fault. This is done in an identical manner to that used for the negative-sequence system.

10.9. GENERAL CONCLUSIONS

This chapter has extended the capabilities of the transient stability analysis program developed in Chapter 9. This has been achieved by producing more advanced models of some basic power-system components and also by introducing models of less frequently simulated equipment.

A transient stability program need not necessarily contain all the models described in order to completely describe a power system. Conversely, a program containing all these refinements is not necessarily adequate for a particular system. It must be anticipated that transient stability programs will be con-tinuously refined as tighter operating constraints coupled with new control strategies are introduced.

Moreover, the h.v.d.c. converter model presented in this chapter has assumed continuous controllability under all circumstances. A much more accurate model of converters is developed in the next chapter and the integration of this model with a multi-machine transient stability program is discussed in Chapter 13.

10.10. REFERENCES

1. S. B. Crary, 1945. *Power System Stability—Steady-State Stability*, Vol. 1. Wiley & Sons Ltd., New York.
2. D. W. Olive, 1966. 'New techniques for the calculation of dynamic response', *IEEE Transactions on Power Apparatus and Systems*, **PAS-85** (7), 767–777.

3. T. J. Hammons, and D. J. Winning, 1971. 'Comparisons of synchronous-machine models in the study of the transient behaviour of electrical power systems', *Proceedings IEE*, **118** (10), 1442–1458.
4. S. Beckwith, 1937. 'Approximating Potier Reactance', *AIEE Transactions*, 813.
5. A. H. Knable, 1956. *Electrical Power Systems Engineering—Problems and Solutions*. McGraw-Hill, New York.
6. H. W. Dommell, and N. Sato, 1972. 'Fast transient stability solutions', *IEEE Transactions on Power Apparatus and Systems*, **PAS-91** (8), 1643–1650.
7. C. P. Arnold, 1976. 'Solutions of the multi-machine power-system stability problem.' Ph.D. thesis, Victoria University of Manchester, U.K.
8. IEEE Committee Report, 1973. 'Dynamic models for steam and hydro turbines in power-system studies', *IEEE Transactions on Power Apparatus and Systems*, **PAS-92** (6), 1904–1915.
9. D. S. Brereton, D. G. Lewis, and C. C. Young, 1957. 'Representation of induction motor loads during power-system stability studies', *AIEE Transactions on Power Apparatus and Systems*, **PAS-76**, 451–461.
10. H. E. Jordan, 1979. 'Synthesis of double-cage induction motor design', *AIEE Transactions on Power Apparatus and Systems*, **PAS-78**, 691–695.
11. C. P. Arnold, and E. J. P. Pacheco, 1979. 'Modelling induction motor start-up in a multi-machine transient stability program.' IEEE PES Summer Meeting. Vancouver, B.C., Canada.
12. E. J. P. Pacheco, 1975. 'Induction motor starting in an electrical power-system transient-stability programme.' M.Sc. dissertation, Victoria University of Manchester, U.K.
13. C. P. Arnold, K. S. Turner, and J. Arrillaga, 1980. 'Modelling rectifier loads for a multi-machine transient-stability programme', *IEEE Transactions on Power Apparatus and Systems*, **PAS-99** (1), 78–85.
14. D. B. Giesner, and H. Arrillaga, 1970. 'Operating modes of the three-phase bridge converter', *Int. J. Elect. Engng. Educ.*, **8**, 373–388.
15. IEEE Working Group on Dynamic Performance and Modeling of DC Systems, 1980. 'Hierarchical Structure'.
16. K. S. Turner, 1980. 'Transient stability analysis of integrated a.c. and d.c. power systems', Ph.D. Thesis. University of Canterbury, New Zealand.
17. CIGRE Working Group 31-01, 1977. 'Modelling of static shunt VAR systems for system analysis', *Electra*, (51), 45–74.
18. E. W. Kimbark, 1956. *Power System Stability: Synchronous Machines*. Wiley & Sons, New York.

11

Dynamic Simulation of
A.C.–D.C. Systems

11.1. INTRODUCTION

During disturbances, the interaction between supply systems and large power converters, especially those used in h.v.d.c. transmission, generally results in large waveform distortion and valve maloperations. Such behaviour cannot be predicted with the quasi-steady-state models described in Chapter 3, because the topological converter changes resulting from the disturbance can not be specified in advance.

The simulation of transients associated with h.v.d.c. transmission is presently carried out in physical models and simulators which provide the required detailed modelling of the converter plant.

However, the pressing need to represent the behaviour of h.v.d.c. converters in conventional a.c. power-system computer studies is exciting the development of digital methods. Such methods are normally limited in their practical application by the restricted size and accuracy of the a.c. system representation and by the cost of running the programs. As a result, programs designed to analyse the transient behaviour of static converters have tended to use a simplified a.c. network representation.

This chapter describes a general dynamic model, formulated in terms of state-space theory. It uses nodal analysis with diakoptical segregation of the plant components undergoing frequent switching, to avoid involving the whole network in unnecessary topological changes. The model accepts varying degrees of a.c. system and converter plant representation to meet the requirements of any particular study, e.g. small disturbances, faults, etc. In each case the individual units of greater relevance are emphasised, whereas the rest of the power system is synthesized into a simpler equivalent circuit.

11.2. FORMULATION OF THE DIFFERENTIAL EQUATIONS

As explained in Chapter 2, a power system can be represented by a network of inductive, resistive and capacitive elements. The behaviour of the network, i.e. the

300

currents in the different branches and the voltages at different nodes, is determined by topological and algebraic constraints.

In cases involving h.v.d.c. links and containing few capacitive branches, mesh analysis results in fewer current differential equations. However, when the a.c. system contains multi-conductor transmission lines with mutual inductive effects and a high level of interconnection, the nodal approach is more efficient. Moreover, the accessibility of individual elements is important in a flexible model, e.g. to include nonlinearities such as saturation, and this is achieved by using the branch current formulation.

Definitions

Let the network contain n nodes interconnecting l inductive branches, r resistive branches and c capacitive branches subject to the following restrictions[1]:

— one end of each capacitive branch (or subnetwork) is the common reference point of the system;
— at least one end of each resistive branch (or subnetwork) is the common reference or a node also having a capacitive connection;
— at least one end of each inductive branch (or subnetwork) is the common reference or a node having a resistive or a capacitive connection.

These restrictions are taken care of in the representation of the system components.

The network nodes can thus be subdivided into three parts according to the types of branches connected to them namely;

— α nodes: which have at least one capacitive connection.
— β nodes: which have a least one resistive connection but no capacitive connections.
— γ nodes: which have only inductive connections.

The topological matrices K_{ln}^t, K_{rn}^t, K_{cn}^t which are the branch-nodes incidence matrices (transposed) of the l, r and c branches, respectively, are defined by their general elements as follows:

$$K_{\rho i}^t = 1, \text{ if node } i \text{ is the sending end of branch } \rho$$

$$= -1, \text{ if node } i \text{ is the receiving end of branch } \rho$$

$$= 0, \text{ otherwise.}$$

These matrices can be partitioned according to node type as follows:

$$K_{ln}^t = [K_{l\alpha}^t K_{l\beta}^t K_{l\gamma}^t]$$

$$K_{rn}^t = [K_{r\alpha}^t K_{r\beta}^t K_{r\gamma}^t] \tag{11.2.1}$$

$$K_{cn}^t = [K_{c\alpha}^t K_{c\beta}^t K_{c\gamma}^t]$$

where $K_{r\gamma}^t$, $K_{c\beta}^t$ and $K_{c\gamma}^t$ are null by definition.

Voltage and current relations

In the absence of current sources, the following nodal equation applies

$$K_{nl}I_l + K_{nr}I_r + K_{nc}I_c = 0 \qquad (11.2.2)$$

where I are the branch current vectors.

Partitioning equation (11.2.2) according to equation (11.2.1) and rearranging:

$$K_{\gamma l}I_l = 0$$

or

$$K_{\gamma l}pI_l = 0 \qquad (11.2.3)$$

$$K_{\beta r}I_r + K_{\beta l}I_l = 0 \qquad (11.2.4)$$

$$K_{\alpha c}I_c + K_{\alpha r}I_r + K_{\alpha l}I_l = 0 \qquad (11.2.5)$$

The following branch equations can be written as matrix expressions for each branch type.

(a) Inductive branches

$$E_l - p(L_lI_l) - R_lI_l + K^t_{l\alpha}V_\alpha + K^t_{l\beta}V_\beta + K^t_{l\gamma}V_\gamma = 0 \qquad (11.2.6)$$

or

$$pI_l = L_l^{-1}(E_l - pL_l \cdot I_l - R_lI_l + K^t_{l\alpha}V_\alpha + K^t_{l\beta}V_\beta + K^t_{l\gamma}V_\gamma) \qquad (11.2.7)$$

(b) Resistive branches

$$- R_rI_r + K^t_{r\alpha}V_\alpha + K^t_{r\beta}V_\beta = 0 \qquad (11.2.8)$$

or

$$I_r = R_r^{-1}(K^t_{r\alpha}V_\alpha + K^t_{r\beta}V_\beta) \qquad (11.2.9)$$

(c) Capacitive branches

$$C_cp(K^t_{c\alpha}V_\alpha) = I_c \qquad (11.2.10)$$

Premultiplying by $K_{\alpha c}$ and substituting equation (11.2.5) yields

$$K_{\alpha c}C_cK^t_{c\alpha}pV_\alpha = - K_{\alpha l}I_l - K_{\alpha r}I_r \qquad (11.2.11)$$

or

$$pV_\alpha = C_\alpha^{-1}J_\alpha \qquad (11.2.12)$$

where

$$C_\alpha^{-1} = K_{\alpha c}C_cK^t_{c\alpha} \qquad (11.2.13)$$

and

$$J_\alpha = - K_{\alpha l}I_l - K_{\alpha r}I_r$$
$$= K_{\alpha c}I_c \qquad (11.2.14)$$

Evaluation of dependent variables

Premultiplying equation (11.2.9) by $K_{\beta r}$ yields

$$V_\beta = -R_\beta(K_{\beta l}I_l + K_{\beta r}R_r^{-1}K_{r\alpha}^t V_\alpha) \qquad (11.2.15)$$

where

$$R_\beta = (K_{\beta r}R_r^{-1}K_{r\beta}^t)^{-1} \qquad (11.2.16)$$

Premultiplying equation (11.2.7) by $K_{\gamma l}$ and substituting equation (11.2.3) yields

$$V_\gamma = -L_\gamma K_{\gamma l}L_l^{-1}(E_l - pL_l \cdot I_l - R_l I_l + K_{l\alpha}^t V_\alpha + K_{l\beta}^t V_\beta) \qquad (11.2.17)$$

where

$$L_\gamma = (K_{\gamma l}L_l^{-1}K_{l\gamma}^t)^{-1} \qquad (11.2.18)$$

Equations (11.2.7) and (11.2.12) can now be integrated by numerical techniques to assess the system behaviour. The vectors V_β (equation (11.2.15)), V_γ (equation (11.2.17)) and I_r (equation (11.2.9)) may be eliminated entirely from the solution so that only I_l, V_α and the input vectors are explicit in the equations to be integrated. However, although this elimination reduces the overall number of equations, it has the following computational disadvantages.

— some of the original matrix sparsity is sacrificed;
— the modified incidence matrices now contain noninteger elements, thereby requiring actual multiplications rather than simple additions and subtractions when performing connection matrix-by-vector products;
— in addition to modifying L_γ and $K_{\gamma l}$ at every converter topological alteration, an auxiliary branch matrix is created in the elimination process, which also need reforming at these times.

It is, therefore, recommended to retain the equations in the above form and to evaluate the dependent variables V_β, V_γ and I_r at each step. Moreover, this formulation provides useful output information directly.

Formulation for variable topology

In Fig. 11.1, the converter transformer windings and the d.c. smoothing reactor together with the bridge valves, which are considered as switches, form a subnetwork of inductive elements whose topology varies periodically with the changes in valve states. Such a network contains only γ-nodes and the variable topology is manifested by changes in L_γ and $K_{\gamma l}$.

To avoid involving the whole network in a topological change affecting only part of it, it is generally convenient to use an extra type of branch (k) and node (δ) to represent the part of the system subject to the frequent topological change, i.e. the converters.

With that addition, the inductive subnetwork contains $(l + k)$ inductive

304

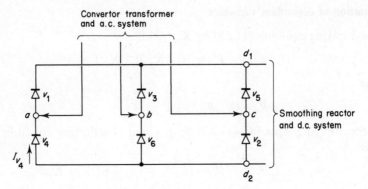

Fig. 11.1. Injected currents

branches; the k branches denoting branches involved in the converter representation, and the l branches denoting all inductive branches external to and not connected directly to the converter. An inductive branch between the two inductive subnetworks will thus be specified as a k branch so that the connection matrix $K^t_{l\delta}$ is null. Therefore, δ nodes connect only inductive branches within a converter subnetwork (k branches) whereas γ nodes also involve external inductive (l) branches. Therefore:

$$K^t_{kn} = [K^t_{k\alpha} K^t_{k\beta} K^t_{k\gamma} K^t_{k\delta}] \tag{11.2.19}$$

and $K^t_{l\delta}$, $K^t_{r\delta}$, $K^t_{c\delta}$ are all null.

Because of the nature of the converter subnetwork, it may be assumed that it contains no e.m.f. sources, or time variant inductances (i.e. $pL_k = 0$). Applying partitioning to Kirchhoff's current law, yields:

$$K_{\delta k} I_k = 0$$

or

$$K_{\delta k} p I_k = 0 \tag{11.2.20}$$

$$K_{\gamma k} I_k + K_{\gamma l} I_l = 0 \tag{11.2.21}$$

$$K_{\beta k} I_k + K_{\beta l} I_l + K_{\beta r} I_r = 0 \tag{11.2.22}$$

$$K_{\alpha k} I_k + K_{\alpha l} I_l + K_{\alpha r} I_r + K_{\alpha c} I_c = 0 \tag{11.2.23}$$

The k branch voltage equations may be written as

$$p I_k = L_k^{-1} (- R_k I_k + K^t_{k\alpha} V_\alpha + K^t_{k\beta} V_\beta + K^t_{k\gamma} V_\gamma + K^t_{k\delta} V_\delta) \tag{11.2.24}$$

along with three equations from the previous section

$$p I_l = L_l^{-1} (E_l - R_l I_l - pL_l \cdot I_l + K^t_{l\alpha} V_\alpha + K^t_{l\beta} V_\beta + K^t_{l\gamma} V_\gamma] \tag{11.2.25}$$

$$I_r = R_r^{-1} (K^t_{r\alpha} V_\alpha + K^t_{r\beta} V_\beta) \tag{11.2.26}$$

$$p V_\alpha = C_\alpha^{-1} K_{\alpha c} I_c \tag{11.2.27}$$

and from equation (11.2.23) the latter can also be written as

$$pV_\alpha = C_\alpha^{-1}(-K_{\alpha l}I_l - K_{\alpha k}I_k - K_{\alpha r}I_r) \tag{11.2.28}$$

The three dependent variables V_β, V_γ and V_δ can now be evaluated directly as follows:

Premultiplying equation (11.2.9) by $K_{\beta r}$, yields

$$V_\beta = -R_\beta(K_{\beta r}R_r^{-1}K_{r\alpha}^t V_\alpha + K_{\beta l}I_l + K_{\beta k}I_k) \tag{11.2.29}$$

Using equation (11.2.20) and multiplying equation (11.2.24) by $K_{\delta k}$, yields

$$0 = K_{\delta k}L_k^{-1}(-R_kI_k + K_{k\alpha}^t V_\alpha + K_{k\beta}^t V_\beta + K_{k\gamma}^t V_\gamma + K_{k\delta}^t V_\delta) \tag{11.2.30}$$

Then defining

$$L_\delta = (K_{\delta k}L_k^{-1}K_{k\delta}^t)^{-1} \tag{11.2.31}$$

yields

$$V_\delta = -L_\delta K_{\delta k}L_k^{-1}(-R_kI_k + K_{k\alpha}^t V_\alpha + K_{k\beta}^t V_\beta + K_{k\gamma}^t V_\gamma) \tag{11.2.32}$$

Differentiation of equation (11.2.21) yields

$$K_{\delta k}pI_k + K_{\delta l}pI_l = 0 \tag{11.2.33}$$

Substituting equations (11.2.24) and (11.2.7) yields

$$K_{\delta k}L_k^{-1}(-R_kI_k + K_{k\alpha}^t V_\alpha + K_{k\beta}^t V_\beta + K_{k\gamma}^t V_\gamma + K_{k\delta}^t V_\delta)$$
$$+ K_{\gamma l}I_l^{-1}(E_l - R_lI_l - pL_l \cdot I_l + K_{l\alpha}^t V_\alpha + K_{l\beta}^t V_\beta + K_{l\gamma}^t V_\gamma) = 0 \tag{11.2.34}$$

or

$$V_\gamma = -L_\varepsilon[K_{\gamma k}L_k^{-1}(-R_kI_k + K_{k\alpha}^t V_\alpha + K_{k\beta}^t V_\beta + K_{k\delta}^t V_\delta)$$
$$+ K_{\gamma l}I_l^{-1}(E_l - R_lI_l - pL_l \cdot I_l + K_{l\alpha}^t V_\alpha + K_{l\beta}^t V_\beta)] \tag{11.2.35}$$

where

$$L_\varepsilon = (K_{\gamma k}L_k^{-1}K_{k\gamma}^t + K_{\gamma l}L_l^{-1}K_{l\gamma}^t)^{-1} \tag{11.2.36}$$

which is not affected by topology changes, but is, in general, not sparse.

Inspection of equations (11.2.32) and (11.2.35) indicates that they are two interdependent equations for the two vectors V_γ and V_δ. To solve for these two vectors would require an iterative process, but the associated computational burden would be prohibitive. However, in a practical interconnected a.c.–d.c. power system, the only interface between converter and nonconverter inductive subnetworks occurs at the a.c. terminals of the convertor transformers. The usual practice is to have a.c. harmonic filters, including a high pass, connected on the a.c. side, which would by definition create a β node at this busbar. Alternatively, the connection of an a.c. line or static capacitors will define it as an α node. Thus, if the restriction $K_{\gamma k} = 0$ is applied, the computational problem is removed without imposing any real system representation restrictions. In the rare case of the converter busbar being a γ node, the formulation without δ nodes would be more economic.

By applying this network restriction, i.e. $K_{\gamma k}$ null, to the above equations, a direct solution is obtained, i.e.

$$V_\delta = -L_\delta K_{\delta k} L_k^{-1}(-R_k I_k + K_{k\alpha}^t V_\alpha + K_{k\beta}^t V_\beta)$$ (11.2.37)

and

$$V_\gamma = -L_\gamma K_{\gamma l} L_l^{-1}(E_l - R_l I_l - pL_l \cdot I_l + K_{l\alpha}^t V_\alpha + K_{l\beta}^t V_\beta)$$ (11.2.38)

Matrix L_δ in equation (11.2.37) is block-diagonal with the blocks corresponding to node clusters which are separated by non-δ-nodes. This implies that a switching operation in a converter will cause a change in only that part of L_δ where the switching takes place.

As already explained, the primary windings of the converter transformers are normally connected to either α or β nodes, and therefore these transformers can be considered short circuited at their primary terminals for the purpose of setting up L_δ or modifying it. For this purpose, the transformer phases can be simply represented by their leakage reactances referred to the secondaries and connected to the corresponding secondary winding ends.

The modification of L_δ to represent the various modes of operation can be effected using a simple procedure which adds or removes single elements from L_δ thereby avoiding re-formation and inversion for each mode.

When an additional link of inductance (L) is connected between nodes i and j in the network, L_δ is modified to the new matrix L'_δ whose general element is given, in terms of the elements of L_δ, by

$$L'_{pq} = L_{pq} - (L_{pi} - L_{pj})(L_{iq} - L_{jq})/S$$ (11.2.39)

where

$$S = L_{ii} + L_{jj} - L_{ij} - L_{ji} + L$$

One of the nodes i and j may be external to the δ-network, in which case the corresponding elements in equation (11.2.39) will be zero. Conducting valves are represented by adding a link of zero inductance and the removal of an existing branch is represented by adding a link of negative inductance equal in magnitude to that of the removed branch and connected in parallel with it. In the latter case, if the branch to be removed provides the only path from one part of the network to the external system or to the common reference point, that part of the network will become isolated. This condition is indicated by a vanishing denominator (S), and it must be avoided.

11.3. TRANSIENT CONVERTER MODEL

The converter bridge contains the three terminals of the transformer secondary windings and the two d.c. terminals. One of the d.c. terminals may be earthed, or may be connected to a d.c. terminal of a second bridge. A reduction in the number of nodes together with a considerable improvement in the efficiency of the treatment can be achieved if the following assumptions are made:

(i) The forward voltage drop in a conducting valve is neglected so that the valve may be considered as a switch. This is justified by the fact that the voltage drop is very small in comparison with the normal operating voltage, quite independent of the current and should, therefore, play an insignificant part in the commutation process, since all the valves commutating on the same side of the bridge suffer similar drops.

(ii) The converter transformer leakage reactances as viewed from the secondary terminals are identical for the three phases. This provides a useful simplification since it enables the use of the same nodal solution matrix (L_δ) for any combination of conducting valves within the same mode. This assumption is not justified if nonlinearities due to saturation are considered but it is perfectly legitimate otherwise.

(iii) The case of a completely isolated transformer secondary is not to be allowed, i.e. at least one of the bridge valves is kept ON when the converter is in a nonconducting state. This assumption is necessary for meeting the restriction on inductive branches imposed by a state variable formulation, i.e. the voltage of one of the terminals must be fixed using a conducting valve. Clearly, this will not influence the remainder of the system since the transformer secondary remains open.

Modes of operation

With reference to the three-phase bridge configuration of Fig. 11.1, the interconnection of the converter transformer secondary windings and the d.c. smoothing reactor, is determined by the number and positions of conducting valves. For the converter bridge to be in a conducting state, two or more valves must be ON in such a way that they provide a closed path for the current as viewed from the d.c. terminals. With six valves in the bridge, conducting states can be categorized into the following modes:

A— (Noncommutating mode): two valves on different sides and arms of the bridge are ON.

B— (Normal commutation): three valves, one on each arm, two commutating and at least one conducting on each side of the bridge.

C— (Noncommutating arm short circuit): both valves on one of the arms are ON, with no other valves conducting.

D— (Commutating arm short circuit): as in C with one or both valves on another arm being ON. With one valve on the second arm conducting (D1), there is a single commutation process. When both valves on the second arm conduct (D2), commutations are simultaneously taking place on both sides of the bridge.

E— (A. C. short circuit): this is a case when the three secondary terminals of the converter transformer are short circuited by the conducting valves.

This occurs when four or more conducting valves involve all three arms of the bridge. With four conducting valves (E1), two commutations occur simultaneously which may be on the same side of the bridge. When five valves conduct (E2), commutations take place on both sides and with six valves conducting (E3), multiple commutations take place on both sides of the bridge.

The bypass valve, used in many existing schemes, has been left out of the above classification since its effect is implicitly included in Modes C, D and E.

Converter topology

The modes of operation stated above are represented by defining a set of three nodes for each converter bridge nominally referred to as the top (δ_t), middle (δ_m) and bottom (δ_b) as shown in Fig. 11.2. δ_t and δ_b correspond to d_1 and d_2, respectively, in Fig. 11.1 during normal operation and the presence of conducting valves determines the connections of the transformer secondary terminals (a, b and c) to any of the three nodes δ_t, δ_m and δ_b.

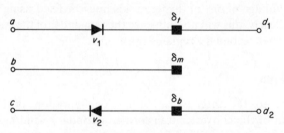

Fig. 11.2. Converter bridge δ nodes

Referring to Fig. 11.2, if valve v_1 is ON then a is set to δ_t and if valve v_4 is ON then a is set to δ_b, otherwise, if both v_1 and v_4 are OFF, a is set to δ_m. When the process is repeated for b and c, all bridge nodes (a, b, c, d_1 and d_2) will have been assigned to the set of δ-nodes defined above.

Since the connections of the transformer secondary windings and the d.c. smoothing reactor are permanently fixed to a, b, c, d_1 or d_2, then the connection matrix $K_{\delta k}$ is fully established by the procedure defined above. There are, however, certain instances when two phases simultaneously have no conducting valves and will, therefore, both be set to δ_m, such as when operating in Mode C. In these cases, the smoothing reactor bridge terminal (d_1 or d_2) is set to the opposite side of the bridge, describing a short circuit, or to the other end of the reactor, describing an open circuit, as the case may be. The latter case is permissible since the smoothing reactor normally has no mutual coupling with any other inductive element in the network. Moreover, if the third phase is also assigned to the node which normally corresponds to the smoothing reactor bridge end, a similar change can be carried out for that phase. This process leaves a free node to which one of the two open phases may be assigned, thereby establishing the independence of these two phases.

The nodal solution matrix L_δ is established for the mode of operation in question, taking into consideration the assumptions and characteristics discussed in the preceding sections, as follows:

Defining a 'basic' L_δ as one corresponding to the noncommutating mode (Mode A), a 'working' L_δ is derived in accordance with the actual mode of operation by adding, removing or changing the connections of the appropriate inductive elements using equation (11.2.32). For example, when the bridge operates in the normal commutating mode (Mode B), with the commutation process taking place in the top half of the bridge, e.g. valves v_1, v_2 and v_3 conduct, a zero inductance link, representing an extra valve conducting in comparison with the 'basic' mode, is inserted between δ_t and δ_m according to equation (11.2.32).

The above treatment is quite general and applies equally well to all the configurations shown in Fig. 11.3. Details of node numbering, formation of the 'basic' L_δ and its modification for the modes of operation are considered in the next two sections.

δ-Nodes and basic L_δ

All the converter configurations shown in Fig. 11.3 assume that one end of the bridge combination is earthed and the other connected to the smoothing reactor. Hence, only two nodes are explicitly defined in six-pulse operation to simulate

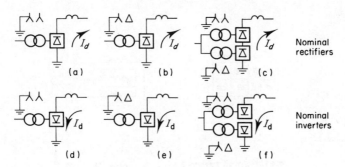

Fig. 11.3. Different converter configurations

Table 11.1

Type of converter		Neutral	Nominal rectifier			Nominal inverter (top-earthed)	δ-range	
			δ_t	δ_m	δ_b		min	max
6-pulse STAR		$m+1$	$m+2$	$m+3$	0	$\delta_t \leftrightarrow \delta_b$	$m+1$	$m+3$
6-pulse DELTA		—	$m+1$	$m+2$	0	$\delta_t \leftrightarrow \delta_b$	$m+1$	$m+2$
12-pulse	STAR	$m+1$	$m+2$	$m+3$	$m+4$	$\delta_t \leftrightarrow \delta_b$	$m+1$	$m+5$
	DELTA	—	$m+4$	$m+5$	0	$\delta_t \leftrightarrow \delta_b$	$m+1$	$m+5$

310

valve switching, with a further node needed to represent the neutral of the transformer secondary in a star bridge. Five nodes altogether are needed to simulate twelve-pulse operation of which one is shared between the two bridges and another representing the neutral of the star-connected secondary. The allocation of new δ-nodes to a new set of converter data, assuming that m such nodes have already been defined, is shown in Table 11.1.

Table 11.1 indicates that a fixed structure is used in representing the variation in topology caused by valve switchings. This implies that, during commutations for instance, redundant nodes are retained to the detriment of computation efficiency. However, the dynamic reordering and relocation of data, which would otherwise be required, is avoided by adopting the fixed structure. The converter bridge, on the other hand, should, under normal conditions, be in a noncommutating state, which requires all the nodes defined, for the greater part of the time. In some special cases, explicitly the noncommutating arm short-circuit and the open-circuit condition, the number of nodes defined above is insufficient to deal with these cases in the direct manner used in the others. Action is taken in these special cases to free one of the redundant nodes and use it to represent the condition as described in the next section. This approach is preferred, to deal with

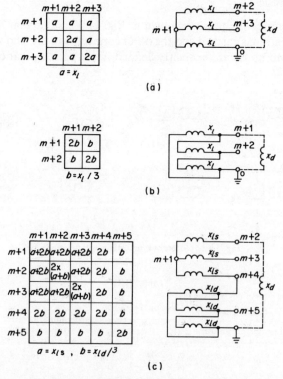

Fig. 11.4. Noncommutating nodal matrices. (a) Star; (b) Delta; (c) Twelve-pulse

the special conditions since the addition of extra nodes would considerably increase the computation time.

The basic noncommutating solution matrix (L_δ) is established at input by a direct procedure in which each secondary winding of the converter transformer is replaced by its equivalent leakage reactance and the smoothing reactor is excluded. The procedure results in the matrix shown in Fig. 11.4. An inductance x_d, representing the smoothing reactor, is then connected between $m + 1$ and 0 for a six-pulse delta converter or between $m + 2$ and 0 for six-pulse star and twelve-pulse converter using equation (11.2.32). The equivalent leakage reactance referred to above is derived from the inductance matrix of the transformer and is given by:

$$x_l = L_{22} - L_{21} \cdot L_{12}/L_{11}$$

The matrices given in Fig. 11.4 can be verified by inspection.

Representation of modes of operation

In normal operation, converter bridges alternate between the noncommutating and commutating modes, i.e. Modes A and B defined earlier. The nodal solution matrices used in these cases are, respectively, the unmodified basic L_δ and the one modified by connecting a zero inductance link between δ_m and either δ_t or δ_b, depending on where the commutation process is taking place. The commutating short-circuit condition (Mode D1) is also easily represented by connecting a single zero inductance link between δ_t and δ_b. The three-phase short-circuit (Mode E1) is represented by connecting a zero inductance between δ_m and δ_b, and another between δ_t and δ_b.

However, in the noncommutating d.c. short circuit (Mode C), due to the limited number of δ-nodes available, special action is taken to represent the conditions. With six-pulse operation the smoothing reactor is removed from its normal δ-node and connected to the common reference point. This action frees a node to which one of the two nonconducting arms may be assigned.

Twelve-pulse converters can be more complicated, depending on which of the two bridges is in Mode C. If the short circuit is in the star bridge, the reactor is removed from its normal δ-node and connected to the node shared between the two bridges ($m + 4$ in Table 11.1), as shown in Fig. 11.5(a). If, however, the delta bridge is in Mode C, then all the branches in the star bridge that are connected to ($m + 4$) are disconnected from this node and connected to the common reference point. The branches switched over are effectively connected in parallel and the program lumps them together and, therefore, one removal and connection process is required. Fig. 11.5(b) shows this condition with the star bridge in a commutating state on the shared-node side. Finally, if both bridges are in Mode C, the smoothing reactor and one transformer branch in the star bridge are both disconnected from their normal δ-nodes and connected to the common reference point as shown in Fig. 11.5(c).

To represent the converter when it is in a nonconducting mode, i.e. the current

312

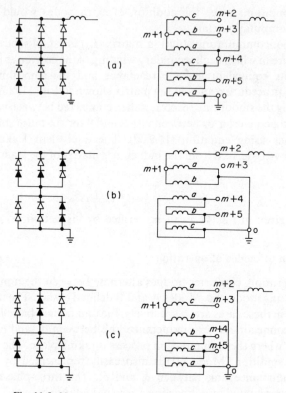

Fig. 11.5. Noncommuting arm short circuit in twelve-pulse
converters

path is interrupted, the modification of L_δ required is to remove the smoothing
reactor from its normal δ-node and connect it to an extra δ-node (δ_x) whose
voltage is set equal to the d.c. line voltage. This frees a node which can be assigned
to any of the transformer secondary terminals. The extra node δ_x is not
represented in L_δ and, therefore, the structure of the converter network is
unchanged by its definition. The condition arrived at describes a true open circuit
on the smoothing reactor side of the converter.

In all the cases described above, the matrix L_δ is modified by the switching
actions indicated according to equation (11.2.31). Moreover, since the modification will only change the elements of L_δ corresponding to the nodes of the
converter in which the switching has taken place, the modification is limited to
the parts of L_δ ranging between the minimum and maximum node numbers
defined in Table 11.1.

The $K_{\delta k}$ incidence matrix is formed for the specific combination of conducting
valves as described briefly in the section on converter topology. Fig. 11.6 shows,
as an example, a twelve-pulse converter in which the Δ-bridge is undergoing
commutation. The converter δ-nodes are assigned to the bridge and transformer
terminal nodes, according to the combination of conducting valves given, as

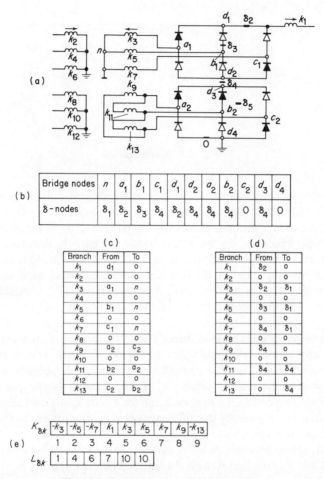

(a)

(b)

Bridge nodes	n	a_1	b_1	c_1	d_1	d_2	a_2	b_2	c_2	d_3	d_4
δ - nodes	δ_1	δ_2	δ_3	δ_4	δ_2	δ_4	δ_4	δ_4	0	δ_4	0

(c)

Branch	From	To
k_1	d_1	o
k_2	o	o
k_3	a_1	n
k_4	o	o
k_5	b_1	n
k_6	o	o
k_7	c_1	n
k_8	o	o
k_9	a_2	c_2
k_{10}	o	o
k_{11}	b_2	a_2
k_{12}	o	o
k_{13}	c_2	b_2

(d)

Branch	From	To
k_1	δ_2	o
k_2	o	o
k_3	δ_2	δ_1
k_4	o	o
k_5	δ_3	δ_1
k_6	o	o
k_7	δ_4	δ_1
k_8	o	o
k_9	δ_4	o
k_{10}	o	o
k_{11}	δ_4	δ_4
k_{12}	o	o
k_{13}	o	δ_4

(e)

$K_{\delta k}$

$-k_3$	$-k_5$	$-k_7$	k_1	k_3	k_5	k_7	k_9	$-k_{13}$
1	2	3	4	5	6	7	8	9

$L_{\delta k}$

1	4	6	7	10	10

Fig. 11.6. Formation $\delta - k$ incidence matrix

shown in Fig. 11.6(b). The permanent branch list shown in Fig. 11.6(c) is transformed to the working list, defined in terms of δ-nodes shown in Fig. 11.6(d). Finally, the $K_{\delta k}$ incidence matrix, shown in Fig. 11.6(e), is derived from the working branch list. The process is limited to the bridge nodes, and the external connections of the converter network, established permanently at input, are indicated by zero entries.

Valve voltages and currents

Referring to Fig. 11.1, the forward voltage across the top and bottom valves of arm p are $(V_p - V_{d1})$ and $(V_{d2} - V_p)$ respectively. Since terminals p, $d1$ and $d2$ correspond directly to the δ-nodes defined in the section on converter topology, whose voltages can be evaluated using equation (11.2.29), the calculation of valve

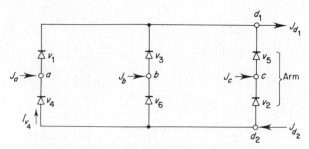

Fig. 11.7. Injected currents

voltages becomes a matter of routine. The relations stated will apply regardless of whether the valve is conducting or not, for, when the valve is conducting, its end will be set to the same δ-nodes and its voltage will correctly be calculated at zero.

The current relations, on the other hand, are entirely determined by two factors: the currents in the converter branches, i.e. the smoothing reactor and the transformer windings, and by the prevailing combination of conducting valves.

Considering the set of injected currents shown in Fig. 11.7, conducting valves provide the paths for these currents to flow in the system and the instantaneous values of valve currents can be obtained from these injected currents. The injected currents themselves are obtained directly from the branch currents. J_{d1} and J_{d2} are equal and numerically equivalent to the d.c. current. The phase injected currents are obtained from the current in the transformer secondary windings, e.g. for a star-connected secondary, J_a is the negative of the current in the secondary winding connected to node a and for a delta-connected secondary, the injected currents are given by the difference between the currents in the windings connected to the node.

The relationship between the valve currents in the top and bottom sides of the bridge (I_{vt} and I_{vb}) on arm p, with the injected current at p, J_p, is

$$J_p = I_{vt} - I_{vb} \tag{11.3.1}$$

when the bottom valve (vb) is OFF:

$$I_{vt} = J_p \tag{11.3.2}$$

and when the top valve (vt) is OFF:

$$I_{vb} = -J_p \tag{11.3.3}$$

In these two cases, an implied supplementary relation, setting the nonconducting valve current to zero, enables the solutions given by equations (11.3.2) and (11.3.3) to be obtained. However, when both valves conduct the supplementary relation must be found elsewhere. The method suggested considers the three valves on the top half of the bridge, whose currents add up to J_{d1}, i.e.:

$$J_{d1} = \Sigma I_{\text{top}} \tag{11.3.4}$$

in this relation, provided p is the only arm with both valves conducting, I_{vt} is implicitly defined.

Equation (11.3.4) may be rewritten, substituting by equation (11.3.2) for the top conducting valves on phases other than p, to give

$$I_{vt} = J_{d1} - \sum_{q \neq p} J_q \qquad (11.3.5)$$

where q are the phases having their top valves conducting.

The current in the bottom valve of phase p may now be calculated using equation (11.3.1), or by a relation similar to equation (11.3.5) giving

$$I_{vb} = J_{d2} + \sum_{r \neq p} J_r \qquad (11.3.6)$$

where r are the arms having their bottom valves conducting.

It can be seen that no solution is possible when two bridge arms have both valves conducting since there is an insufficient number of independent equations. A further relation would be provided by considering the current which may circulate in the closed loop formed by the two arms but this entails assumptions which may not be justifiable. In practice, the case is likely to arise only in the special circumstance when two valves are fired simultaneously at a time when another pair of valves, which would close the loop, are conducting. Otherwise, there is insufficient voltage across a further incoming valve, if the three other valves are already conducting and a firing pulse is applied to the fourth, for it to conduct. For this reason the operating Modes D2, E2 and E3, defined in the earlier section have been left out.

Initial conditions

Conducting valves in each bridge are determined from knowledge of phase 'a' voltage phase angle θ_a, measured with respect to the steady-state load-flow reference, and taking into consideration the inherent phase shift in the bridge transformer θ_s (0 or $+30°$) and the delay angle α.

Representing the a.c. voltage as a cosine function, the voltage crossover for valve $v1$ in the bridge, at time $t = 0$, will have occurred 'A' degrees earlier. Referring to Fig. 11.8, A is given by:

$$A = 60 + \theta_a + \theta_t$$

Valve $v1$ will, therefore, have started conducting B degrees earlier, where

$$B = A - \alpha$$

In this condition valves v_1 and v_6 are conducting and valve v_2 is due to be fired at time F_{v2}, which is 60 degrees later than the firing of v_1, hence,

$$F_{v2} = 60 - B$$

However, if B is less than zero then v_1 is yet to be fired. Conversely, If B is

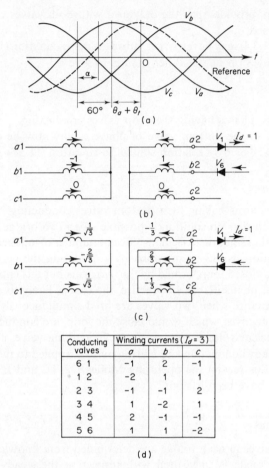

Fig. 11.8. Initial conditions in converter bridges. (a) Determination of conducting valves; (b) Currents in star bridge; (c) Currents in delta bridge; (d) Delta winding currents

greater than 60 degrees then v_1 will have been conducting for more than one firing period indicating that v_2 is already conducting. In either case, the sequence is changed until the correct pair of conducting valves has been defined. This approach ignores the commutation angle which, if included, would probably complicate the issue without offering a substantial advantage since the initial system conditions are approximate in the first place.

A refinement is to adjust the load-flow reference angle so that commutation periods are avoided at the converter. This can be effected by choosing the firing time of the next valve, $k + 1$, to be

$$F_{v_{k+1}} = C$$

where in the above analysis $k = 1$. Then if the steady-state commutation angle is μ.

$$C = 60 - \mu \qquad \text{for six-pulse operation}$$

and

$$C = 60 - \theta_s - \mu \qquad \text{for twelve-pulse operation}$$

This will ensure maximum period exists before any alteration in the converter topology is effected by a valve commutation. Therefore

$$B = k \cdot 60 - C$$

and

$$A = k \cdot 60 + \alpha - C$$

and the required value for θ_v will be

$$\theta_a^{sp} = (k - 1) \cdot 60 + \theta_t + \alpha - C$$

which can be achieved by altering the load-flow reference angle by $(\theta_a^{sp} - \theta_a)$

Having established the pair of conducting valves in the bridge, the currents in the transformer windings must be established on both sides. For a star bridge, shown in Fig. 11.8(b), the secondary winding currents of each phase are set equal to the given d.c. current if its bottom valve is conducting and to the negative of that if the top valve is conducting, otherwise it is set to zero. Primary (a.c. side) winding currents are set equal to the corresponding secondary currents with the signs reversed.

Current is forced to flow in a secondary winding of a delta bridge even though both valves corresponding to its phase are not conducting. In the case shown in Fig. 11.8(c), the secondary winding of phase (c) carries a third of the d.c. current with valves v_2 and v_5 both not conducting. In general, one winding carries two-thirds of the d.c. current in one direction while the other two carry one-third each, flowing in the opposite direction. The table shown in Fig. 11.8(d) gives the current distribution in the delta winding for all combinations of conducting pairs of valves. The primary winding currents are set to $(-\sqrt{3})$ times the corresponding secondary currents.

Converter control system

In order to illustrate the representation of control schemes in the context of full dynamic simulation a model based on the equidistant firing[2] control system is described here.

A valve is switched to the conducting state when there is adequate positive voltage between anode and cathode and a firing pulse is present. The next valve to fire is then set by a ring counter in the sequence indicated in Fig. 11.1 and its firing instant is given by:

$$F_i = F_{i-1} + 60 + \Delta P_i \tag{11.3.7}$$

where ΔP_i is the corrective action produced by the control system which in the steady state is zero.

With rectifier operation under constant current control (C.C.)

$$\Delta P_i = K_1(I_d - I_{ds}) + K_2\frac{dI_d}{dt} \tag{11.3.8}$$

where K_1 and K_2 are adjustable constants.

In practice, the C.C. action is somewhat more complex, e.g. the constant current amplifier in the phase-locked oscillator system is provided with a feedback damping circuit. In essence the factors K_1 and K_2, corresponding to the effective gain of the system, would be complex time functions representing the overall effect of damping and delay circuits.

As indicated by equations (11.3.7) and (11.3.8) the firing instant is independent of the delay angle α. However, the excursions of the firing instant must be limited so that valves are not triggered outside a certain range dictated by hardware limitations or operating requirements and to prevent loss of synchronism between the firing system and the supply. Hence, limit stops are provided to prevent the firing from occurring before a minimum delay (α_{min}) of about 7 degrees or trigger the valve prematurely if a maximum delay (α_{max}) of about 120 degrees occurs before F_i.

Valve extinctions are detected when their currents become negative. By linear interpolation the actual zero-current crossings instants are determined and it is at this instant that the valves are switched to the nonconducting state. Since it is not possible to integrate through discontinuities, the integration steps must start with the conditions prevailing immediately after the change in a valve state and these are obtained by the interpolation process.

A valve extinction may produce a nonconducting condition in the converter, i.e. the current path is interrupted due to there being no conducting valves in one half of a bridge. Action is taken in such a case to ensure that the valves nearest to the smoothing reactor are OFF and that the minimum number of valves remain in the conducting state to provide a path to the common reference point, thus avoiding the need for defining new network nodes.

Optimal extinction angle control (E.A.) is normally exercised with inverter operation. This is achieved by comparing the smallest of the extinction angles per cycle (γ_{min}) with a reference optimum (γ_0) and producing a correction given by

$$\Delta P = K_3(\gamma_{min} - \gamma_0) \tag{11.3.9}$$

If the firing is retarded by the optimum control, the following commutation angles will increase, hence the correction must be smaller in magnitude than the error, i.e. K_3 must be smaller than unity. If the extinction angle of a particular valve i goes below γ_0 a safety control produces a firing advance

$$\Delta P_i = K_4(\gamma_i - \gamma_0) \tag{11.3.10}$$

where K_4 is a constant of similar value to K_3, though normally chosen to be greater to ensure safe operation.

The inverter is provided with both E.A. and C.C. control. The value of the C.C. setting being lower than for the rectifier. E.A. is the normal control and C.C. is only called into action under transient conditions, when rectifier and inverter currents may be different, or when the rectifier is unable to increase the link current beyond the inverter C.C. setting. The inverter may also be provided with a minimum delay angle limit which is usually greater than 90°.

11.4. TRANSIENT GENERATOR MODEL

A phase-variable formulation has to be used in order to develop a general model capable of handling symmetrical, unsymmetrical, nonlinear and distorted waveform conditions with a single basic program.[3] As compared with the single-phase 'd, q' model, the computational burden increases as a result of the time-varying inductive coefficients.

For the purpose of sign conventions, in writing the differential equations the machine is treated here as a motor, i.e. positive current flows into the positive terminal of a circuit branch.

Direct and quadrature axes representation is retained for the rotor circuits as it fits well into the actual geometry and winding arrangements. The position of the rotor is specified with reference to phase a and the angle, θ, is measured from the axis of phase a in the positive direction of rotation.

The following terminal voltage relationships can be written in matrix notation:

$$[V_t] = p[L \cdot i] + [R \cdot i]$$

$$= [L] \cdot \frac{d}{dt}[i] + \omega \frac{d}{d\theta}[L] \cdot [i] + [R] \cdot [i] \qquad (11.4.1)$$

where:

$$\omega = d\theta/dt \text{ is the angular velocity}$$

$$[L] = \left[\begin{array}{c|c} L_{ss} & L_{sr} \\ \hline L_{rs} & L_{rr} \end{array}\right]$$

and

$$L_{ss} = \begin{bmatrix} \begin{array}{c} L_{aa} + L_{a2}\cos 2\theta \\ + L_{a4}\cos 4\theta \end{array} & \begin{array}{c} -L_{ab} - L_{ab2}\cos 2(\theta+30) \\ -L_{ab4}\cos 4(\theta+30) \end{array} & \begin{array}{c} -L_{ac} - L_{ac2}\cos 2(\theta+150) \\ -L_{ac4}\cos 4(\theta+150) \end{array} \\ \begin{array}{c} -L_{ab} - L_{ab2}\cos 2(\theta+30) \\ -L_{ab4}\cos 4(\theta+30) \end{array} & \begin{array}{c} L_{bb} + L_{b2}\cos 2(\theta-120) \\ + L_{b4}\cos 4(\theta-120) \end{array} & \begin{array}{c} -L_{bc} - L_{bc2}\cos 2(\theta-90) \\ -L_{bc4}\cos 4(\theta-90) \end{array} \\ \begin{array}{c} -L_{ac} - L_{ac2}\cos 2(\theta+150) \\ -L_{ac4}\cos 4(\theta+150) \end{array} & \begin{array}{c} -L_{bc} - L_{bc2}\cos 2(\theta-90) \\ -L_{bc4}\cos 4(\theta-90) \end{array} & \begin{array}{c} L_{cc} + L_{c2}\cos 2(\theta+120) \\ + L_{c4}\cos 4(\theta+120) \end{array} \end{bmatrix}$$

$$L_{sr} = \begin{bmatrix} L_{afd}\cos\theta & L_{akd}\cos\theta & -L_{akq}\sin\theta \\ L_{bfd}\cos(\theta-120) & L_{bkd}\cos(\theta-120) & -L_{bkq}\sin(\theta-120) \\ L_{cfd}\cos(\theta+120) & L_{ckd}\cos(\theta+120) & -L_{ckq}\sin(\theta+120) \end{bmatrix}$$

$$L_{rs} = L_{sr}^t$$

$$L_{rr} = \begin{bmatrix} L_{ffd} & L_{fkd} & 0 \\ \hline L_{fkd} & L_{kd} & 0 \\ \hline 0 & 0 & L_{kq} \end{bmatrix}$$

Torque matrix:

$$\frac{d}{d\theta}[L] = \begin{bmatrix} \dfrac{\partial L_{ss}}{\partial \theta} & \dfrac{\partial L_{sr}}{\partial \theta} \\ \hline \dfrac{\partial L_{rs}}{\partial \theta} & \dfrac{\partial L_{rr}}{\partial \theta} \end{bmatrix} \tag{11.4.2}$$

where

$$\partial L_{rr}/\partial \theta = 0 \quad \text{and} \quad \partial L_{rs}/\partial \theta = \left(\frac{\partial L_{sr}}{\partial \theta}\right)^t$$

The model described contains two damper circuits aligned with the direct and quadrature axes respectively. The machine inductance matrix is thus 6×6 square. The fourth harmonic of inductance variation, although not normally known, is considered to be of relevance and has been included.

The AVR and governor representation when included, should be treated as a separate set of equations. This strategy is recommended due to the wide variation of the characteristic eigenvalues of such subsystems allowing a consequential saving in computation time.

A detailed analysis of the electric network without generator control is perfectly adequate for a number of cycles after a disturbance or for disturbances lasting only for a few cycles, like a temporary valve mal-functioning of a static converter. These are carried out over a few cycles of fundamental frequency only and the generator controls (AVR and governor) can either be left out or approximated by constrained algebraic relations. When specially required, both the AVR and governor representations can be added to the present model as separate subsystems taking the applied field voltage and the rotor angular speed as the interface variables, or in the case of thyristor-controlled excitation it is possible and desirable to represent the dynamic behaviour of the thyristor bridge.

11.5. TRANSMISSION LINES

The response of a transmission line to its terminal conditions can be analysed by wave propagation techniques. The modal components used in the wave equations require a sampling interval chosen so that travel times are integral multiples of this sampling time. When the line forms part of an extensive power-system such as shown in Fig. 11.9, with multiple a.c. and d.c. lines of varying lengths, the travel times may be very different. The problem of different velocities

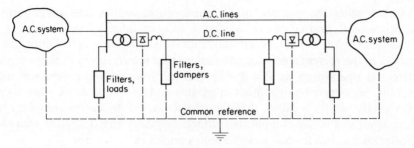

Fig. 11.9. H.v.d.c.–a.c. power system

of propagation, associated with the different modes of a multiconductor line, compounds this so that a satisfactory sampling inverval which will coincide with the converter switching operations will be too small for efficient computation.

A network approach is therefore adopted as described in Chapter 2 whereby the line is segmented into an appropriate number of short line pi-segments, according to the highest frequency of interest.

H.v.d.c. and h.v.a.c. multiconductor lines or cables are defined by their series impedance $(R + jX)$ and shunt susceptance (B_c) matrices per unit length. The pi-segment elements are derived from the physical geometry of the conductors, earthing and conductor–earth resistivities.

Defining the B matrix, whose order is the number of conductors, including earth wires, and whose general element is given by:

$$B_{ij} = \log_e (D_{ij}/d_{ij})$$

where

D_{ij} = distance between the ith conductor and the image of the jth conductor;
d_{ij} = distance between the ith and jth conductors $(i \neq j)$; or

d_{ij} = radius of ith conductor $(i = j)$.

The capacitance matrix per unit length is $C = 2\pi\varepsilon \cdot B^{-1}$.

The useful part of C corresponds to the 'live' conductors since the earth wires are at earth potential and may, therefore, be eliminated. This is achieved by discarding the rows and columns in C corresponding to the earth wires.

The resistance and inductance matrices (series impedance) are given by:

$$Z = R_c + R_e + j\omega(L_g + L_c + L_e)$$

where suffixes g, c and e signify, respectively, the effects of geometry, conductor and earth path.

The contribution of the geometry to the total inductance is obtained directly from the B matrix as $L_g = \mu B/2\pi$.

The effect of the conductor on the resistance at low frequencies is limited to the d.c. resistance per unit length and, where skin effect is significant, a corrected power frequency value may be used. The internal inductance L_c is constant and is calculated using the concept of geometric mean radius. The earth return contributions R_e and L_e can be calculated using an infinite series.

The earth wires cannot be entirely eliminated from the series impedance matrices in the present study since the impedance matrix, given in terms of the differential operator p as $Z = R + Lp$, has an inverse whose elements are, generally, high degree polynomials in (p). It is only in special cases, when these polynomials can be reduced, that the elimination of the earth wires may be achieved without undue complication. Single frequency analysis, when p may be replaced by $j\omega$, and lossless analysis are examples of such cases.

Therefore the equivalent circuit of a typical transmission line segment, and the corresponding inductance and capacitance matrices, are shown in Fig. 11.10. The elements of the capacitance matrix are related to the equivalent circuit by, typically,

$$C_{aa} = C_1 + C_4 + C_5$$

and

$$C_{ab} = -C_4$$

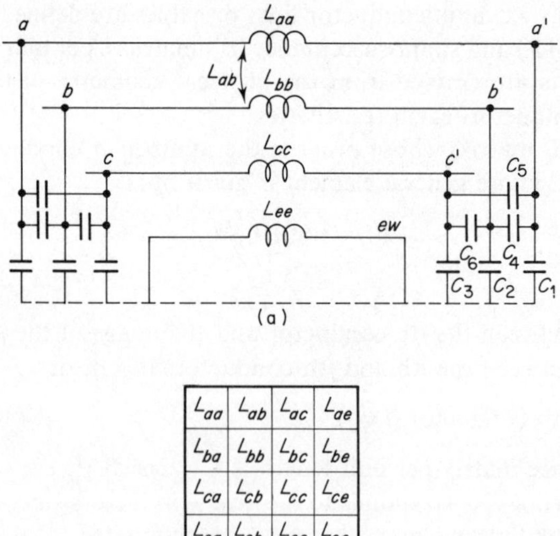

(a)

L_{aa}	L_{ab}	L_{ac}	L_{ae}
L_{ba}	L_{bb}	L_{bc}	L_{be}
L_{ca}	L_{cb}	L_{cc}	L_{ce}
L_{ea}	L_{eb}	L_{ec}	L_{ee}

(b)

	a'	b'	c'
a'	C_{aa}	C_{ab}	C_{ac}
b'	C_{ba}	C_{bb}	C_{bc}
c'	C_{ca}	C_{cb}	C_{cc}

(c)

Fig. 11.10. Transmission line representation (typical pi-section). (a) Equivalent pi-segment model; (b) Inductance matrix; (c) Capacitance matrix

11.6. TRANSIENT TRANSFORMER MODEL

The transformer offers different impedances to the different components of current depending on the type of magnetic circuit and on the connections of the terminals and the neutral. Moreover, in the presence of static converters, phase shifts due to different connections, must also be represented to cater for twelve-pulse converter operation. It is, therefore, inaccurate to represent the transformer by a simple series impedance.

Neglecting core losses, a single-phase transformer can be modelled in terms of self and mutual inductances as:

$$\begin{bmatrix} V_1 \\ V_2 \end{bmatrix} = \begin{bmatrix} L_{11} & L_{12} \\ L_{21} & L_{22} \end{bmatrix} \begin{bmatrix} pI_1 \\ pI_2 \end{bmatrix} + \begin{bmatrix} R_1 & 0 \\ 0 & R_2 \end{bmatrix} \begin{bmatrix} I_1 \\ I_2 \end{bmatrix} \tag{11.6.1}$$

where I_1 and I_2 are taken as positive when they flow into their respective windings.

Numerically (in p.u.) the self inductances are equal, as are the mutuals. Their relationship is approximately

$$L_{12} = L_{21} = \sqrt{L_{11}(L_{22} - L_{1a})} \tag{11.6.2}$$

where L_{1a} is the total leakage inductance of the transformer. The value of the self-term in modern transformers is very high (e.g. 1 percent magnetizing current) and the leakage reactance is very small (e.g. 5 percent) so that in p.u.

$$L_{11} = L_{22} = 100 \text{ p.u.}$$

and thus

$$L_{12} = L_{21} = 99.975 \text{ p.u.}$$

The numerical value of the off-diagonal terms is therefore close to that of the diagonal terms. Since the formulation requires inversion of the transformer inductance, numerical instabilities could be created by the use of inaccurate data.

If there are further windings on the same core the matrix equation can be extended to include them.

Three-phase transformers can be treated in the same way with the possible simplification resulting from ignoring interphase mutuals and reducing the equations to three independent sets for the three phases. This is accurate when banks of single-phase units are used and acceptable with five-limbed units.

Effects like phase-shifts, zero-sequence circulating currents, neutral earthing etc. are automatically catered for by the terminal connections. Off-nominal taps can be represented by suitable modification of the self and mutual elements of the tap-side winding, as described in Chapter 2, noting that the current in the tap-winding is identical to that in the main winding and that the terminal voltage is the algebraic sum of the main and tap-winding voltages.

If the converter transformers are subject to off-nominal tap change, it is necessary to recalculate the elements of the inductance matrix. When the nonlinear relationship between the derived equivalent circuit and the magnetizing characteristics are known, the elements of the coefficient matrix can be continuously updated to correspond with the instantaneous values of the currents, voltages or flux linkages. In such cases, the formulation using δ_k and L_δ suggested in Section 11.2 increases the programming complexity and it is easier to use only the L_γ formulation.

Transformer saturation

In the coupled-circuit model used to represent the transformers, the magnetizing current may be directly formed by summing the instantaneous primary and secondary currents, with the appropriate sign. Multiplication by the magnetizing reactance will result in the instantaneous magnetizing flux linkage, i.e.

$$i = i_1 + i_2 \tag{11.6.3}$$

$$\psi_m = L_m i \tag{11.6.4}$$

where in per unit, the magnetizing reactance L_m is equivalent to the self-reactance (L_{11}, L_{22}) of the transformer.

A typical transformer magnetization curve[4] is indicated, with hysteresis neglected, in Fig. 11.11(a). As can be seen, the knee point of the curve is well defined and the curve may be closely described by a two-slope linear approximation as in Fig. 11.11(b), since the slope in the saturated region above the knee is almost linear, and the slope in the unsaturated region is the linear unsaturated value of magnetizing reactance.

In cases where the instantaneous magnetization curve is not defined for a particular transformer, it may be derived from the r.m.s. characteristics usually supplied by the manufacturers, by a conversion technique, which ignores hysteresis, eddy current losses and winding resistances.[5]

When a transformer saturates, its operating point on the transformer magnetization curve, which may be expressed as magnetizing flux linkage versus magnetizing current in instantaneous values, is shifted into the nonlinear region. In the dynamic analysis formulation, instantaneous magnetizing current is directly available, whilst magnetizing flux is expressed by the time integral of the voltage impressed at the transformer terminals. The state-space formulation, however, requires constant coefficients over each integration time step, with no discontinuities occurring in flux linkages.

The product of the instantaneous magnetizing current, obtained at time t_j, using equation (11.6.3), and the constant value for magnetizing reactance over the previous interval $(t_{j-1}$ to $t_j)$, will provide an instantaneous magnetizing flux linkage (ψ_m). This value may then be located on the known magnetizing characteristic to give an updated value for the magnetizing current (i_m). The following relationships apply with reference to Fig. 11.11(b):

if

$$\psi_m < \psi_1$$

$$i_m = \psi_m / L_1 \tag{11.6.5}$$

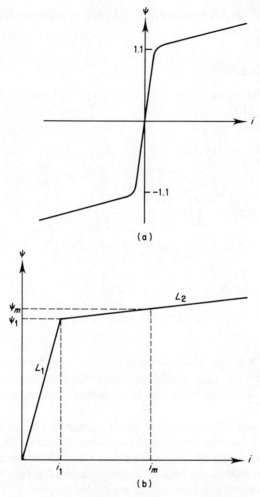

Fig. 11.11. (a) Typical transformer magnetization characteristic; (b) Linearized transformer magnetization characteristic

if

$$\psi_m > \psi_1 \qquad\qquad (11.6.6)$$

$$i_m = \psi_1/L_1 + (\psi_m - \psi_1)/L_2 \qquad\qquad (11.6.7)$$

where L_1 and L_2 are the gradients of the dual slope approximation.

To provide a generalized model for cases where the knee point is less well defined, a multislope linearized approximation can be used, as an extension of the two-slope approximation.

Since saturation will affect each phase independently, data for each phase of the transformer bank must be stored separately. Reconstruction of the system matrix L_I^{-1} can be performed directly, due to its block diagonal structure, by simply inserting L_t^{-1} into the relevant location. If L_y is dependent on L_t, it must also be reformulated.

11.7. STATIC ELEMENTS

Provision has to be made to represent various types of shunt-connected circuits as shown in Fig. 11.12.

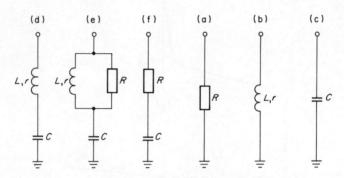

Fig. 11.12. Static shunt elements

Circuits (a) and (b) provide representation of local loads at the converter terminals in order to investigate their damping effect and harmonic penetration.

Static capacitors (c) may be present for compensation and also to represent a d.c. cable.

Circuits (d), (e) and (f) are typical of h.v.d.c. harmonic filters and line dampers.

Owing to their large-time constant, the converter harmonic filters have a major influence on the converter waveforms following a disturbance and they have to be accurately represented in the small-step dynamic simulation.

Static series elements consist of linear, uncoupled elements and are represented by series inductances and capacitances. The latter are restricted, in that they must be part of a capacitive network which provides a path to the common reference point of the system; in practice, these capacitors are connected in series with transmission lines and the restriction is met.

11.8. A.C. SYSTEM

Modelling of each component in an a.c. power system comprising many generators, transmission lines, etc., is possible, but for a large system this becomes computationally uneconomic due to the size and complexity of the system.

Inevitably some assumptions must be made to obtain a less complex, yet sufficiently accurate, model for the system. System components of particular interest must be represented in detail; e.g. when a synchronous generator dominates an a.c. system, the machine may be represented by its full transient equations as in Section 11.4. With static converter plant, it is necessary to explicitly model a.c. harmonic filters and the converter transformers. The combined effects of other components may be treated by a form of network reduction. Various problems occur on reduction of (parts of) an a.c. system to an equivalent form. Not the least of these is establishing a model accurate for all harmonics of the power frequency. Use of a simple fundamental frequency model, in general, gives inaccurate and conservative results.

The a.c. system forms an oscillatory system with the a.c. harmonic filters which may have one or more natural resonant frequencies of a low order. Disturbances may excite these natural frequencies and subsequent waveform distortion, with possible overvoltages, may occur in the a.c. system. These, in turn, will interact with the converters' controllers and affect their responses.

Correct a.c. system representation on a reduced basis would require knowledge of the impedance-frequency locus of the a.c. network. From this information, an a.c. system representation such as proposed by Hingorani[6] and illustrated in Fig. 11.13(a) may be used by the dynamic analysis program. However, this information is subject to operational changes, such as line switching, which can alter the shape of the locus quite drastically. As a result, the availability of a locus for a particular system operating point is somewhat suspect. To avoid this problem, some assumptions must be made which allow for the use of a less complex model.

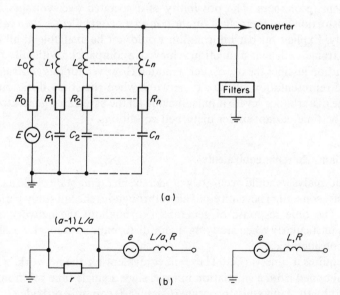

Fig. 11.13. (a) nth order network equivalent—to represent the harmonic impedances of n resonant frequencies; (b) (Modified) Thevenin equivalents of network

Usually a locus would show a tendency for the system to be inductive for the important lower harmonics (up to fifth). The a.c. filters will be capacitive below the fifth harmonic and so a parallel resonant condition *may* exist around the third or fourth harmonic, depending on the strength of the a.c. system. The stronger the system the higher the resonant frequency will be in general. An equivalent circuit which maintains a constant impedance angle over the lower frequency range and provides realistic damping of harmonic voltages has been proposed by Bowles.[7] Giesner[8] proposed an adapted version of this (Fig. 11.13(b)) in which the source impedance (which may be calculated from the short-circuit level at the converter transformer busbars) is split so that there are two, not necessarily equal, reactive elements with the resistive elements designed to give the required impedance angle. The representation of a simple e.m.f. behind a series (source) impedance is also provided for, although as explained, this will only be accurate at a single frequency. The impedance angle varies markedly from system to system. Its correct representation is important as its effect on damping of resonant conditions such as overvoltages is critical. Typical values of 75 and 85 degrees for receiving and sending end a.c. systems respectively have been suggested.[7] For frequencies at or above the resonance of the high pass filter, the a.c. system impedance will be swamped. Thus the only assumption made is in the mid-range of frequencies where the likelihood of a resonant condition in realistic strength systems is small.

The use of an equivalent representation for an a.c. system therefore assumes there are no network resonances yet gives an adequate treatment to possible a.c. filter-system resonances. The possibility of associated over-voltages and waveform distortion and their effect on the converter controllers is then catered for adequately. Explicit busbar information would not be available at all busbars, but the extremely efficient simulation achieved in comparison with full a.c. system representation justifies the use of such a model. However, for this equivalent to be a realistic representation of the a.c. network when the study duration is long and/or the disturbance severe it must include all the associated generator effects, i.e. must be time-variant under disturbed conditions.

Time variant Thevenin equivalents

A dynamic analysis should accurately model the changing state of the a.c. system. The use of a constant Thevenin equivalent throughout the full study period is not realistic. The time response of generators throughout the network, must be included, particularly when studying major disturbances, as machine effects will then be considerable.

This requires a time-variant Thevenin equivalent for the network, which can only be obtained from a simulation in time, since a single time solution, such as performed by the fault study program (Chapter 8) can only provide equivalents valid at that single instant of time.

The multimachine transient stability models described in Chapters 9 and 10 include a quasi-steady-state approximation of the d.c. link behaviour. This model can be used to provide a Thevenin equivalent for each d.c. link terminal at each integration step. This equivalent is thus a reduced representation of the a.c. system as seen from the converter's a.c. busbar, and will include the associated generator effects.

It will be shown later that interactive coordination between the Transient Convertor Simulation and Transient Stability program can improve the accuracy of the a.c. system dynamic equivalents.

In cases where the full a.c. system is represented, i.e. an equivalent representation is not used, each machine may be represented by a time varying equivalent, or in the case of a system-dominating machine, by its full transient equations.

11.9. D.C. SYSTEM

The d.c. line parameters vary from scheme to scheme and the natural frequency of oscillation ranges between 10 and 60 Hz in existing schemes.

In power-frequency related studies requiring mainly the behaviour of the d.c. link at the two terminals and the fault point, it is sufficiently accurate to model the line by a number of π equivalents, depending on distance, and including the smoothing reactors. The behaviour of the smoothing reactors is reasonably linear for currents up to two or three times the rated value and their presence limits considerably the magnitude and rate of rise of current.

By way of illustration, Fig. 11.14 shows a monopole equivalent of the New Zealand four-bridge bipole h.v.d.c. configuration described in Chapter 3 (Fig. 3.8). It includes a π-model for every 50 miles of overhead line and a shunt capacitance representation of the sub-marine crossing. Such a model is considered sufficiently accurate for the purpose of analysing bipole d.c. faults and a.c.

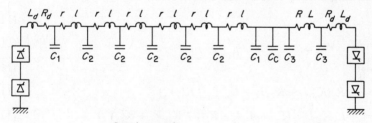

Data in per unit

Smoothing Reactors : L_d = 0.65807 R_d = 0.00042

S.I. overhead line : l = 0.09686 r = 0.00509 $C_1 = C_2/2$
     = 0.064752

N.I. overhead line : L = 0.040362 R = 0.00212 C_3 = 0.026983

Submarine cable : C_c = 1.599743

Fig. 11.14. Monpole equivalent for New Zealand h.d.v.c. link in d.c. line fault studies

disturbances. Of course, the analysis of a single-pole fault on a bipole system would require explicit representation of each pole and any mutual effects.

11.10. SOLUTION OF THE NETWORK EQUATIONS

Numerical integration

The choice of integration method is limited to the so-called single-step methods which are self-starting. This is because multistep methods require information store from previous steps and they are normally started by establishing a few points using a single-step procedure. Taking into consideration the frequency of the occurrence of discontinuities in converter bridges, it can be seen that a multistep method would hardly start to operate before a discontinuity occurs and the single-step procedure called upon again to restart the process.

Two methods for the solution of the differential equations namely, the fourth-order Runge–Kutta method and the implicit integration procedure based on a trapezoidal approximation are explained in Appendix I.

Choice of state variables

By defining

$$\psi_l = L_l I_l, \quad \psi_k = L_k I_k \quad \text{and} \quad Q_\alpha = C_\alpha V_\alpha$$

and using ψ_l, ψ_k, Q_α as the state variables, the magnitudes of the derivatives are of the same order as the voltages and currents of the network. The vectors I_l, I_k and V_α become dependent variables related to the state variables by algebraic relations.

The relevant network equations developed in Section 11.2 are transformed to the following (noting that E_k and E_r are zero):

$$I_l = L_l^{-1} \cdot \psi_l \tag{11.10.1}$$

$$I_k = L_k^{-1} \cdot \psi_k \tag{11.10.2}$$

$$V_\alpha = C_\alpha^{-1} \cdot Q_\alpha \tag{11.10.3}$$

$$V_\beta = -R_\beta(K_{\beta l} I_l + K_{\beta k} I_k + K_{\beta r} R_r^{-1} K_{r\alpha}^t V_\alpha) \tag{11.10.4}$$

$$V_\gamma = -L_\gamma K_{\gamma l} L_l^{-1}(E_l - pL_l I_l - R_l I_l + K_{l\alpha}^t V_\alpha + K_{l\beta}^t V_\beta) \tag{11.10.5}$$

$$V_\delta = -L_\delta K_{\delta k} L_k^{-1}(-R_k I_k + K_{k\alpha}^t V_\alpha + K_{k\beta}^t V_\beta) \tag{11.10.6}$$

$$I_r = R_r^{-1}(K_{r\alpha}^t V_\alpha + K_{r\beta}^t V_\beta) \tag{11.10.7}$$

$$p\psi_l = E_l - R_l I_l + K_{l\alpha}^t V_\alpha + K_{l\beta}^t V_\beta + K_{l\gamma}^t V_\gamma \tag{11.10.8}$$

$$p\psi_k = -R_k I_k + K_{k\alpha}^t V_\alpha + K_{k\beta}^t V_\beta + K_{k\delta}^t V_\delta \tag{11.10.9}$$

$$pQ_\alpha = -K_{\alpha l} I_l - K_{\alpha k} I_k - K_{\alpha r} I_r \tag{11.10.10}$$

Per unit system

Normal steady-state power-system analyses use a per unit (p.u.) representation of voltages, currents and impedances in which these quantities are scaled to the same relative order, thereby treating each quantity to the same degree of accuracy. In dynamic analyses, instantaneous phase quantities are used in both a.c. and d.c. networks and at the same time derivatives of relatively fast changing variables are evaluated.

The usual cosinusoidal relation between a state variable and its derivative indicates that the relative difference in magnitude is ω, which may be high, so that the degree of accuracy will not be the same for each. If the state variables are changed by a factor ω_0 (the fundamental a.c. system angular frequency), and reactances and susceptances are used, instead of inductances and capacitances, the integration process may be carried out as a function of the angular displacement of a reference vector rotating at velocity ω_0, with the result that all variables and coefficient matrices, and hence state variable derivatives, are of the same relative orders of magnitude.

Therefore, by defining a base angular frequency ω_0, the values to be used for the coefficient matrices are impedances and admittances based on this frequency, for all parts of the network, including the d.c. link. The voltages and currents may then be scaled to instantaneous phase quantities. Any convenient power-invariant set of bases may be chosen for the network parameters. The integration step will then be an angle rather than a time increment.

Solution of the transformed network equations

Equations (11.10.1) to (11.10.10) are solved for one integration step, starting from given initial conditions and using the integration process specified.

Both integration methods used require the evaluation of equations (11.10.1) to (11.10.10) several times per step. The Runge–Kutta method requires this evaluation four times per step regardless of the step length. The optimal step for the Runge–Kutta method will thus be the largest that gives a stable solution.

The implicit method requires the evaluation of the changes in dependent variables and state variables during the iterative process of the implicit integration algorithm. The smallest total number of evaluations of equations (11.10.1) to (11.10.10) occur with a step length of about one degree for the particular system under consideration.

Discontinuities

The valve-switching instants must be determined as accurately as possible, so that the relevant network changes may be carried out.

Valve-firing discontinuities are decided beforehand by the control system and the integration step-length can be changed so that the firing coincides with the end of a step.

Valve extinctions, on the other hand, can only be predicted at the expense of slowing down the computation. It is preferred to detect extinctions after they have occurred, when valve currents become negative. At this point, the time of extinction can be assessed by linear interpolation. The time thus obtained is then used to interpolate all the other quantities in the network. This is a first-order approximation assuming that the variables change linearly over the step. If a more accurate assessment is required then a backward integration to the estimated time of discontinuity may be carried out with a further interpolation.

Voltage crossovers do not result in discontinuities but their occurrence must be detected to provide a reference for the firing control system. This is achieved by linear interpolation in a manner similar to the calculation of valve extinction instants.

11.11. COMPUTER IMPLEMENTATION

Figure 11.15 shows the general flow diagram of a computer program developed to implement the Transient Converter Simulation derived in previous sections. It is important to separate, as far as possible, the processes indicated in the flow diagram in order to simplify the program and its verification.

Input and establishing network equations

The data for a dynamic simulation study can be very considerable even for a relatively simple system. For example, a system including a six-pulse d.c. link, a single-circuit a.c. line, and the usual a.c. filters, can easily have upwards of 100 inductive branches, a dozen resistive branches and about 90 nodes of different types. If the data had to be input in branch-list form, it would involve a great deal of preparation and almost certainly some error, particularly since the inductive and nodal matrices contain mutual elements which can increase the volume considerably. Furthermore, any small change in the data, such as changing the number of a.c. line segments, would involve a whole lot of data modification. A more general approach should be adopted in which the data input is as simple as possible and the job of expanding this data into the full network model is left to the computer.

For each system component, a set of control parameters (composed of four numbers) is read in, giving all the information required for expanding the given component data and converting it to the form required by the program. As well as these control parameters, a representative sample of the circuit constants of the component is read in together with the one-line currents and other parameters, such as control system constants and settings of converter bridges. This data is systematically expanded into individual branches and nodes, the one-line currents and voltages are converted into instantaneous phase quantities and individual branch currents and node voltages are calculated and assigned. The new network nodes, created by the expansion of the data, are assigned a type in accordance with the types of branches connected to them as defined in Section

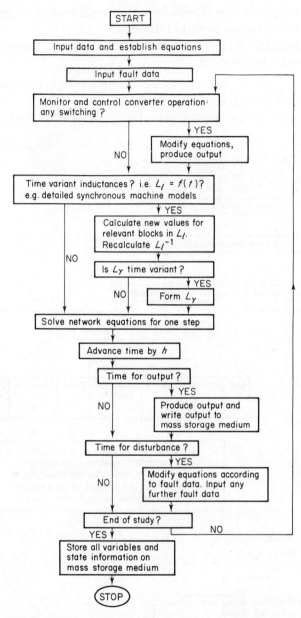

Fig. 11.15. Flow diagram of digital computer program control

11.1. A set of cross-referencing information is created relating the system busbars to the new network nodes, ordering the different types of nodes into their respective sequences and relating them to the network nodes. If the component is a converter, bridge valves are set to their conducting states from knowledge of the a.c. busbar voltages, the type of converter transformer connection and the initial delay angle which is given as part of the data.

The procedure described above, when repeated for all the components of the system, generates the following:

(i) L_l, R_l, L_k, R_k, C_α; in compact form, together with their corresponding indexing information;
(ii) the branch lists for inductive and resistive branches in terms of the set of

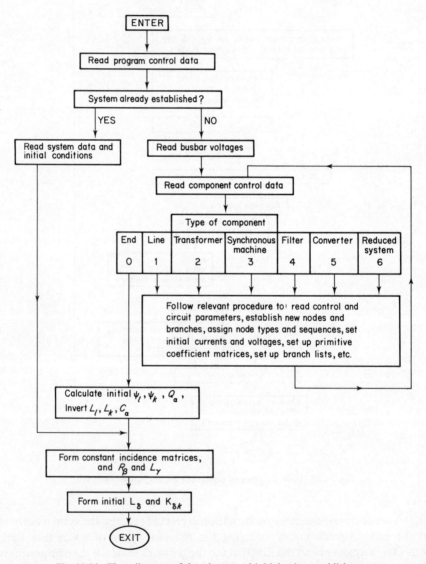

Fig. 11.16. Flow diagram of data input and initial value establishment

new network nodes created by the procedure, together with the type designation and cross-referencing information;

(iii) the initial values of branch current and network node voltages;

(iv) for converter bridges, the basic (noncommutating) nodal solution matrix L_δ and its indexing information, and the initial valve states in each bridge.

Alternatively, network data and initial conditions, stored from a previous (initializing) run may be input in its final expanded form.

The network equations are established by defining all the coefficient values of the variables ψ_l, ψ_k and Q_α. This process involves the inversion of the matrices L_l, L_k, R_r and C_α, the formation of permanent incidence matrices, including an auxiliary connection matrix used in the calculation of valve currents, and the formation of R_β, L_γ, L_δ, and $K_{\delta k}$, i.e. all the variable matrices. All dependent variables and the derivatives of state variables are then calculated so that the solution of the network equations may commence.

Figure 11.16 shows a general flow diagram of the process.

Monitoring and control of converter bridges

At each step of the integration process, converter bridge valves are tested for extinction, voltage crossover, and conditions for firing. If indicated, changes in valve states are made and the control system is activated to adjust the phase of firing. Moreover, when a valve switching takes place, the network equations and the connection matrix $K_{\delta k}$ are modified. The flow diagram in Fig. 11.17 illustrates the monitoring and control procedure.

External disturbances

Externally applied disturbances, as opposed to those inherent in the network as a result of valve switchings, can be identified by a set of data defining the disturbance and indicating the time of application. Besides the application of a.c. and d.c. faults other typical requirements for the dynamic simulation are:

(i) a.c. fault clearance;

(ii) d.c. line charge/restart/reversal;

(iii) if a d.c. fault is to occur, detection parameters need to be established.

Simultaneous application of disturbances should be possible.

Output

The main output stream dumps the calculated results onto a mass storage medium at specified intervals, say 10–20 degrees of fundamental frequency. However, output immediately before and after switching operations is also necessary. A separate program is recommended to retrieve the dumped results and produce time-response graphs of the required variables. It is also advisable to produce graphs of variables which are not explicitly defined in the program, such

336

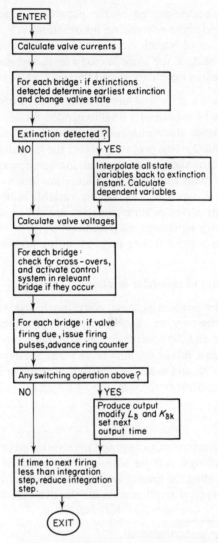

Fig. 11.17. Flow diagram of monitoring and
control of h.d.v.c. converters

as commutation voltages. Finally, it should be possible to magnify selected parts
of the plottings to show more detail.

Interactive coordination with transient stability analysis

The role of a multimachine transient stability program in obtaining time-variant
a.c. system representation has been discussed in Section 11.8.

The Transient Converter Simulation (TCS) results need to be processed to provide information on fundamental power and voltage at the converter terminals for each integration step of the transient stability study.

In this way, the quasi-steady-state approximation to d.c. link performance can be replaced by a more accurate representation of the link performance and further iterations can be made if necessary to improve the interaction between the two programs as detailed in Fig. 11.18. To prevent iterating between the two

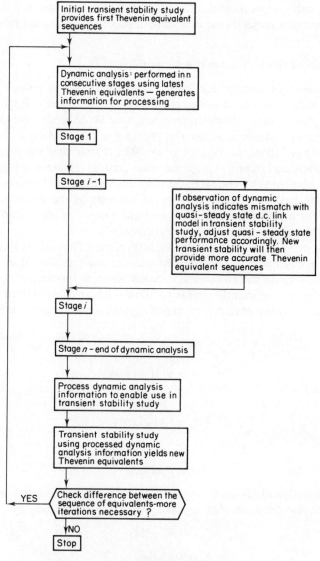

Fig. 11.18. Flow diagram of iterative interation between dynamic analysis and transient stability computer programs

programs, an expensive exercise (indicated by the LHS loop in Fig. 11.18), the following alternative approach is suggested.

It is possible to observe the trends as the TCS progresses through consecutive stages, and if necessary, to interactively modify the performance of the quasi-steady-state d.c. link model over the period already dynamically simulated, so as to match more closely the dynamic simulation results. This is shown by the RHS loop connected between stages (i–1) and (i) in Fig. 11.18. An updated sequence of time-variant network equivalents are then provided which can be used from stage (i) on. Although several short duration transient stability runs may be required with this approach, these are less expensive than the alternative iteration scheme.

Incorporation of time variant network equivalents[(9),(10)]

In the presence of multibridge h.v.d.c. converters, the transient converter simulation requires a very small integration step (of the order of 0.1 ms). On the other hand, the transient stability program, used to determine the generators' response during disturbances, usually provides reliable information with integration steps of 10 ms, i.e. one for every 100 dynamic simulation steps.

In the transient stability program, the gradual changes of a.c. system parameters from the subtransient to the transient, as obtained from the generator response, are used to update the injected currents at the terminals of each generator. The updated current injections yield a new Thevenin source voltage, while the Thevenin impedance remains constant.

The r.m.s. magnitude (E) and phase angle (θ) of each Thevenin voltage at times t_j and t_{j+1} as calculated by the transient stability study are stored in a data file. The rates of change of these Thevenin voltages over each time step of the transient stability study are sufficiently similar to allow a linear interpolation to find an appropriate equivalent at each time step of the transient converter simulation, i.e. at time t_x

$$E_x = E_j + (E_{j+1} - E_j)\Delta t \qquad (11.11.1)$$

$$\theta_x = \theta_j + (\theta_{j+1} - \theta_j)\Delta t \qquad (11.11.2)$$

where

$$\Delta t = (t_x - t_j)/(t_{j+1} - t_j)$$

and

$$t_j \le t_x \le t_{j+1}$$

The dynamic analysis then calculates instantaneous phase values for each Thevenin equivalent source voltage at t_x from these quantities, by

$$e_a = E_x \cos(\theta_x + \omega t)$$

$$e_b = E_x \cos(\theta_x + \omega t - 120°) \qquad (11.11.3)$$

$$e_c = E_x \cos(\theta_x + \omega t + 120°)$$

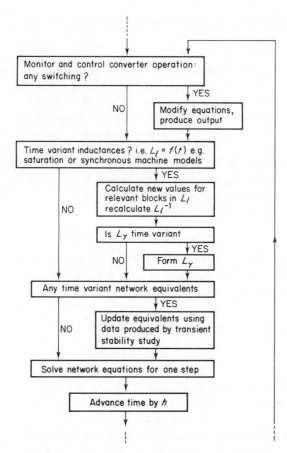

Fig. 11.19. Modification to program control flow diagram
to accommodate time variant Thevenin equivalents and
transformer saturation

The necessary modifications to the overall program control, illustrated in Fig. 11.15, to incorporate a facility for time variant network equivalents, are illustrated in Fig. 11.19.

Careful alignment of the transient stability data file with the dynamic analysis is necessary to correctly identify fault application and removal times. An initial solution at $t = 0$ by the transient stability program which yields information in the form of a load flow, provides the initial Thevenin equivalents. To shorten the establishment of prefault initial conditions, the reference angle in the a.c. system(s) is adjusted so that the run begins within a commutation-free period. Thus, all future angles obtained from the transient stability data file must be similarly adjusted.

The dynamic initial conditions are obtained using the constant initial Thevenin equivalents. Fault application initiates the process of obtaining the time variant Thevenin equivalents from the transient stability data file, as described above.

A consequence of using a time-variant equivalent is that frequency deviations occur as the a.c. system(s) swing. Provided control information in the form of I_d, dI_d/dt (both nonzero) and δ_i (from crossovers) is still available, this will not cause any problems. However α must be measured from the crossover of voltages (as for δ), if a correct indication of the absolute value of firing angle is required.

Alternatively, a reference to the system frequency may be formed so that distortion of the a.c. voltage waveforms does not confuse the firing control pattern when viewed as an absolute firing angle (α). The system swing is then used to modify the interval between firings (ΔF_i), which are $60°$ in the steady state. Under disturbed conditions the interval will be $(60° + \Delta P_i)$. To compensate for frequency shift the interval will become

$$\Delta F_i = (60° + \Delta P_i)\cdot(1 + \Delta\theta/180°) \tag{11.11.4}$$

where $\Delta\theta = \theta_{j+1} - \theta_j$ is the change in Thevenin source angle over the relevant transient stability step. This may also be used to yield a value of α relative to the actual a.c. voltage waveforms.

11.12. PROGRAM OPTIMIZATION AND USE

Scope

The computer program must handle a combined a.c.–h.v.d.c. system of reasonable size and flexible configuration. Hence it must be capable of referring to a h.v.d.c. converter not only in the special context of a rectifier-line-inverter arrangement, but also in terms of a multi-terminal h.v.d.c. system, for example. The number of transmission lines, transformers, etc., are limited only by the storage allocated to the combined total of their equivalent circuit elements. Provision should be made for different transmission line configurations, for both a.c. and d.c. operation, and the number of segments used to represent the lines should be variable within the limits of the total storage available.

Sparsity storage

The coefficient and incidence matrices in equations (11.10.1) to (11.10.10) contain a large proportion of zero elements (e.g. Fig. 11.20). Only the nonzero elements need be stored in order to save computer storage and time in performing matrix operations. Further reduction in storage is achieved by storing only a sample of any constant sets which are repeated, i.e. assuming linearity and balance, all three-phase matrix elements, e.g. multiple-segment transmission line series inductance matrix, and each phase of a bank of single-phase transformers. This will not apply to transmission line nodal capacitances nor to resistances, the latter (R_k, R_l, R_r, R_β) are stored as simple diagonal vectors whose indexing information is inherently given by the relative position of the elements.

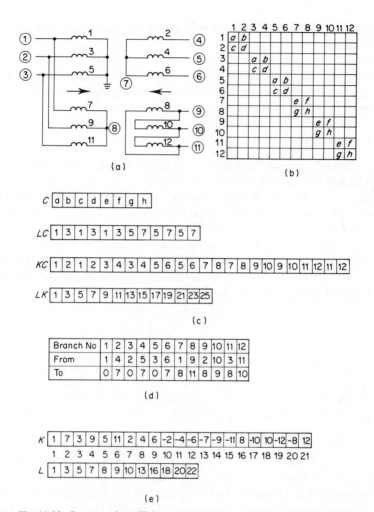

Fig. 11.20. Storage of coefficient and incidence matrices. (a) Equivalent circuit; (b) Full inductance matrix; (c) Compact inductance matrix; (d) Branch list; (e) Node-branch incidence matrix

Coefficient matrices

The representative samples of nonzero matrix elements can be stored in a vector of real constants C, and the indexing information in three integer vectors LC, KC, LK, as in the example in Fig. 11.19(c). C contains the nonzero elements of the full matrix, scanned row-wise, ignoring repeated sets. The vector LC gives the relative address in C of the start of information of each row that is defined in the full matrix. KC contains the column numbers of all nonzero elements in the full matrix while LK gives the relative address in KC of the start of information for each row.

The branch and node coefficient matrices $(L_l, L_k, L_\delta, C_\alpha)$ are also ordered in such a way that they are composed of block-diagonal submatrices of low order. Each block is a square matrix representing a set of mutually coupled branches with few, if any, null elements. The full matrix inverses are therefore identical to the original matrix in structure. This makes it relatively easy to calculated inverse and matrix-by-vector products in a systematic manner.

Incidence matrices

Incidence matrix information can be stored in two integer vectors: the coefficient vector (K) and the associated index vector (L).

Taking a node-branch incidence matrix as an example, L gives the relative address of the start of information in K for each node. The coefficient vector K contains the numbers of branches that are connected to each node, preceded by a minus sign if the node happens to be the receiving-end of the branch in question. This is effectively storing the matrix by rows, replacing $a + 1$ by its column number, $a - 1$ by the negative of the column number, and completely ignoring zeros. Figure 11.20(e) shows the node-branch incidence matrix for the example given.

Implementation

It is recommended to use subroutines to perform the tasks of forming, storing and indexing the coefficient and incidence matrices. As well as these tasks, which are basically required at input, the solution of the network equations is best achieved by using special routines to perform the coefficient matrix inversion, the evaluation of matrix-by-vector products, etc.

The product of a compactly-stored incidence matrix D and a vector V can be obtained by algebraically adding the elements of V whose relative addresses and signs are given by the elements of K pertaining to each row in D. The product of the transpose of D and a vector can be obtained by a similar modified procedure, and it is therefore only necessary to store one form of incidence matrices. The real coefficient matrix-by-vector product can be achieved by summing up the products of the nonzero elements in each row and the elements of the vector V, whose relative addresses are given by the column numbers stored in KC.

Multiple products, such as those used in equations (11.10.5) and (11.10.6), can be formed by scanning the expressions from right to left, progressively carrying out single matrix-by-vector multiplications until the final product is reached.

11.13. GENERAL CONCLUSIONS

A general method has been developed for the computation of a.c.–d.c. system disturbances which combines the efficiency of transient stability modelling and the accuracy of transient simulation of converter behaviour.

The transient converter simulation uses time-variant Thevenin equivalents, formed by linear interpolation, from the discrete values obtained at every step of the transient stability study. Similarly, transient stability compatible equivalents are formed from the transient converter simulation during the disturbance. Such interaction permits the use of reduced equivalent circuits and results in moderate computing requirements. Every cycle of real-time simulation requires about 14 seconds of CPU time on an IBM-3033 computer. As an indication of computing cost, a complete run of a typical disturbance required 3 minutes of CPU time. Moreover, the transient converter simulation is only required for short runs, i.e. during and immediately after the fault period. The extra computing time required for stability studies is relatively insignificant.

The use of a mass storage medium during simulations allows postprocessing of results. Any variable or combination of variables can be plotted on any scale desired, without the transient converter simulation being re-run.

The interactive model developed provides a useful tool in assessing converter controllability during disturbances and the effect of alternative control and protection schemes. It also provides accurate information of fault current and voltage waveforms in the vicinity of large converter plant, the subject of Chapter 12. Such information is essential for the rating of switchgear and the setting of protective relays.

Finally the accuracy of the model used to simulate the disturbance is critical in the assessment of transient stability, a subject discussed in Chapter 13.

11.14. REFERENCES

1. J. Arrillaga, H. J. Al-Khashali and J. G. Campos Barros, 1977. 'General formulation for dynamic studies in power systems including static converters', *Proc. IEE*, **124** (11), 1047–1052.
2. J. D. Ainsworth, 1968. 'The phase-locked oscillator—A new control system for controlled static converters', *Trans. IEEE*, **PAS-87** (3), 859–865.
3. J. Arrillaga, J. G. Campos Barros and H. J. Al-Khashali, 1978. 'Dynamic modelling of single generators connected to h.v.d.c. converters', *Trans. IEEE*, **PAS-97** (4), 1018–1029.
4. H. W. Dommel, 1971. 'Nonlinear and time varying elements in digital simulation of electromagnetic transients', *Trans. IEEE*, **PAS-90**, 2561–2567.
5. H. W. Dommel, 1975. 'Transformer models in the simulation of electromagnetic transients.' Paper 3–1/4, 5th Power System Computation Conference (PSCC), Cambridge, England.
6. N. G. Hingorani and M. P. Burberry, 1970. 'Simulation of a.c. system impedance in h.v.d.c. system studies', *Trans. IEEE*, **PAS-89** (5/6), 820–828.
7. J. P. Bowles, 1970. 'A.c. system and transformer representation for h.v.d.c. transmission studies', *Trans. IEEE*, **PAS-89** (7), 1603–1609.
8. D. B. Giesner and J. Arrillaga, 1971. 'Behaviour of h.v.d.c. links under balanced a.c. fault conditions', *Proc. IEE*, **118** (3/4), 591–599.
9. M. D. Heffernan, K. S. Turner, J. Arrillaga and C. P. Arnold, 1981. 'Computer modelling of a.c.–d.c. disturbances—Part I. Interactive Coordination of generator and converter transient models', *Trans. IEEE*. Winter Power Meeting, Atlanta.
10. K. S. Turner, M. D. Heffernan, C. P. Arnold and J. Arrillaga, 1981. 'Computer modelling of a.c.–d.c. disturbances—Part II. Derivation of power frequency variables from converter transient response', *Trans, IEEE*. Winter Power Meeting, Atlanta.

12

A.C.–D.C. System Disturbances

12.1. INTRODUCTION

Conventional faulted system analysis, as described in Chapter 8, cannot represent the highly nonlinear response of h.v.d.c. converters to fault conditions. Such response and subsequent effects are often neglected without proper justification and fault studies are carried out by ignoring the presence of the d.c. link or representing it by quasi-steady-state models.

While this simplification may be justifiable in particular cases, the power ratings of h.v.d.c. links are generally too large to be ignored and their speed of controllability has a profound effect on the entire system behaviour during and immediately after the fault. There is, therefore, a need for detailed transient converter simulation with realistic representation of both the a.c. and d.c. systems as explained in Chapter 11. According to their origin, a.c.–d.c. disturbances can be divided into converter faults, a.c. faults and d.c. faults.

Converter faults, though relatively frequent, are normally self-recovering and have little influence on the system as a whole. Moreover, from the modelling point of view, the converter operating modes resulting from internal converter disturbances are not different from those produced by converter faults consequential to the occurrence of external disturbances.

This chapter describes the application of the dynamic simulation program developed in Chapter 11 to the analysis of a.c. and d.c. fault conditions. The New Zealand a.c.–d.c. power system is used as a test model for the results.

A.C. FAULT MODELLING

12.2. REDUCED NETWORK REPRESENTATION

To determine an adequate equivalent network, the areas of specific interest must first be identified. For a.c. fault studies, the most important points of interest for the observation of overcurrents, overvoltages and waveform distortion will be the converter terminals and the fault location. Therefore, in general, explicit information will be required at three locations, viz. at the a.c. terminals of each converter, and the fault location.

The New Zealand Primary Transmission system consists of two separate 220-kV a.c. networks of comparable capacity, containing 21 buses in the North Island, and 19 buses in the South Island, which are interconnected by a ± 250-kV h.v.d.c. transmission link. As the system is not dominated by any particular component (other than the d.c. link), it is appropriate to consider an equivalent network representation, with just the three locations detailed above being modelled explicitly. Effectively then, at each node of interest a Thevenin Equivalent must be obtained for the a.c. system, as seen looking into the system from the relevant node.

Approximate equivalent networks can be obtained from an a.c.–d.c. fault study programme with quasi-steady-state representation of the d.c. link.[1] The program uses nodal system analysis, and the steady-state prefault nodal voltages of the a.c.–d.c. system are described by the following matrix equation.

$$V = ZI + ZJ \qquad (12.2.1)$$

where

Z is the bus impedance matrix,

I is a vector of nodal injected currents due to a.c. generators,

and

J is a vector of nodal currents injected by the d.c. converters.

The model of the d.c. link assumes that the converters retain controllability during the disturbances, and thus maintain the current setting. When a three-phase fault occurs in one of the a.c. systems, represented by a current I^f flowing out of the faulted bus p, the quasi-steady-state model resolves the d.c. link equations. With the d.c. link on constant-current control (at one of the two converter stations), the magnitude of the current injections, J^f, will remain fixed, but the respective phase angles of the injections will depend on the terminal voltages and the d.c. link operating conditions (α_r, δ_i).

From the matrix equation (12.2.1), the following expressions are extracted for the nodal voltages at the converters and fault buses (Fig. 12.1).

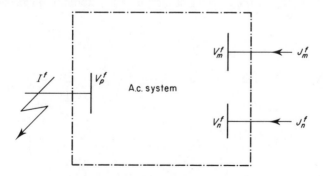

Fig. 12.1. Reduced equivalent network

$$V_m^f = V_m + Z_{mm}(I_m^f + J_m^f) + Z_{mn}(I_n^f + J_n^f) + Z_{mp}I^f \qquad (12.2.2)$$

$$V_n^f = V_n + Z_{nm}(I_m^f + J_m^f) + Z_{nn}(I_n^f + J_n^f) + Z_{np}I^f \qquad (12.2.3)$$

$$V_p^f = -Z^f I^f = V_p + Z_{pm}(I_m^f + J_m^f) + Z_{pn}(I_n^f + J_n^f) + Z_{pp}I^f \qquad (12.2.4)$$

where

Z^f is the fault impedance,
I^f is the fault current, and
V_m, V_n, V_p are the prefault voltages at the rectifier, inverter and fault buses respectively, in the absence of d.c. transmission.

Equations (12.2.2) to (12.2.4) describe a matrix set containing three interrelated Thevenin equivalents. Each equation contains two terms which couple that equivalent to the other two equivalents, viz.

$$Z_{kj}(I_j^f + J_j^f) \quad \text{for } k \in m, n, p$$

$$j \in m, n, p$$

$$j \neq k$$

where

$$J_p^f = 0 \quad \text{and} \quad I_p^f = I^f$$

This equation set can be easily incorporated into the transient converter simulation of Chapter 11 by modifying equations (11.10.1), (11.10.5) and (11.10.8) to allow for the mutual coupling term, which has both real and imaginary components of impedance, i.e.

(i) R_l now contains off diagonal terms, and
(ii) L_l contains additional off diagonal terms.

The initial conditions are obtained using the process described in Section 11.11. The quasi-steady-state fault study, which provides the Thevenin equivalents, also provides the steady-state prefault data necessary for the dynamic analysis to begin the establishment of correct dynamic initial conditions.

The applicability of the equivalent circuit at harmonic frequencies has been

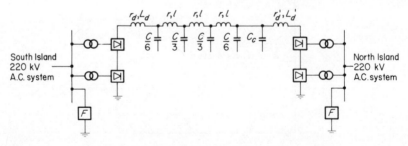

Fig. 12.2. Equivalent h.v.d.c. representation for a.c. faults

discussed in Section 11.8. The test results included here relate to a purely fundamental frequency a.c. network impedance. However, whenever the impedance-frequency locus of the system is known it can be modelled in the program by appropriate equivalent networks as illustrated in Fig. 11.13(a).

As explained in Section 11.9 for a.c. fault studies, the d.c. line representation may be simplified to a double-bridge monopole equivalent, with three π-sections for the South Island overhead line, a large shunt capacitance for the submarine cable and a series impedance for the short overhead line in the North Island. The latter is incorporated into the smoothing reactor data and the complete d.c. side representation is shown in Fig. 12.2.

12.3. FAULT APPLICATION AND REMOVAL

The following typical sequence is used to model the behaviour of the a.c. fault.

  (i) detection of fault with relevant discrimination;
 (ii) operation of the relevant circuit breakers to isolate the fault within 3–10 cycles;
(iii) after a period sufficient for the fault path to deionize, the breaker is reclosed thus completing the original circuit;
(iv) if the fault re-establishes, the sequence is repeated for up to two to three times and then if the fault still persists, the breakers will permanently isolate the fault.

The node specification at the fault location is determined by the degree of a.c. system representation, i.e. equivalent circuit or full representation.

Equivalent a.c. system representation

A fault branch (Z^f), which may be resistive or inductive, is established in the original branch list. Any value of Z^f which is not infinite in the prefault period would create an incorrect system loading. However, very large values of Z^f affect the rate of convergence of the numerical integration of the state variable equations.

This can be avoided by including in the Thevenin equivalent, as seen from the fault point, a generating source exactly equal to the effective loading created by the noninfinite shunt impedance to ground at the fault point. By the principle of superposition, the dynamic model will then see the fault bus as an open-circuited terminal which will have the correct voltage.

At fault application time, the shunt impedance is altered to form the appropriate faulted network.

In the case of a remote fault with a purely resistive fault branch of zero fault resistance, R_r^{-1} will be infinite and R_β will be zero. This would necessitate a very short integration step to achieve a reasonable rate of convergence for the numerical integration process. Instead, a resistive-inductive fault is selected to avoid convergence problems. In the case of a converter terminal fault (with high-pass filters connected), either a resistive (R_r) or inductive (jX_l) fault of any magnitude can be simulated.

With a three phase-to-earth fault, each phase will have an equivalent which is grounded through a fault branch. For a single-phase fault the two healthy phases will retain their prefault configurations. Similarly, a phase–phase fault can be simulated by connecting the relevant phases together and if the fault involves earth they may be connected to ground as well.

The actual time of fault clearance is determined by the following sequence of events. At a prescribed time following fault application, for example, five cycles, a

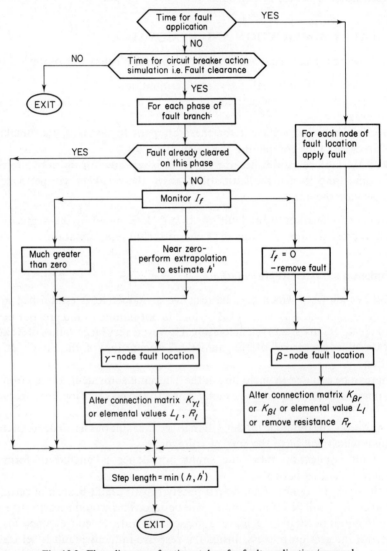

Fig. 12.3. Flow diagram of actions taken for fault application/removal

flag is set which indicates that fault clearance is to be initiated on the relevant fault branch(es) at the next zero crossing of current in the fault branch(es). Extrapolation is used to predict the time when a zero crossing will occur; i.e. if t_j is the present time step, and h the step length, then

$$t_j = t_{j-1} + h \qquad (12.3.1)$$

Therefore, if the current in the fault branch at time t_j is I_j, then

$$h' = hI_j/(I_j - I_{j-1}) \qquad (12.3.2)$$

so that if h' is less than the next integration step (the nominal value as calculated in the integration routine), the next time step will be

$$t_{j+1} = t_j + h' \qquad (12.3.3)$$

which, assuming linear extrapolation, will coincide with the zero crossing of fault current. Fault clearance will then be effected as described above by alteration of an elemental impedance value to re-establish the prefault circuit.

This process is repeated for each of the phases which are faulted. A flow diagram depicting the actions taken for fault application and removal is illustrated in Fig. 12.3.

Full a.c. system representation

An example of fault application and removal with full a.c. system representation for a (three-phase-to-earth) a.c. line fault is now described with reference to Fig. 12.4(a):

(i) Insertion of the relevant fault impedance at the fault point in parallel with the shunt capacitance of the π model corresponding to that location, by alteration of a pre-established connection matrix element.

(ii) Subsequent to the specified circuit-breaker operating time, the line is isolated by reforming the connection matrix elements associated with the line ends (breaker locations), on detection of current zeros there.

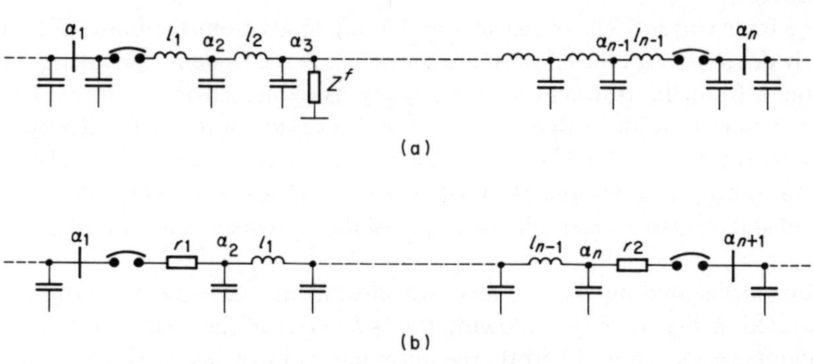

(a)

(b)

Fig. 12.4. (a) Full system representation of a line fault and circuit breakers; (b) Inclusion of circuit breaker resistances

(iii) After a specified time, the breakers are reclosed and the fault to ground is removed by reforming the original connection matrices.

Each of the breaker operations can simulate the inclusion of resistance in the make/break sequences. A series resistive element may be included at the relevant line end (r_1, r_2 in Fig. 12.4(b)), the value of which may be practically zero without affecting convergence patterns, since the line ends are generally connected to other lines, and are therefore in general α nodes.

Opening of the circuit breaker thus involves

(i) increasing r_1 and r_2 at a rate defined by the breaker contact resistance vs time behaviour;
(ii) at a certain resistance, setting elements $\alpha_1 r_1$ and $\alpha_{n+1} r_2$ to zero in the connection matrix $K_{\alpha r}$.

Reclosing the circuit breaker involves:

(i) removing Z^f by setting element $\alpha_3 Z^f$ to zero in $K_{\alpha r}$;
(ii) reinstating elements $\alpha_1 r_1$ and $\alpha_{n+1} r_2$ to their original values $(1, -1)$ in $K_{\alpha r}$; and
(iii) decreasing r_1, r_2 to comply with the preinsertion resistor values.

In cases where breaker contact resistance is not modelled, only step (ii) in the opening action is simulated. Similarly in cases where no preinsertion resistance is used, only steps (i) and (ii) are performed, with r_1 and r_2 equal to zero.

12.4. TEST SYSTEM AND RESULTS

The test studies illustrated are primarily concerned with the fault and immediate postfault periods. Extensions of the study-periods to investigate system stability are discussed in Chapter 13.

As an example of a severe disturbance, the case of a three-phase fault at the inverter a.c. busbar has been selected. Some fault impedance is included to retain sufficient terminal voltage for the commutations (approximately 10 percentage of nominal).

The fault current, illustrated in Fig. 12.5(a), apart from the initial d.c. offset, clearly demonstrates the effect of the time-variant a.c. system equivalent voltage obtained from the transient stability study as explained in Sections 11.8 and 11.11. From an initial value of 1 p.u. the Thevenin source at the inverter end reduces to 0.5 p.u. after five cycles. This is depicted in Fig. 12.5(b). The phase angle at the inverter terminal is shown in Fig. 12.5(c) and the combination of Figs. 12.5(b) and (c) also indicate the variation of the equivalent source at the rectifier end.

The corresponding a.c. voltage waveforms at the inverter terminal are illustrated in Fig. 12.5(d). Following the behaviour of the source voltage in the postfault period (Fig. 12.5(b)), the terminal voltage levels show a gradual recovery.

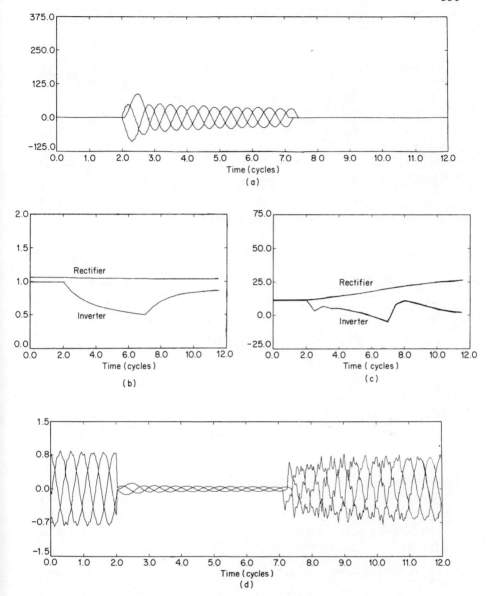

Fig. 12.5. Typical results for a three-phase short circuit at the inverter end. (a) A.c. fault currents; (b) Thevenin voltages; (c) Thevenin angles; (d) Inverter a.c. voltages

The behaviour of the d.c. link is indicated in Fig. 12.6, which shows the d.c. powers at the rectifier and inverter ends.

A second example, involving a less severe disturbance, is the case of a three-phase fault sufficiently distant from the rectifier end to maintain about 65 percent of the nominal voltage at the rectifier terminal. Results for the a.c. current and

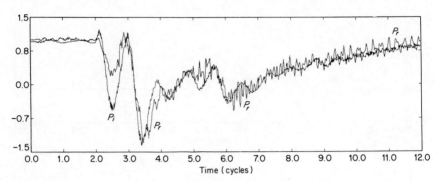

Fig. 12.6. Converter powers for a three-phase short circuit at the inverter end

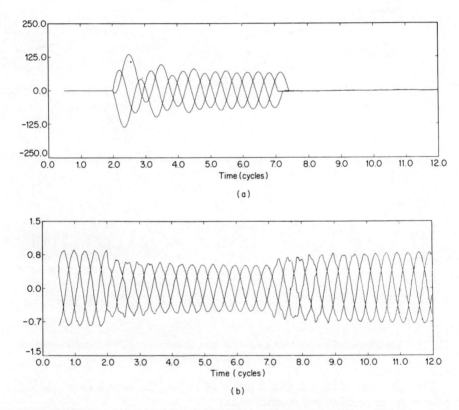

(a)

(b)

Fig. 12.7. Typical a.c. results for a three-phase short circuit at the rectifier end. (a) A.c. fault currents; (b) Rectifier a.c. voltages

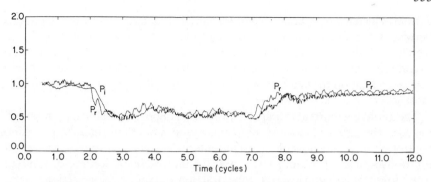

Fig. 12.8. Converter powers for a three-phase short circuit at the rectifier end

voltage waveforms are illustrated in Figs. 12.7. The variation of d.c. powers is plotted in Fig. 12.8.

D.C. FAULT MODELLING

12.5. OVERCURRENT CHARACTERISTICS OF D.C. LINE FAULTS

In the absence of circuit breakers, d.c. transmission faults can not be cleared by temporary switching followed by autoreclosing. On the other hand, the inherent C.C. control of h.v.d.c. converters prevents large and sustained fault currents.

The levels of rectifier and fault overcurrents resulting from a d.c. line short circuit depend on system and fault parameters, fault timing and location, mode of converter operation, and subsequent control action. Peterson[2] investigated the level of transient rectifier overcurrent by isolating several parameters to determine their effects. His analysis was based on a quasi-steady-state d.c. representation, and ignored the effects of the a.c. system.

However, the assumption of a sinusoidal voltage beyond the commutation reactance is not generally valid. The increased rectifier current during the fault period will produce harmonic current imbalance in the a.c. system, which will always result in a.c. waveform distortion for an a.c. system with finite short-circuit capacity. Moreover, the extra alternating current required as I_d increases must be supplied from the a.c. network, thus causing further voltage regulation at the converter terminals. The use of an infinite a.c. system representation will in general yield pessimistic levels of rectifier overcurrent.

The presence of the smoothing reactor limits the magnitude and rate of rise of overcurrents. Under high overcurrent, reactor saturation would increase substantially the peak fault current and a special design of reactors is required to retain their nominal inductance with up to two to three times rated current. The effect of resistances in the converter transformers and smoothing reactors on peak currents is small.

Since the d.c. system contains only passive elements, the main source of fault current is the a.c. system, with d.c. line discharge contributing additional fault current. The maximum fault current magnitude is determined by the impedance between a theoretical a.c. infinite bus and the fault location. Therefore, strong rectifier a.c. systems will result in higher fault and rectifier overcurrents than would a d.c. line fault with a weak (i.e. low S.C.R.) rectifier a.c. system. A zero resistance fault on the line side of the rectifier smoothing reactor is therefore the most severe in terms of location.

The initial overcurrent is limited by the surge impedance of the d.c. line (which includes the line capacitance, inductance and smoothing inductance). Since control action is only possible at discrete intervals (every 60° for six-pulse, and 30° for twelve-pulse operation), even with instantaneous detection, a transient overcurrent will occur. However, control action, initiated by a protection scheme, which retards the firing into inversion, will limit the maximum overcurrent level when compared to normal regulator action (C.C.), since the latter has time delays (circuit time constants) and/or gain limits incorporated in their design to ensure stable operation in the steady state.

Severe oscillations of d.c. voltage and current can occur following normal control action, due to possible resonance effects between the d.c. line and the control system.[3] This highlights the need for an overriding protective action to limit overcurrent stresses and ensure rapid fault clearance and recovery. Normally, no more than one commutation occurs in twelve-pulse operation after a line fault, before protective action is taken by the current control system.

Rectifiers operating with minimum firing angles at fault inception will exhibit higher overcurrents than those with larger firing angles, since the effective voltage driving the fault is then greater. Figure 12.9(a) indicates the effect of the rectifier delay angle on the maximum level of rectifier overcurrent, at the time of the fault inception. These curves assume instantaneous fault detection, and the fault is applied immediately after a valve firing so that no control action can occur until the next firing instant. The levels are pessimistic in that they are obtained with an infinite a.c. source,[2] so that for six-pulse operation:

$$I_{dr}^{max} = (2 + \cos \alpha_r - \sqrt{3} \sin \alpha_r) \pi / 6X_c \qquad (12.6.1)$$

and for twelve-pulse operation:

$$I_{dr}^{max} = (\sqrt{2} + \sqrt{6} - (2 + \sqrt{3}) \sin \alpha_r + \cos \alpha_r) \pi / 12X_c \qquad (12.6.2)$$

To correct for nonzero overlap angle, α is replaced in these equations by $\alpha + \gamma'$, where γ' is the overlap angle during the fault, which is larger than the prefault overlap angle. Thus for normal operation, where delay angles are usually less than 30 degree, twelve-pulse operation will have smaller overcurrents than six pulse operation.

Figure 12.9(a) also applies when the abscissa angle depicts the time between fault inception and the previous valve firing. The twelve pulse curve will, therefore, only be applicable from 0 to 30 degree, and thus twelve pulse operation always has a lower peak overcurrent than does six-pulse operation. The negative slope of the curves is indicative of the time which must elapse before protective

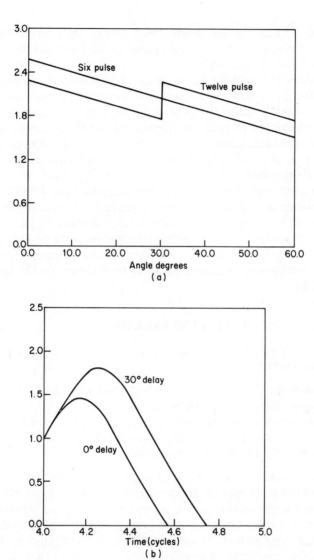

Fig. 12.9. (a) Effect of instant of fault (and α_0) on I_{dr}^{max}; (b) Effect of detection delay on I_{dr}^{max}

retard action of the rectifier firing angle can be initiated. Again, instantaneous detection is assumed, this time with a constant firing angle of α^{min}. The cumulative effects of initial firing angle and elapsed time before protective action becomes effective can be obtained from these curves.

The effect of noninstantaneous detection on the level of overcurrent is shown in Fig. 12.9(b) for a zero resistance earth fault on the line side of the rectifier end terminal. The nominal firing angle was 25 degrees with the fault applied 10 degrees after the last firing. Calculations based on the above idealized formulae

(12.6.1) and (12.6.2) indicate levels much higher than those obtained by the dynamic simulation, for both instantaneous detection and for detection 30 degrees after fault application.

This indicates the need for the more accurate simulation in which the a.c. system, filters, d.c. system parameters and converter controls are accurately modelled, in order to assess the effect of voltage regulation, commutation, and waveform distortion. Unless the simulation can accurately represent the actual timing of valve firings in relation to fault occurrence, and the subsequent effects of rapid control action associated with normal or protective action, inaccurate results will be obtained.

The interactive use of the state variable dynamic analysis and a transient stability program described in Chapter 11 provides accurate simulation of the a.c.–d.c. system behaviour over the entire fault and recovery period.

The dynamic analysis requires suitable representation of practical d.c. line fault detection and recovery schemes. Similarly, accurate representation of the expected fault behaviour, including the arc characteristics, must be incorporated into the model.

12.6. DETECTION OF D.C. LINE FAULTS

Based on the characteristics discussed in Section 12.5 a d.c. line fault detection and protection scheme must:

(i) respond rapidly to d.c. line faults independent of converter or plant parameters, and of fault location;
(ii) not respond to converter faults, a.c. system(s) faults, or reduced d.c. voltage due to operational changes (e.g. step changes in d.c. voltage);
(iii) determine if converter restart has been successful.

Since, as described in Section 12.5, the protection must be fast acting, the detection scheme must be able to identify the travelling waves set up on the transmission lines by each type of fault, and to discriminate between them.

The polarity of these travelling wave gradients, with respect to line voltage, holds enough information to determine whether the fault is bipolar or monopolar, and to indicate the pole(s) involved. In the case of a monopolar fault, travelling waves are also induced on the healthy pole by virtue of mutual coupling action. Investigations into possible overvoltage thereon, by Hingorani[4] and Kimbark,[5] have shown that in certain circumstances the coupling action as a result of a monopolar fault may be sufficient to cause a flashover on the healthy pole. The magnitudes of overvoltages induced on the line by coupling action are dramatically influenced by the d.c. line terminations. Fault location and type, as well as line characteristics, will also affect the overvoltage magnitude at the terminals. As indicated in Section 12.5, converter control will have no effect on the first-wave reflections at the terminals, which in general involve the peak overvoltages.

Because of the d.c. link smoothing reactors, inverter faults and a.c. system changes will cause considerably lower gradients of line voltage and current at the rectifier than will d.c. line faults, since there is much more inductance between the detection unit and the fault location. Similar discrimination is possible with current gradients, since the two types of travelling waves initiated by the fault are interrelated.

The fault resistance will also affect the travelling wave gradients. In the case of a high-resistance ground fault close to the inverter end of a long line, information obtained at the rectifier end, based exclusively on voltage magnitude and gradients, may not be sufficiently reliable, since it may be indistinguishable from voltage information resulting from operational changes or a.c. faults. Resorting to comparison of current at either end is also unreliable, and would require signal transmission. The use of current gradient, however, is a reliable signal. The computer model uses the weighted sum of the direct voltage and direct current gradients at the converters, for each pole of the d.c. line,[6] i.e.

$$\eta = K_5(dV_{dc}/dt) + K_6(dI_{dc}/dt) \qquad (12.6.1)$$

which will be directly related to the travelling waves initiated by the fault, and contains information relating to the fault type, size and location. Since the terminations of the d.c. line will affect the gradients of the travelling waves, this must be considered in setting suitable detection levels.

By suitable choice of K_5 and K_6, restraints may be put on η to ensure that a monopolar fault is not seen as a fault on the healthy pole of a bipole system, due to the voltage which may be induced thereon by mutual coupling action. Of course, if this induced voltage is sufficient to initiate a flashover on the initially unfaulted pole, an extremely undesirable condition, then η must be such as to initiate the detection/protection scheme associated with that pole as well. On detection of a line fault, the converter bridges associated with the faulted pole(s) will then undergo the relevant protective action as described below.

12.7. FAULT CHARACTERISTICS

Arc extinction and deionization

Converter blocking is normally used only for internal station faults, and is not necessary for d.c. line faults, where a method of fast retardation of firing angles is most effective.

On detection of a line fault, by driving the rectifier into inversion (say $\alpha_r = 120$ to 135 degrees), the line voltage will rapidly drop appreciably below the line side value, forcing the rectifier current to extinguish more quickly than it would under normal C.C. control action. Similarly, the drop in line voltage at the inverter due to the line fault will extinguish the inverter current quickly.

However, this condition of zero energy transfer into the link will not immediately extinguish the fault. In the case of a d.c. link such as in the New Zealand scheme, the combination of a long overhead line (with smoothing reactors) and a relatively long submarine cable represents a lightly damped oscillatory RLC circuit with a natural frequency of approximately 50 Hz. This circuit has a significant energy storage capacity. With no external sources, i.e. once both converters have shut down, the time for the energy in this circuit to dissipate is determined by the relative magnitude of the circuit resistances, which will be made up from three components:

  (i) d.c. line resistance;
  (ii) earth fault resistance; and
  (iii) arc resistance.

The value of line resistance will be known for a particular line and therefore to decide upon the interval necessary for complete arc extinction, the properties of the arc, as well as the specific earth resistance, must be established. Tests carried out by Kohler[7] determined the following arc voltage and the arc path deionization times of a free burning arc:

  (i) The arc voltage drop for arc lengths of 0.4 to 2.0 metres seemed to be independent of fault (rectifier) current and of arc length, and had an average value of 2 kV per metre of arc.
  (ii) The deionization time, measured from the instant when the arc current goes below 1A, must be from 20–50 ms to prevent reignition of the arc on converter restart. Higher line recharge voltages were found to require slightly longer deionization periods.
  (iii) A minimum interruption time of 100–150 ms may be sufficient to restore full d.c. transmission if the fault is of a temporary nature.

In these tests by Kohler, the d.c. line was not simulated, so that the only storage element involved was the d.c. smoothing reactor. For this reason, relatively quick extinction was achieved—of the order 20–30 ms. However, when the line storage elements are represented, the extinction time will be longer, depending on the decay of the energy stored in it.

The values used for earth resistance will then greatly affect the time necessary to achieve this current extinction, since as discussed above, the magnitude of resistance in the fault circuit will determine how quickly the naturally oscillating fault voltage (and fault current) envelope decays. In general, earth resistances of at least the magnitude of the tower footing resistance must be included, except that the towers may be connected in parallel by earth wire. Typically values between 5 and 20 Ω can be expected.

Since, as detailed above, the arc voltage drop is relatively constant, the arc resistance is dependent on the instantaneous arc current. Therefore, the arc resistance must be calculated as a function of the instantaneous arc current.

The fault impedance is predominantly resistive and therefore the fault voltage and current will be in phase. With a naturally oscillating current waveform, the period of which is dependent on the line parameters (normally in the range 10–60 Hz), there will be several instances at which the arc may extinguish. However, if the voltage impressed across the still ionized arc path has sufficient peak amplitude to sustain further arc current, i.e. the voltage at the fault location is greater than 2 kV per metre of arc, then the arc path will continue to conduct current. Once the maximum value of voltage (line–earth value) is insufficient to re-ignite the still ionized arc path after current has passed through zero, then the current will remain at zero. The deionization time, representing the period taken for the arc path to deionize, will then be measured from this time on.

The majority of h.v.d.c. schemes, including the New Zealand scheme considered here, have higher direct voltage and direct current ratings than the test system used by Kohler. Therefore, Uhlmann's[8] more conservative recommendation of 100 ms is used as a safe deionization interval in this chapter.

After this time, the arc path will generally be sufficiently deionized to withstand normal voltage. During the arc extinction and deionization periods, the rectifier maintains a firing angle in the range 120 to 135 degrees. Normal inverter C.C. control action would have driven the inverter delay angle towards its minimum in an attempt to maintain link current at K_{di}^{sp}.

In the New Zealand scheme, the presence of a large capacitance (submarine cable) at the end of a long overhead line can create line voltage oscillations when the inverter firing angle is advanced (by C.C. control) towards 90 degrees and these oscillations can force the line voltage to change polarity following line fault occurrence. Furthermore, resonance effect between the d.c. line and the control system may sustain the line voltage oscillations, thus extending arc extinction times. A protection scheme which overrides the inverter C.C. control and maintains E.A. control reduced this problem considerably.

Converter restart

On completion of the deionization period, the program initiates the restart procedure for the link. In most cases, the line is restored to normal voltage and prefault power. If re-energization at full voltage is not acceptable (e.g. due to wet or dirty insulators), then a lower voltage may be used by bypassing one or more of the converter bridges, or by the use of constrained control action, i.e. larger minimum limits on α_r and γ_i.

Transmission cannot restart itself, since operation at the completion of deionization is with both converters in inversion, i.e. $\alpha_r = 125°$ and $\gamma_k = \gamma_0$. A starting order is therefore needed to remove the protective control action and to release the normal control systems of the converters. The actual time required for the d.c. link to regain nominal voltage and current will depend on the properties of the d.c. line and the converter controls. Generally, fast restart with a speed of power recovery conducive to stable operation, but with avoidance of overswings in both a.c. voltage and current, is required. Oscillations of these parameters may otherwise occur at the natural frequency of the line, and may be sufficient to cause arc restrike or to put additional stress on other plant. For this reason, restart by an instantaneous release of the normal controllers is not practical.

To avoid overswings in d.c. voltage (and current), it is recommended to increase the direct voltage exponentially to the nominal value, with minimum overshoot, rather than as a sudden step. This unit usually operates only for start-up or re-energization following a line fault. The time constant of the exponential increase must be long compared to the natural frequency of the line (e.g. 100 ms, cf. 20 ms), to ensure stable, controlled recovery of the d.c. system from the fault condition is obtained. A similar method is used in practice on the New Zealand scheme with an equivalent time constant near 100 ms.

If the line fault persists, or the arc restrikes due to inadequate deionization, the above process may be repeated a few times with increased deionization periods and/or reduced restart voltages, and then, if it still persists, the faulted pole removed from service.

12.8. RESULTS

Fault application

Once satisfactory initial conditions for the test system are achieved (requiring about two cycles of dynamic simulation) a pole-to-pole short circuit adjacent to the line side of the rectifier smoothing reactor is applied, immediately after a valve firing in one of the rectifier bridges. Thus, for twelve pulse operation the fault occurs 30 degrees before any retard action becomes effective. Initiation of the fault is implemented by alteration of the relevant connection matrix element ($K_{\alpha r}$ or $K_{\alpha l}$) for an already established branch element, or by insertion of a new branch element (R_r) and reformation of $K_{\alpha r}$.

Fault detection

In the state-variable formulation, the detection principle is implemented by monitoring the rates of change of the state variables ψ_l and Q_α on each pole of the d.c. line at each converter terminal, where l is the smoothing reactor branch and α is the node relating to the converter's line side termination. Equation (12.6.1) may then be rewritten as

$$\eta = K_5 p\psi + K_6 pQ_\alpha \tag{12.8.1}$$

An indication of the relative detection times for the two converter units is shown in Figs. 12.10(a) (converter direct currents) and (b) (line side converter voltages). An indication of the travelling wave initiated by the fault can be obtained from this graph, as can the natural frequency of the d.c. line. The rectifier detection unit is seen to respond almost instantaneously to the fault, as the voltage gradient is very high. The protective action initiated, with the next firing pulse retarded, is seen to be effective after 30 degrees, as predicted. This is indicated by a change in the rate of rise of rectifier current. The time delay associated with the travelling waves on the line is indicated by the inverter detection delay of around 5 ms between fault initiation and identification. Inverter current is seen to extinguish after approximately 8 ms, and the rectifier current 6 ms later. The oscillation of inverter line voltage is seen to be of about 50

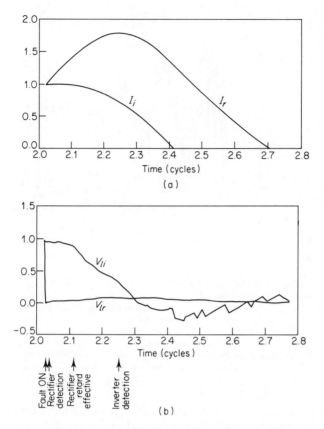

Fig. 12.10. Coordination of d.c. line fault detection action-detection responses

Hz frequency. The envelope of the fault voltage is seen to have a peak of around 0.15 p.u. or about 30 kV, which will promote a healthy arc.

Arc extinction and deionization

Prior to rectifier shut-down (which takes about 14 ms), the earth resistance has little influence on the fault voltage and energy state of the d.c. line.

Subsequent to rectifier turn-off, the only fault current source is the energy stored in the d.c. line, and the fault current will thus decay exponentially with a time constant dependent on the total circuit resistance.

Studies for earth resistances of 2, 5, 10 and 20 ohms were conducted; a typical fault current is illustrated in Fig. 12.11 and the extinction times achieved are summarized in Table 12.1.

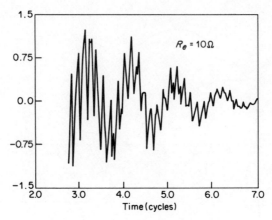

Fig. 12.11. Earth fault current

Table 12.1 Arc extinction times as a function of earth resistance

	Earth resistance (ohms)			
	2	5	10	20
Arc extinction times (ms)	110	100	90	80

It is apparent that each halving of the earth resistance increases arc extinction by approximately a half cycle. The arc extinctions occur at the zero crossings of arc current, provided the peak arc voltage is below the arc sustain level (approximately 2 kV/metre). The arc path resistance then increases effectively to infinity, since no current flows, and provided restriking does not occur, the fault branch may be removed or disconnected. This action is invoked at the completion of the specified deionization time, when it is assumed the arc path is completely deionized.

D.C. line recharge and power transmission restart

To recharge a d.c. line from $\alpha_r = 125°$ and $\gamma_i = \gamma_0$, α_r is reduced by a exponential control function, while maintaining $\gamma_i = \gamma_0$. This ensures the rectifier recharges the line fully before the inverter begins to draw power. This will mean that both the rectifier and inverter constant current controllers will be inoperative, since the inverter is on E.A. control and the rectifier on the start control function.

Following the completion of the deionization period, α_r is stepped from 125 to 90 degrees over one firing instant. Subsequent firing action is then controlled by the recharge function

$$\alpha_r = \alpha_0 + (90° - \alpha_0)\exp(-k\,\Delta t) \tag{12.8.2}$$

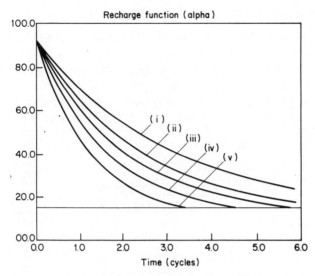

Fig. 12.12. Characteristics of exponential recharge k values: (i) 2.0; (ii) 2.4; (iii) 3.0; (iv) 4.0; (v) 5.0

where

α_0 is the control angle which will give nominal line voltage, and
Δt is the elapsed time since transfer of the converter control to the restart unit (i.e. from when $\alpha_r = 90°$).

The constant k will determine the response rate.

Figure 12.12 illustrates the variation of α_r for various values of k, as a function of elapsed time (Δt). In this case $\alpha_0 = 10°$. For a d.c. line which has a time constant of approximately 20–25 ms, it is desirable to have no more than 63 percent of nominal voltage when $\Delta t = 25$ ms, in order to obtain critical damping of the no-load voltage rise. Curve (ii) exhibits this characteristic, and actually achieves nominal voltage when Δt is approximately 100 ms.

The recharge process is thus implemented by adjusting the firing instant F_{j+1} by

$$\Delta P_{j+1} = \alpha'_{j+1} - \alpha_j \qquad (12.8.3)$$

where

$$\alpha'_{j+1} = 10 + 80 \exp(-2.4\,\Delta t) \qquad (12.8.4)$$

and where Δt is measured to F_{j+1}.

When the d.c. line is fully recharged to nominal voltage at the correct polarity, conditions are conductive to resumption of normal power transfer. At this time α_r will be close to nominal and δ_i will be on its specified optimum value. The rectifier then switches from the line recharge controller to C.C. control, whilst the inverter continues on E.A. control.

Full simulation of a d.c. line fault

The temporary bipole fault under consideration is now followed from the instant of fault initiation through to full fault recovery and normal power transmission, using the TCS and TS programmes described in previous chapters according to the protection philosophy discussed in the above sections. A flow diagram indicating the coordination of the special control functions used for protection and recovery of d.c. line faults is shown in Fig. 12.13. The results are indicated in Figs. 12.14(a)–(g).

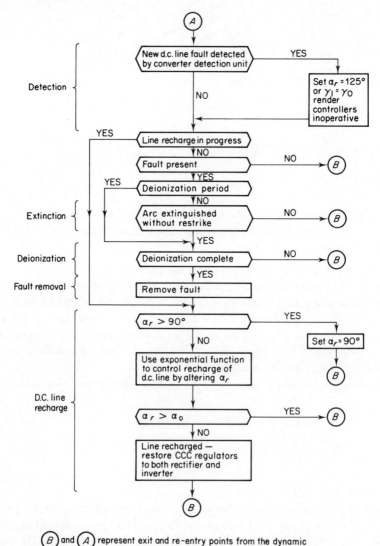

$\textcircled{B}$ and $\textcircled{A}$ represent exit and re-entry points from the dynamic simulations main controlling subroutine

Fig. 12.13. Flow chart of coordination of special control characteristics for d.c. line fault protection

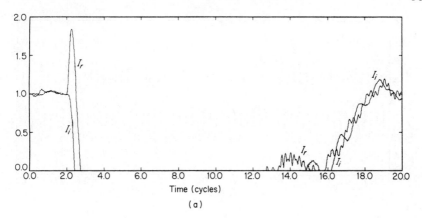

(a)

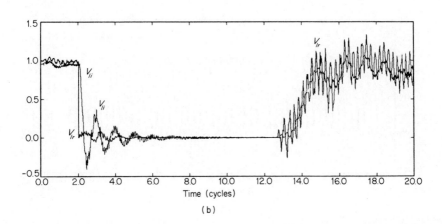

(b)

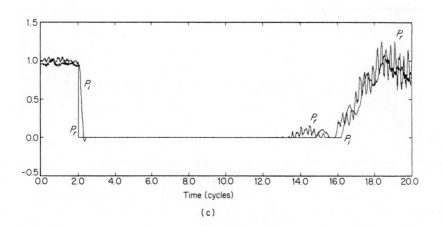

(c)

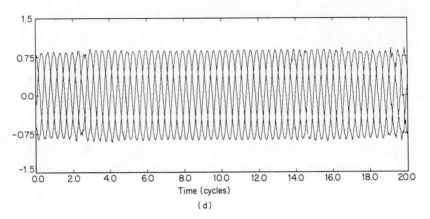

(d)

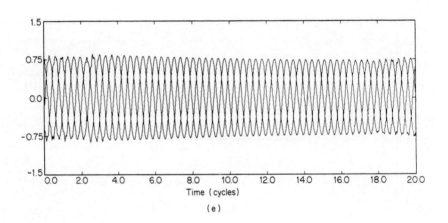

(e)

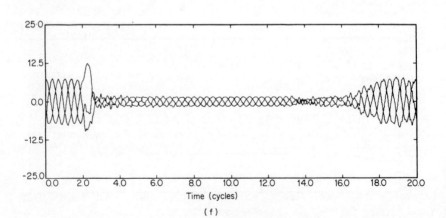

(f)

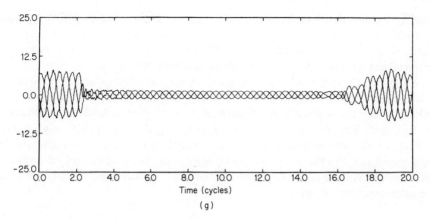

Fig. 12.14. (a) Converter d.c. currents; (b) Converter d.c. voltages; (c) Converter d.c. powers; (d) Rectifier a.c. voltages; (e) Inverter a.c. voltages; (f) Rectifier a.c. currents; (g) Inverter a.c. currents

The fault was applied and detected using the process described in Section 12.8. As α_r is retarded to 125°, Fig. 12.14(f) shows that two phases at the rectifier end continue to conduct current for an extended period. The initial rectifier current peak, which occurs 4 ms after the fault is applied, is only 80 percent above the setting and decays to zero, 15 ms after fault initiation. The inverter has already shut down by this time. Figure 12.14(b) illustrates the existence of voltage oscillations, with the inverter voltage reaching 50 percent of the rated voltage with reverse polarity. The oscillations die out within approximately 80 ms. Due to the continued presence of the harmonic filters, a residual current level remains throughout the fault (Figs. 12.14(f) and (g)), with considerable initial harmonic content causing corresponding a.c. voltage waveform distortion (Figs. 12.14(d) and (e)). This is more evident at the rectifier end due to the large initial increase in direct current and also due to the large change in control angle, which does not occur at the inverter end under E.A. control. The voltage distortion is seen to reduce quickly as the residual (filter) currents become settled. In each case this takes about five cycles, giving an indication of the relatively slow time response of the harmonic filters.

During the same period the line voltages are seen to decay, and as indicated in Fig. 12.14(b), extinction of the arc occurs some 80 ms after fault initiation.

During the deionization period, which extends for approximately 100 ms, the a.c. voltage waveforms become sinusoidal but some amplitude variation is noticeable in Figs. 12.14(d) and (e) as a result of the time-variant Thevenin sources.

Some 200 ms after fault initiation, the arc is expected to be completely deionized, the fault branch is removed and fault recovery commences. The rectifier direct current and both line voltages indicate the effect of the exponential recharge function. The current involved in line recharge is seen to be relatively small. Because the inverter is on E.A. control, the inverter does not conduct current until the line is charged to about 90 percent of nominal. The recharge time, measured from deionization until the line is recharged and converter current has begun to increase, is seen to be about 80 ms.

Following the recharge period, which is completed 280 ms after fault initiation, the converter currents are seen to increase quickly and steadily to the set values. This increase takes place with the rectifier on C.C. control, and the inverter on E.A. control. The current is seen to reach its setting after 60 ms with only a minor initial overshoot, indicating that the C.C. control constants used are adequate. At this time (cycle 19), the d.c. power levels are not at their prefault values since the inverter a.c. voltage is still depressed, although it is increasing steadily with the reinjection of link power.

12.9. REFERENCES

1. J. Arrillaga, M. D. Heffernan, C. B. Lake and C. P. Arnold, 1980. 'Fault studies in a.c. systems interconnected by h.v.d.c. links', *Proc. IEE*, **127**, Pt. C (1), 15–19.
2. H. A. Peterson, A. G. Phadke and D. K. Reitan, 1969. 'Transients in e.h.v.d.c. power systems: Part I—rectifier fault currents', *Trans. IEEE*, **PAS-88** (7), 981–989.
3. M. D. Heffernan, J. Arrillaga, K. S. Turner and C. P. Arnold, 1980. 'Recovery from temporary h.v.d.c. link faults', *Trans. IEEE*, Paper 80SM 675–679. PES Summer Meeting, Minneapolis.
4. N. G. Hingorani, 1970. 'Transient overvoltages on a bipolar h.v.d.c. overhead line caused by d.c. line faults', *Trans. IEEE*, **PAS-89** (4), 592–610.
5. E. W. Kimbark, 1970. 'Transient overvoltages caused by a monopolar ground fault on bipolar d.c. line: theory and simulation', *Trans. IEEE*, **PAS-89** (4), 584–592.
6. A. Erinmez, 1971. 'Direct digital control of h.v.d.c. links.' Ph.D. thesis, University of Manchester.
7. A. Kohler, 1967. 'Earth fault clearing on an h.v.d.c. transmission line with special consideration of the properties of the d.c. arc in free air', *Trans. IEEE*, **PAS-86** (3), 298–304.
8. E. Uhlmann, 1960. 'Clearing of earth faults on h.v.d.c. overhead lines', *Direct Current*, **5** (2), 45–57, 65–66.

Transient Stability Analysis of A.C.–D.C. Systems

13.1. INTRODUCTION

Having developed a detailed transient converter simulation program (TCS) suitable for analysing the behaviour of an h.v.d.c. link just after a disturbance, it is only necessary to merge this model with a transient stability (TS) program to complete the suite of programs. It is then possible to examine the response of all system components over a period of several seconds.

Many of the problems associated with the merging of the two programs have been discussed in Chapter 11. This was done to ensure that acceptable a.c. system information was made available to the TCS programme. This chapter firstly deals with the processing of the information, as obtained from the TCS, into a form suitable for inclusion in the TS programme.

The second part of this chapter is concerned with three transient stability analysis studies of the New Zealand a.c.–d.c. power systems. These studies are used to illustrate principles as well as to justify the use of the integrated programs.

13.2. REQUIREMENTS OF THE TRANSIENT STABILITY PROGRAM

Transient stability (TS) studies are normally carried out on the assumptions of balanced and sinusoidal waveforms and in the presence of h.v.d.c. links such assumptions need to be extended to the voltage and current waveforms at the converter terminals. As explained in Chapters 11 and 12, part of the study requires detailed transient converter simulation, the results of which cannot be directly incorporated in a TS study.

It is therefore necessary to derive fundamental frequency quantities from TCS voltage and current waveforms before information can be transferred from the TCS to the TS study.

Two variables are needed to transfer the converter information from the TCS to the TS study. A number of alternatives are available, i.e. real power, reactive power, voltage magnitude and current magnitude. Various definitions have been offered for reactive power in the presence of distorted waveforms,[1][2][3] but a meaningful value of reactive power can only be obtained from sinusoidal components of voltage and current of the fundamental frequency. This fundamental frequency reactive power can only be derived from spectral analysis of the voltage and current waveforms and since both these variables can be used on their own, reactive power becomes redundant. The above problems do not exist in the definition of real power and further justification for its use as a variable is given later.

The choice of the second variable is between voltage and current magnitude. Since the converter acts as a current harmonic source, the current waveform, particularly during and immediately after a disturbance, exhibits greater distortion and modulation effects than the voltage waveform. This is illustrated in the typical waveforms of Fig. 13.1.

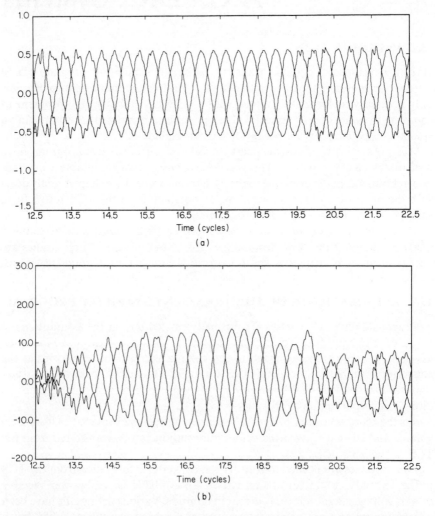

Fig. 13.1. Comparison between voltage and current waveforms at the rectifier terminal during current setting changes (from a TCS study). (a) Rectifier voltages; (b) Rectifier currents

The spectral analysis of the voltage waveform is, therefore, less prone to error and the voltage magnitude is a better choice of variable.

A specification of voltage and real power is required at each integration step of the TS study when the quasi-steady-state d.c. link model is replaced by the results of TCS.

To obtain fundamental components of voltage and power, the TCS waveforms must be subjected to Fourier processing over a discrete number of fundamental cycles of the waveforms. A component of voltage and power can be obtained at any sample point using a complete cycle (an aperture size of period T) spanning $T/2$ both sides of the sample point as illustrated in Fig. 13.2(a). The sampling points must be synchronized with the TS study integration steps and, in order to ensure that sufficient samples are used,[4] the TS program step length should not exceed half of one fundamental cycle. It is convenient, therefore, to fix the step length and sample rate at half a cycle of the fundamental frequency as illustrated in Fig. 13.2(b).

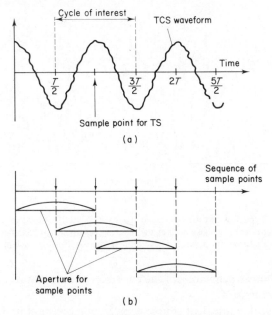

Fig. 13.2. Interaction between TS sample points and a TCS waveform. (a) TCS waveform; (b) Sample points

13.3. OBTAINING INFORMATION FROM THE TCS PROGRAM

The voltage waveforms at the terminals of a d.c. link during typical disturbances, such as those illustrated in Fig. 13.3 cannot be interpreted in terms of simple periodic waveforms.

The reasons for this are:

(i) Converter control action introduces deviations in the equidistant firing regime and valve firings do not necessarily occur periodically.

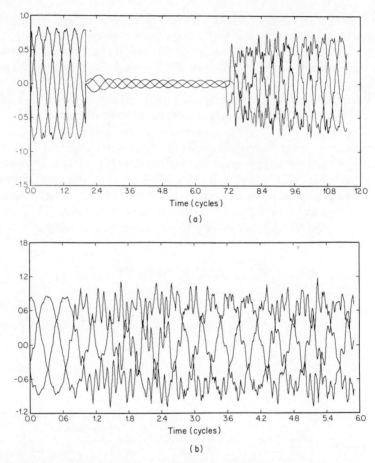

Fig. 13.3. Typical waveforms obtained from a TCS study. (a) Inverter voltage waveform for a three-phase fault at the inverter Terminal; (b) Rectifier voltage waveforms for a single-phase fault in the inverter a.c. system

(ii) The nonlinear behaviour of the converter excites disturbances which are not periodic in T.

(iii) The network equivalent source is given in a constant frequency frame of reference but the TS study deviates from constant frequency after a disturbance, effectively altering the fundamental period.

Spectral analysis

The above effects can be classified in terms of modulation, frequency mismatch and nonperiodic noise. Their presence means that a simple fundamental component of voltage (or current) cannot be accurately obtained merely by using a standard discrete Fast Fourier Transform (FFT),[5] on the samples which make up the TCS waveforms of Fig. 13.3.

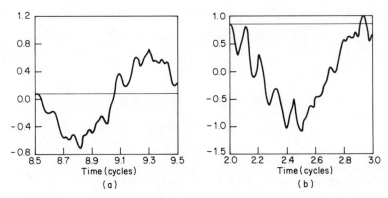

Fig. 13.4. Single cycles of one-phase obtained from TCS waveforms of Fig. 13.3 (a)
Inverter voltage; (b) Rectifier voltage

Figure 13.4 illustrates single cycles of distorted waveforms. The discontinuities at the boundaries of these individual cycles show that considerable levels of nonperiodic frequencies exist in the waveforms due to the nonlinearities associated with control action, filter response and converter behaviour. If these waveforms are assumed to be periodic and are subjected to Fourier transformation to obtain the spectral components of the signals (in particular the fundamental component), the phenomenon known as spectral leakage[6] occurs. Spectral leakage results in the nonperiodic noise contributing to each of the periodic spectral components present, introducing uncertainty in their identification by the Fourier transformation.

Work[7] has been done to determine the magnitude of the contribution of the non periodic noise. The results of over 100 different cycles of TCS waveforms indicate differences of up to 0.7 percent for very distorted waveforms. The maximum difference observed is 1 percent for the cycle illustrated in Fig. 13.4(b).

The differences are small and it can therefore be concluded that spectral leakage does not contribute significant errors when identifying the fundamental components of TCS waveforms.

It is therefore possible to obtain fundamental powers and voltages from TCS waveforms by simple application of an FFT, in the presence of modulation, frequency mismatch and non periodic noise. This can be done with the same degree of accuracy expected from other input data in transient stability studies.

Simplification in obtaining results

An approximation[7] can be used to reduce the effort of performing spectral analysis of the TCS waveform.

Apparent power is defined as:

$$s(t) = v(t) \cdot i(t) \tag{13.3.1}$$

The real component of $s(t)$ is the average value of $s(t)$ taken over a period, T, i.e.

$$P_{\text{rms}} = \frac{1}{T} \int_0^T v(t) \cdot i(t) \, dt \qquad (13.3.2)$$

This is readily evaluated from discrete signals using:

$$P_{\text{rms}} = \frac{1}{N} \sum_{n=1}^{N} v_n i_n \qquad (13.3.3)$$

where N is the number of samples.

The real power evaluated in this way is the sum of fundamental and harmonic real powers and the results[8] from many TCS's show that, in general, the harmonic real power is small and that the rms approximation for real fundamental power is a good one. Figure 13.5 shows two representative curves. In the case of Fig. 13.5(a), although the maximum error is 4 percent, the cummulative error over the simulation period is small. The maximum error occurs for a short period when highly distorted signals exist.

The harmonic power flow can only result from in-phase components of

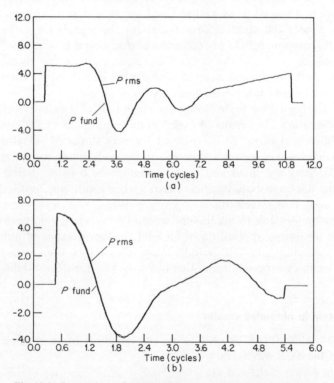

Fig. 13.5. Comparison of RMS and fundamental powers for the cases shown in Fig. 13.3. a) Three-phase fault at the inverter end b) Single-phase fault at the inverter end

harmonic currents and voltages. Considering the relatively low resistive component of the converter transformer and the large X/R ratio of the a.c. system, the in-phase component of the harmonic currents and voltages are very small. Therefore the use of rms power is a good approximation and spectral analysis is not needed to derive the power variable.

The rms value of a voltage, $V(t)$, is defined over a period T as

$$V_{rms} = \sqrt{\frac{1}{T} \int_0^T v(t)^2 \, dt} \qquad (13.3.4)$$

For discrete data this becomes:

$$V_{rms} = \sqrt{\frac{1}{N} \sum_{n=1}^{N} (V_n)^2} \qquad (13.3.5)$$

where N is the number of samples.

Figure 13.6 shows a comparison of fundamental voltage (positive sequence) and rms voltage for the TCS waveforms of Fig. 13.3. Figure 13.6(b) represents a

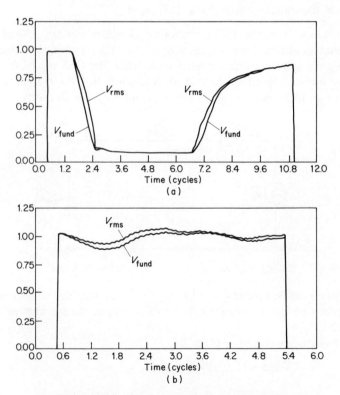

Fig. 13.6. Comparison of RMS and fundamental voltages for the TCS waveforms of Fig. 13.3. (a) Inverter voltage; (b) Rectifier voltage

severely distorted case with a 5 percent difference observed during the period immediately after disturbance initiation. This coincides with the period of maximum distortion in Figure 13.3(b) at which time harmonic content is high. This error is too large to be considered insignificant when transferring data from the TCS to the TS programme.

It is therefore necessary to retain the spectral analysis to determine a fundamental voltage magnitude from the transient converter simulation waveforms but the rms approximation for power avoids the need for spectral analysis of current.

13.4. INCORPORATING TRANSIENT CONVERTER SIMULATION IN A TRANSIENT STABILITY STUDY

When a detailed transient converter simulation has been performed on a system, the TS compatible equivalents derived from the results can be used to replace the QSS model during the most disturbed part of the study. When the response of the d.c. link has settled the TCS is terminated. The TS study then reverts back to using the QSS d.c. link model for the remainder of the study.

Solution at the converter node using TCS input

The information from the TCS is embodied in real power (P_{dyn}), and voltage, (E_{dyn}), variables obtained from processing of the TCS waveforms. These become the specifications at the link terminal node when the QSS model is removed. Using the unified algorithm developed in Section 10.5 the system equivalent shown in Fig. 13.7 is solved, given the P_{dyn} and E_{dyn} specifications.

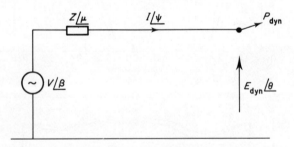

Fig. 13.7. Equivalent system for including TCS data

This system can be used to solve directly for the a.c. currents which can then be superimposed on the a.c. network in the same way as the a.c. current from the QSS model.

The complex power $S\underline{/\phi}$ is given by

$$S\underline{/\phi} = (E_{\text{dyn}}\underline{/\theta})(I\underline{/-\psi}) \tag{13.4.1}$$

or

$$S\underline{/\phi} = (E_{\text{dyn}} \cdot V\underline{/\theta + \mu - \beta})/Z - (E_{\text{dyn}}^2\underline{/\mu})/Z \tag{13.4.2}$$

Taking real parts of equation (13.4.2) and rearranging, θ can be obtained directly by:

$$\theta = \beta - \mu + \cos^{-1}[(P_{dyn}Z)/(E_{dyn} \cdot V) + (E_{dyn} \cdot \cos\mu)/V] \qquad (13.4.3)$$

The a.c. current can now be obtained using:

$$I\underline{/\psi} = (V\underline{/\beta} - E_{dyn}\underline{/\theta})/Z\underline{/\mu} \qquad (13.4.4)$$

This solution requires little computational effort when compared to the QSS link model. With very low converter terminal voltages, the first term of equation (13.4.3) becomes very large and any mismatch between E_{dyn} and P_{dyn} can cause the cosine function to exceed unity. This difficulty is overcome by releasing the power specification and allowing it to change until the equation can be solved. The small adjustment of power setting has little effect on the transient stability analysis, since the converter power involved is small under very low voltage operating conditions.

Alignment of TCS input with TS

The TCS results may be reduced to a sequence of data points, spaced half a cycle apart and each representing the fundamental component over the cycle spaced about that point. The integration step of the TS study should then be fixed at half a cycle to coincide with the data points from the TCS.

In the TS study, at a disturbance initiation, a new solution is obtained for the system immediately after the disturbance application. However, because the TCS results are essentially waveform orientated, a correct data point cannot be obtained at the new solution after disturbance application. Figure 13.3(a) shows sudden discontinuities at approximately two and seven cycles and Figure 13.6(a) demonstrates the averaging effect of processing the waveform results.

It is therefore necessary to extrapolate the processed TCS data points to provide, as closely as possible, a representation of the link performance immediately before or after a disturbance application. This effect is demonstrated in Fig. 13.8 in which the discrete integration steps of the TS study are related to the data points obtained from TCS.

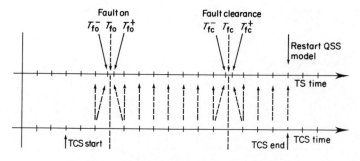

Fig. 13.8. Data alignment between TCS and TS program

Fault clearance requires special consideration as the mechanism differs from the two models. In the TS model the inherent assumption is made that at the time of fault clearance, T_{fc}, all three phases clear simultaneously and a new solution is obtained immediately (i.e. at T_{fc}^+). For the TCS, although the decision is made to clear the fault at T_{fc}, only one phase will open within 60 degrees of T_{fc} (i.e. at the first current zero), the remaining phases opening in the following 120°. This would appear to distort the alignment of the two models for the TS solution at T_{fc}^+.

However, the disturbance caused by the first phase clearance has the greatest influence on d.c. link control, and in spite of the TCS clearing delay, the post fault TCS data, obtained by processing one full cycle after T_{fc}, can be used to superimpose the immediate effect of the fault clearance on the TS solution at T_{fc}^+.

At the first and last points of the TCS data sequence, the QSS model is solved as well, to provide a check on the accuracy of matching between the two programs and the two link models.

13.5. MODIFICATION OF TRANSIENT STABILITY PROGRAM FOR A.C. FAULTS

The modelling of a.c. faults takes different forms in the TS and TCS programmes. For the TS program described in Chapter 9, the fault admittance is included directly in the admittance matrix of the network when the fault is to be present, i.e. the fault bus is included implicitly in the equation

$$[\bar{V}] = [\bar{Y}]^{-1}[\bar{I}] \tag{13.5.1}$$

Alternatively, the fault can be separated from the admittance matrix to obtain a solution for the fault current using the Thevenin equivalent at the faulted node. The fault current can then be superimposed onto the network in the same way as the injected current obtained at the d.c. link nodes.

It is necessary to solve both the fault and the d.c. link simultaneously because of their mutual coupling shown in equations (12.2.2) to (12.2.4).

The simultaneous solution can be achieved by reformulating the d.c. link

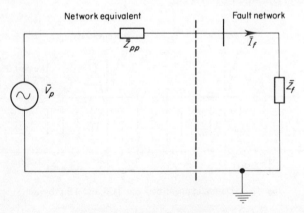

Fig. 13.9. Fault bus representation

model completely so that the fault equations are directly included in the Newton–Raphson algorithm. However, this special formulation would only be required to provide the time variant equivalent for TCS during the fault period. Thus the resulting extra computational demand should be avoided if possible.

Algorithm for simultaneous fault and link solution

The two different models can be retained by using a simple algorithm in the TS program which involves the simultaneous solution of the d.c. link model and the fault during the period when it is applied.

The solution of the fault is obtained by using the simple network of Fig. 13.9

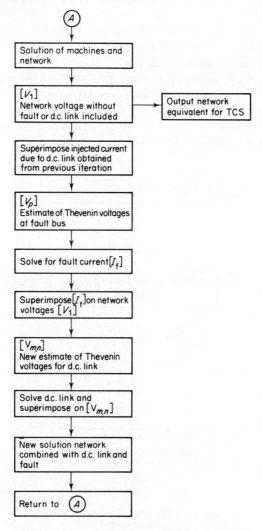

Fig. 13.10. Flow chart for simultaneous solution of fault and d.c. link

380

where $\bar{V}_p$ and $\bar{Z}_{pp}$ are the Thevenin voltage and impedance at the fault bus. The fault current is obtained from:

$$\bar{I}_f = \bar{V}_p/(\bar{Z}_{pp} + \bar{Z}_f) \qquad (13.5.2)$$

For this to be a correct solution, the Thevenin theorem states that all other voltage and current sources must be present, i.e., the network equivalent at the fault bus must be obtained with the d.c. link injected currents included. The converse also applies in obtaining the correct solution for the d.c. link. The solutions of both link and fault are combined in the algorithm of Fig. 13.10.

The algorithm relies on a good estimate of the d.c. injected currents at the first iteration of each step if the speed of convergence of the TS program is not to be impeded. However, the link solution obtained at the previous step is suitable except for the first iteration after fault application. In this case, since the fault has a dominant effect on the network, it is solved alone at the first iteration. A new d.c. link solution can then be obtained for entry into the algorithm (at A) of Fig. 13.10. This starting process is illustrated in Fig. 13.11.

The provision of a time variant system from the TS program is much simpler with a d.c. fault. For this case, it is not necessary to provide explicit fault bus representation since the fault is contained within the d.c. system being simulated. The equivalent provided for the TCS program in this case is exactly the same as that obtained for the QSS d.c. link model.

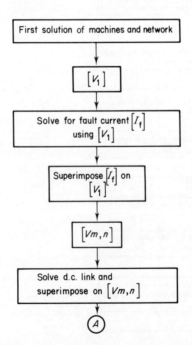

Fig. 13.11. Starting process after discontinuity

13.6. RESULTS USING THE COORDINATED PROGRAM

In this section, the a.c. and d.c. system responses to typical disturbances, using the coordinated program, are compared with those obtained using the QSS converter model. From the results, it is possible to assess the validity of the QSS model used in the TS program. Techniques adopted to ensure accuracy and economy during the running of the coordinated program are also discussed wherever necessary.

The New Zealand a.c. power system used for these studies consists of the South Island primary system as shown in Fig. AII.2 and a simple Thevenin equivalent for the North Island primary system in parallel with the synchronous condensers at the inverter terminals. The QSS model represents each d.c. link converter by a single bridge equivalent while for the TCS the d.c. link is represented using a twelve-pulse configuration as illustrated in Fig. 13.12.

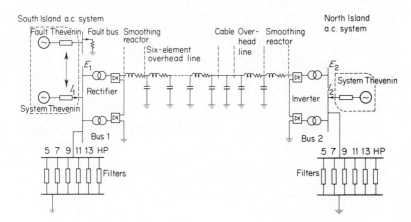

Fig. 13.12. D.c. link representation for transient converter simulation

In this representation, the transient stability program, using a QSS d.c. link model, provides the time variant system equivalents for the TCS. In turn the instantaneous values of I_1, I_2, E_1 and E_2 of Fig. 13.12 are obtained from the TCS to provide r.m.s. power and fundamental frequency voltage at the converter terminal bus, for use in the transient stability programme.

Three worst case studies consisting of a rectifier a.c. system (South Island) fault, an inverter a.c. system (North Island) fault and a d.c. system fault are evaluated.[8]

Rectifier a.c. system fault

A three-phase fault of five cycle duration is applied at the Twizel bus of the South Island a.c. system (see Fig. AII.2). The fault produces a 25 percent drop in a.c. voltage at the rectifier terminal immediately after fault application but represents a severe disturbance to the South Island system because of its position.

The voltage and reactive power profiles during the study are given in Fig. 13.13.

382

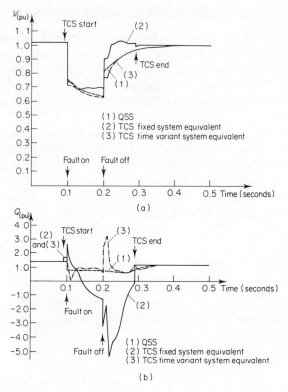

Fig. 13.13. Voltage and reactive power at the rectifier for the rectifier a.c. system fault. (a) Terminal a.c. voltage; (b) Reactive power

Three sets of results are shown, the first utilizes the QSS model alone without any use of TCS. The second uses the TCS together with the TS simulation but there is no coordination between the two parts of the program and thus fixed a.c. system equivalents are used in the TCS. The third set of results is obtained when the two simulations are coordinated and the TCS has a time variant a.c. system equivalent.

For the second case, the power factor of the injected current is forced to change from lagging to leading as a result of specifying voltage and real power at the converter node. This results in the impossible condition of the converter supplying reactive power to the system. The unsuitability of this form of modelling is further emphasized during the postfault period where the TCS recovery voltage considerably exceeds that of the dynamically responsive system.

The third set of results using interactive coordination gives a much better match between the QSS model performance and the TCS results. The departure between the two models immediately after fault inception and fault removal is due mainly to the different filter models. In the QSS model, filters are represented by a shunt admittance while the TCS reflects the full dynamic response of the filters. For the TCS representation, the combination of the dynamic filter response and control action results in an initial drop in reactive power demand from the a.c. system. The mismatch is more pronounced following fault clearance.

A further cause for mismatch is the requirement to align the TCS results with the TS integration steps, at the disturbance times, as shown in Fig. 13.8. This distorts the behaviour of the TCS results immediately after fault removal.

Although the reactive power profile is useful in indicating the accuracy of matching between the two programes, it is the real power transferred by the d.c. link which affects transient stability analysis. Fig. 13.14 shows the real power flow at the link terminals. The inverter power (Fig. 13.14(b)) is shown with positive power flow out of the link terminal to facilitate a comparison between the two ends of the link.

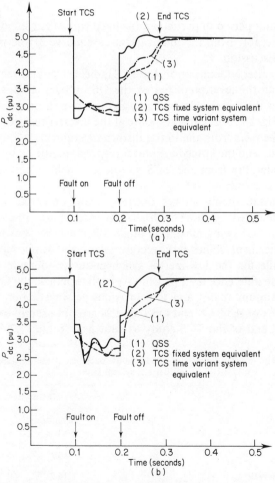

Fig. 13.14. Real power flows at d.c. link terminals for the rectifier a.c. system fault. (a) Rectifier; (b) Inverter

384

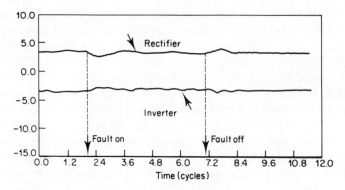

Fig. 13.15. Converter d.c. currents for the rectifier a.c. system fault

In the postfault period in particular, the fixed equivalent simulation (curve 2), departs considerably from the time variant case (curve 3) and may be omitted from further discussion.

Immediately after fault inception, the TCS indicates that the rectifier d.c. power flow is less than the level predicted by the QSS model. This is caused by the sudden drop in the rectifier d.c. voltage, giving rise to a drop in d.c. current as illustrated in Fig. 13.15. Control response at the inverter restores the d.c. current and the link recovers from the initial disturbance quickly.

At the inverter end the immediate power response predicted by the QSS model deviates considerably from the TCS solution, which reflects the delay effect caused by the d.c. line.

Although instantaneous power differences are observed, the nett energy transfer as calculated by the TCS and QSS models are similar, thus causing relatively small first swing maximum angle differences as shown in Table 13.1. For the high-inertia machines at the rectifier terminals (Benmore) the difference is 3 percent, while for the low-inertia synchronous condensers at the inverter (Haywards), the difference is 12 percent. At both terminals the QSS model gives the larger maximum angles, and thus provides pessimistic but safe results.

The solutions of the TCS and QSS models are not exactly coincident at the beginning and end of the TCS study as shown in Table 13.2. The results are

Table 13.1. Maximum swing angles of synchronous machines at the d.c. link terminals for the rectifier a.c. system fault

Case	Maximum swing angle	
	Benmore (rectifier)	Haywards (inverter)
(1) QSS	16.9°	− 8.1°
(2) TCS fixed equivalent	14.7°	− 5.9°
(3) TCS variant equivalent	16.4°	− 7.1°

Table 13.2. Coincidence of solutions at start and end of TCS for the rectifier a.c. system fault

| | | Link power (p.u.) | | Link a.c. voltage (p.u.) | |
		Rectifier	Inverter	Rectifier	Inverter
Start	QSS	5.00	−4.76	1.025	0.980
	TCS	4.99	−4.78	1.022	0.979
End	QSS	4.63	−4.42	0.964	0.965
	TCS	4.51	−4.28	0.955	0.963

obtained at the last integration step of the TS program for which a TCS data point is available. A QSS solution of the link is obtained at this step but the injected a.c. currents are derived from the final TCS data point (refer to Fig. 13.8). In this case, the mismatches are small and maximum mismatch occurs for the inverter power and is approximately 3 percent of the QSS value.

Inverter a.c. system fault

Faults at inverter terminals provide a severe test for the d.c. link controls and often result in repeated commutation failures causing considerable disruption to the normal valve firing sequence at the inverter. This form of converter malfunction is of particular importance because it cannot be represented by the QSS model and a transient converter simulation is essential.

To provide a worst case study, a three-phase fault of five cycles duration is applied very close to the inverter terminals. Referring to Fig. 13.12, the fault is applied to Bus 2 and a small fault resistance (1 ohm) is included to provide sufficient source voltage for commutation to be maintained. The fault causes an 80 percent drop in voltage at Bus 2 immediately after fault inception.

Since the fault is applied at Bus 2, and the transient converter simulation models it explicitly, the TCS currents (I_2 of Fig. 13.12), include the fault current and the fault need not be applied in the a.c. system. Because the fault is resistive,

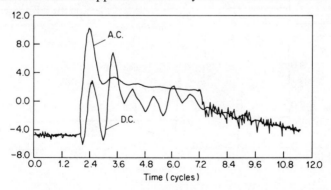

Fig. 13.16. A.c. and d.c. apparent power at inverter terminal for the inverter a.c. system fault

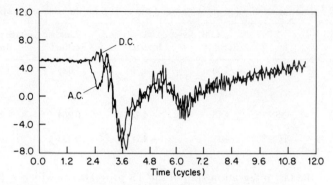

Fig. 13.17. A.c. and d.c. apparent power at rectifier terminal for the inverter a.c. system fault

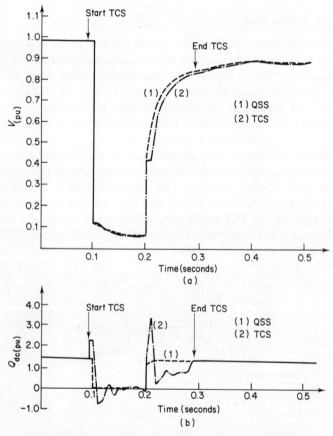

Fig. 13.18. Voltage and reactive power at the inverter for the inverter a.c. system fault. (a) Terminal voltage; (b) Reactive power flow

the real power obtained in deriving the TCS equivalents, includes the real power flowing into the fault. The difference between the a.c. power flowing into Bus 2, and the d.c. power flow at the inverter is illustrated in Fig. 13.16.

Immediately after the fault is applied, the inverter experiences repeated commutation failures which persist during the fault period. The commutation failures produce oscillations in the d.c. power at the inverter terminal, as shown in Fig. 13.16, illustrating the effect which they have on the d.c. link performance. A comparison between Fig. 13.16 and 13.17 shows that the d.c. power at the rectifier terminal is smoothed by the d.c. line response. The oscillations in Fig. 13.16 are therefore not due to the line.

Figure 13.18 illustrates the difference between the QSS and TCS a.c. voltage and reactive power at the inverter terminals. This case has a similar filter response to that for the rectifier a.c. fault study. In the post fault period the reactive power demands of the link are lower in the TCS results because of the slower link recovery for this model.

A good match is obtained for the voltages, during the fault period because of the dominance of the fault in the TCS results. In the postfault period the match is not so accurate because of the slower link response determined by transient converter simulation.

The active power flows at the two terminals of the link are shown in Fig. 13.19. The QSS model shows significant departures from the results obtained using TCS. The differences between the results are caused by three factors:

(a) Due to the d.c. line response, power continues to be fed into the rectifier end of the line for one cycle after the fault occurs at the inverter bus.

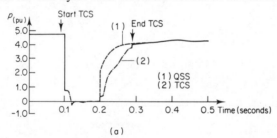

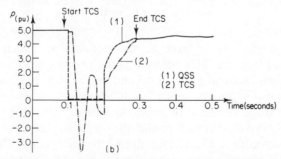

Fig. 13.19. Real power flows at the d.c. link terminals for the inverter a.c. system fault. (a) Inverter; (b) Rectifier

388

(b) At the inverter terminal, immediately before fault clearance takes place (i.e. the first detected current zero after breaker operation), a commutation failure occurs. This converter disturbance delays the clearance time of the fault. It is impossible for a QSS model to detect this sequence of events.

(c) In addition to the delay due to the commutation failure, inverter recovery is also affected by the response of the d.c. line and the controller at the rectifier end. As shown in Fig. 13.19(b) the TCS rectifier response is slower than the QSS response.

The accurate response obtained from the transient converter simulation affects the results of the TS analysis to a greater degree than for the rectifier a.c. system fault study, as shown by the swing curves in Fig. 13.20. In this case, the maximum swing angle obtained using the QSS model is less than that obtained using the TCS model (i.e. the results are optimistic) and this is especially noticeable for the low-inertia machines at the inverter terminal.

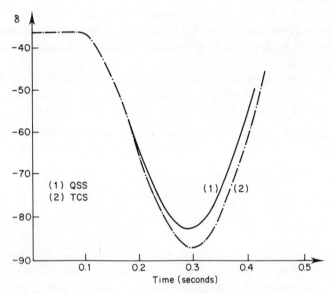

Fig. 13.20. Swing curves of Haywards (inverter) synchronous machines for the inverter a.c. system fault

The QSS model would have to be modified to match the response rate determined from the TCS before it could produce more accurate results. However, accurate matching of the QSS and TCS responses cannot be generally applied to this disturbance because of the indeterminate nature of consequential converter disturbances and their effect on normal valve firing sequences. In addition, the QSS model cannot predict the oscillatory response at the rectifier terminal, as illustrated in Fig. 13.19(b) during the fault period. This is due to the rapid rise in d.c. current in response to the collapse of d.c. voltage at the rectifier terminal. The oscillation, however, does not involve any significant nett energy transfer and thus does not affect the swing angle at the Benmore (Rectifier) generator significantly.

The peak swing angle at the Haywards (Inverter) synchronous machines occurs at 0.29 sec. and the voltage waveforms are close to sinusoidal at this time. This is thus a suitable time to terminate the TCS. The maximum difference between the TCS and QSS results at termination is 7 percent for the inverter power. Table 13.3 records the differences between the TCS and QSS variables.

Table 13.3. Coincidence of solutions at start and end of TCS for the inverter a.c. system fault

| | | Link power (p.u.) | | Link a.c. voltage (p.u.) | |
		Rectifier	Inverter	Rectifier	Inverter
Start	QSS	5.00	− 4.76	1.025	0.980
	TCS	5.02	− 4.79	1.024	0.979
End	QSS	4.27	− 4.04	0.981	0.840
	TCS	4.12	− 3.76	0.990	0.838

D.C. line fault

This study is of a short-circuit fault applied on the d.c. line just beyond the rectifier smoothing reactor and initiated immediately after a valve firing. There is thus no power transfer during the fault period. It is a worst case study since control action influence is delayed for 30° until the next firing.

The QSS d.c. link model has severe limitations in its ability to represent this type of disturbance. A d.c. fault, using this model, is represented by instantaneous shutdown of the link at fault initiation and restart when the fault is removed. Moreover, it yields unrealistic current peaks as it cannot accurately represent the timing of valve firings in relation to the fault occurrence, or the subsequent effects of rapid control action. It is also inadequate in representing the transient imbalance of harmonic currents between the converters and the d.c. system caused by the sudden fault current and subsequent control action.

The TCS results show that the a.c. fault current flows for half a cycle after fault initiation. During this period, the rectifier delay angle is retarded into the inverting region (between 120 and 130 degrees) to accelerate the d.c. line discharge. Since the fault is close to the rectifier, this effect is not observable on the d.c. power trace of Fig. 13.21, but is clearly shown in the a.c. power response.

Once control action has isolated the d.c. line, fault arc extinction is dependent on the natural oscillatory response of the d.c. line itself. This period combined with that required for deionization gives a minimum shutdown time of 220 ms or 11 cycles.[10]

At the end of the extinction and deionization period, transmission recovery begins and in the absence of prior knowledge of the TCS response, the QSS model produces instantaneous power recovery at this time, as illustrated by curve 1 of Fig. 13.22.

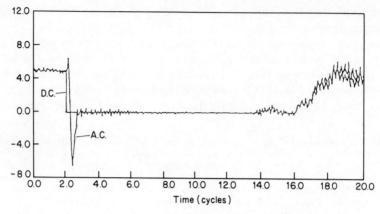

Fig. 13.21. A.c. and d.c. apparent power at rectifier terminal for d.c. line fault

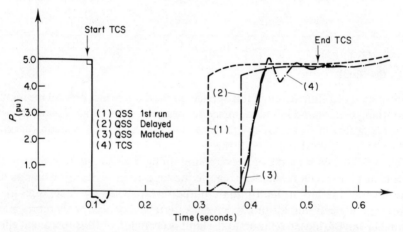

Fig. 13.22. Power flow to rectifier terminal for the d.c. line fault

However, as the TCS progresses, it becomes clear that transmission will not be re-established at this time and matching between the two programs will be poor. A number of subsequent equivalents may be obtained. The one adopted in this case has the restart delayed and is ramped to nominal d.c. current linearly over a two-cycle period. This equivalent provides excellent matching between the two models as illustrated by curve 3 in Fig. 13.22.

As shown in Fig. 13.22, there is a 7 percent overshoot in the TCS link power at full recovery. This is due to the finite rate of response of the controller. The simulation is continued after recovery for a further four cycles to ensure that the oscillation is damped and converges to the same conditions as the QSS model.

Throughout the d.c. fault there are no consequential converter disturbances and because of this, and the accurate matching achieved with the QSS model, the differences between the QSS and the TCS models are small.

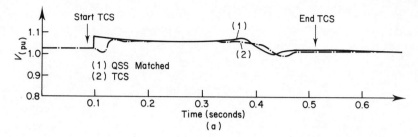

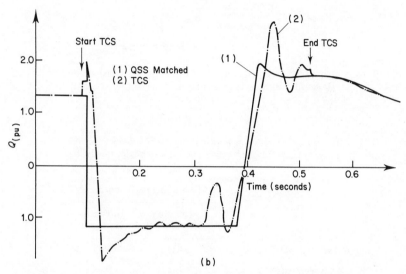

Fig. 13.23. Voltage and reactive power at the rectifier for the d.c. line fault. (a) A.c. terminal voltage; (b) Reactive power flow

Immediately after fault application, extra reactive power is required by the rectifier. This is due to the half cycle of fault current which flows and the associated commutation angle increase as shown in Fig. 13.23(b). This causes a small depression of voltage at the rectifier terminals, which is illustrated in Fig. 13.23(a).

Once the fault current is cleared by control action, this depression changes to a rise in voltage due to the extra reactive power available to the network from the filters.

Towards the end of the fault period, the link draws reactive power from the filters as the line is recharged. The oscillatory response after fault recovery rapidly decays as control action settles. Differences at the inverter terminal are minor.

The difference in swing angles of the machines near the d.c. link terminals for the two models is shown in Fig. 13.24. The results correspond to curves 1 and 4 of Fig. 13.22. In this case, there is a considerable error in the TS assessment with a 16 percent error in the rectifier a.c. system machine first swing peak of the QSS model.

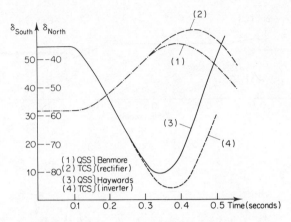

Fig. 13.24. Swing curves of Benmore (rectifier) and Haywards (inverter) synchronous machines for the d.c. line fault

Table 13.4. First swing maximum angles of Benmore (rectifier) and Haywards (Inverter) synchronous machines for the d.c. line fault

	First swing maximum angle	
	Benmore (Rectifier)	Haywards (Inverter)
QSS (1st run)	24.1°	45°
QSS (delayed)	29.1°	50.0°
QSS (matched)	29.0°	49.7°
TCS	28.7°	49.2°

However, as the matching between the QSS and TCS models is improved, the differences in peak swing angle are reduced. For curve 2 of Fig. 13.22, the difference is 2 percent and for curve 3 it is less than 1 percent. The maximum swing angles for these cases are included in Table 13.4. These differences are particularly small and are due in part to the improved matching between models and also to the fact that the peak swing of both machines occurs very close to the d.c. fault clearance time (0.43 sec. and 0.41 sec. respectively). Hence, any differences during the link restart period, i.e. ramped or instantaneous in the case of the QSS model, have little effect on the peak swing angle.

Conclusion

The three case studies presented in this section demonstrate the feasibility and effectiveness of combining both the TCS and TS programmes. Together they provide more accurate modelling of a.c.–d.c. systems disturbances than either programme can provide on its own. Detailed information can be obtained from the TCS regarding fault development, d.c. link response, converter performance, over voltages, over currents, etc.[9],[11]

The different results obtained from the TS analysis of the three cases investigated, indicates that a QSS model cannot be classified as giving either pessimistic or optimistic results for peak swing angles. This emphasizes the need to perform worst case transient converter simulations during investigations of a system so that the QSS link model performance can be modified to more accurately represent d.c. link performance for further studies. The system used for these case studies is relatively strong and a weaker system, where stability is more marginal, may produce much greater differences between the two models.

13.7. OVERALL CONCLUSIONS

In this final chapter, a compromise between the accurate but computationally prohibitive, transient converter simulation and the computationally efficient, but less detailed, transient stability program has been described. The advantage of both these methods can be retained by simultaneously performing transient stability and transient converter simulations with periodic coordination of the results. In this way, the detailed converter model is provided with a time-variant Thevenin equivalent for correct a.c. system representation as described in Chapter 11. Similarly, from the transient converter simulation, accurate power frequency information can be derived to represent the d.c. system behaviour at the converter terminals. The computational requirements are minimized by restricting the use of the transient converter simulation to the most disturbed part of the study. As the d.c. link regains normal controllability and attains predictable behaviour, this detailed simulation is replaced by the quasi-steady-state model.

The results using the coordinated program have demonstrated its viability and usefulness. The New Zealand primary transmission system used as a test, is a strong system with few stability problems. The results clearly illustrate the need for the more realistic modelling proposed in this chapter.

13.8. REFERENCES

1. W. Shepard and P. Zakikhani, 1972. 'Suggested definition of reactive power for nonsinusoidal systems', *Proc. IEE*, **119** (9), 1361–1362.
2. D. Sharon, 1973. 'Reactive power definitions and power factor improvement in nonlinear systems', *Proc. IEE*, **120** (6), 704–706.
3. E. Micu, 1973. 'Suggested definition of reactive power for nonsinusoidal systems', *Proc. IEE*, **120** (7), 796–797.
4. B. P. Lathi, 1968. *Communication Systems*. Wiley and Sons, New York.
5. W. T. Cochran, *et al.*, 1977. 'What is the fast fourier transform?' *Proc. IEEE*, **55** (10), 1664–1674.
6. F. J. Harris, 1978. 'On the use of windows for harmonic analysis with the discrete Fourier transform', *Proc. IEEE*, **66** (1).

7. K. S. Turner, M. D. Heffernan, C. P. Arnold and J. Arrillaga, 1981. 'Computation of a.c–d.c. system disturbances—Part II. Derivation of power frequency variables from converter transient response.' IEEE Winter Power Meeting, Atlanta, Ga.
8. K. S. Turner, 1980. 'Transient stability analysis of integrated a.c. and d.c. power systems.' Ph.D. thesis, University of Canterbury, New Zealand.
9. J. Arrillaga, M. D. Heffernan, C. B. Lake and C. P. Arnold, 1980. 'Fault studies in a.c. systems interconnected by h.v.d.c. links', *Proc. IEE*, **127**, Pt. C (1), 15–19.
10. M. D. Heffernan, K. S. Turner, J. Arrillaga and C. P. Arnold, 1981. 'Computation of a.c–d.c. system disturbances. Part I—Interactive coordination of generator and converter transient models.' IEEE Winter Power Meeting, Atlanta, Ga.
11. M. D. Heffernan, J. Arrillaga, K. S. Turner and C. P. Arnold, 1980. 'Recovery from temporary h.v.d.c. line faults.' IEEE Summer Power Meeting, Minneapolis, Ma.

Numerical Integration Methods

I.1. INTRODUCTION

Basic to the computer modelling of power-system transients is the numerical integration of the set of differential equations involved. Many books have been written on the numerical solution of ordinary differential equations, but this appendix is restricted to the techniques in common use for the dynamic simulation of power system behaviour.

It is therefore appropriate to start by identifying and defining the properties required from the numerical integration method in the context of power system analysis.

I.2. PROPERTIES OF THE INTEGRATION METHODS

Accuracy

This property is limited by two main causes, i.e. round-off and truncation errors. Round-off error occurs while performing artithmetic operations and is due to the inability of the computer to represent numbers exactly. A word length of 48 bits is normally sufficient for scientific work and is certainly acceptable for transient stability analysis. When the stability studies are carried out on computers with a 32-bit word length, it is necessary to use double precision on certain areas of the storage to maintain adequate accuracy.

The difference between the true and calculated results is mainly determined by the truncation error, which is due to the numerical method chosen. The true solution at any one point can be expressed as a Taylor series based on some initial point and by substituting these into the formulae, the order of accuracy can be found from the lowest power of step length (h) which has a nonzero coefficient. In general terms, the truncation error $T(h)$ of a method using a step length h is given by

$$T(h) = O(h^{p+1}) \tag{I.2.1}$$

where superscript p represents the order of accuracy of the method.

The true solution $y(t_n)$ at t_n is thus

$$y(t_n) = y_n + O(h^{p+1}) + \varepsilon_n \tag{I.2.2}$$

395

where y_n is the value of y calculated by the method after n steps, and ε_n represents other possible errors.

Stability

Two types of instability occur in the solution of ordinary differential equations, i.e. inherent and induced instability.

Inherent instability occurs when, during a numerical step by step solution, errors generated by any means (truncation or round-off) are magnified until the true solution is swamped. Fortunately transient-stability studies are formulated in such a manner that inherent instability is not a problem.

Induced instability is related to the method used in the numerical solution of the ordinary differential equation. The numerical method gives a sequence of approximations to the true solution and the stability of the method is basically a measure of the difference between the approximate and true solutions as the number of steps becomes large.

Consider the ordinary differential equation:

$$py = \lambda y \tag{I.2.3}$$

with the initial conditions $y(0) = y_0$ which has the solution

$$y(t) = y_0 e^{\lambda t} \tag{I.2.4}$$

Note that λ is the eigenvalue[1] of the single variable system given by the ordinary differential equation (I.2.3). This may be solved by a finite difference equation of the general multistep form:

$$\sum_{i=0}^{k} \alpha_i y_{n-i+1} - h \sum_{i=0}^{k} \beta_i p y_{n-i+1} = 0 \tag{I.2.5}$$

where α_i and β_i are constants.

Letting

$$m(z) = \sum_{i=0}^{k} \alpha_i (z)^i \tag{I.2.6}$$

and

$$\sigma(z) = \sum_{i=0}^{k} \beta_i (z)^i$$

and constraining the difference scheme to be stable when $\lambda = 0$, then the remaining part of (1.2.5) is linear and the solutions are given by the roots z_i (for $i = 1, 2, \ldots, k$) of $m(z) = 0$. If the roots are all different, then

$$y_n = A_1(z_1)^n + A_2(z_2)^n + \ldots A_k(z_k)^n. \tag{I.2.7}$$

and the true solution in this case ($\lambda = 0$) is given by

$$y(t_n) = A_1(z_1)^n + O(h^{p+1}) = y_0 \tag{I.2.8}$$

where superscript p is the order of accuracy.

The principal root z_1, in this case, is unity and instability occurs when $|z_i| \geq 1$ (for $i = 2, 3, \ldots, k, i \neq 1$) and the true solution will eventually be swamped by this root as n increases.

If a method satisfies the above criteria, then it is said to be stable but the degree of stability requires further consideration.

Weak stability occurs where a method can be defined by the above as being stable, but because of the nature of the differential equation, the derivative part of (I.2.5) gives one or more roots which are greater than or equal to unity. It has been shown by Dalquist[2] that a stable method which has the maximum order of accuracy is always weakly stable. The maximum order or accuracy of a method is either $k + 1$ or $k + 2$ depending on whether k is odd or even, respectively.

Partial stability occurs when the step length (h) is critical to the solution and is particularly relevant when considering Runge–Kutta methods. In general, the roots z_i of (I.2.7) are dependent on the product $h\lambda$ and also on equations (I.2.6). The stability boundary is the value of $h\lambda$ for which $|z_i| = 1$, and any method which has this boundary is termed conditionally stable.

A method with an infinite stability boundary is known as A-stable (unconditionally stable). A linear multistep method is A-stable if all solutions of (I.2.5) tend to zero as $n \to \infty$ when the method is applied with fixed $h > 0$ to (I.2.3) where λ is a complex constant with $\text{Re}(\lambda) < 0$.

Dalquist has demonstrated that for a multistep method to be A-stable the order of accuracy cannot exceed $p = 2$, and hence the maximum k is unity, that is, a single-step method. Backward Euler and the trapezoidal method are A-stable, single-step methods. Other methods not based upon the multistep principle may be A-stable and also have high orders of accuracy. In this category are implicit Runge–Kutta methods in which $p < 2r$, where r is the number of stages in the method.

A further definition of stability has been introduced recently,[3] i.e. Σ-stability which is the multivariable version of A-stability. The two are equivalent when the method is linear but may not be equivalent otherwise. Backward Euler and the trapezoidal method are Σ-stable single-step methods.

The study of scalar ordinary differential equations of the form (I.2.3) is sufficient for the assessment of stability in coupled equations, provided that λ are the eigenvalues of the ordinary differential equations. Unfortunately, not all the equations used in transient stability analysis are of this type.

Stiffness

A system of ordinary differential equations in which the ratio of the largest to the smallest eigenvalue is very much greater than one, is usually referred to as being stiff. Only during the initial period of the solution of the problem are the largest negative eigenvalues significant, yet they must be accounted for during the whole solution.

For methods which are conditionally stable, a very small step length must be chosen to maintain stability. This makes the method very expensive in computing time.

398

The advantages of Σ-stability thus become apparent for this type of problem as the step length need not be adjusted to accommodate the smallest eigenvalues.

In an electrical power-system the differential equations which describe its behaviour in a transient state, have greatly varying eigenvalues. The largest negative eigenvalues are related to the network and the machine stators but these are ignored by establishing algebraic equations to replace the differential equations. The associated variables are then permitted to vary instantaneously.

However, the time constants of the remaining ordinary differential equations are still sufficiently varied to give a large range of eigenvalues. It is therefore important that if the fastest remaining transients are to be considered and not ignored, as so often done in the past, a method must be adopted which keeps the computation to a minimum.

I.3. PREDICTOR—CORRECTOR METHODS

These methods for the solution of the differential equation

$$pY = F(Y, X) \quad \text{with} \quad Y(0) = Y_0 \tag{I.3.1}$$

$$\text{and} \quad X(0) = X_0$$

have all been developed from the general k-step finite difference equation:

$$\sum_{i=0}^{k} \alpha_i Y_{n-i+1} - h \sum_{i=0}^{k} \beta_i F_{n-i+1} = 0 \tag{I.3.2}$$

Basically the methods consist of a pair of equations, one being explicit ($\beta_0 = 0$) to give a prediction of the solution at t_{n+1} and the other being implicit ($\beta_0 \neq 0$) which corrects the predicted value. There are a great variety of methods available, the choice being made by the requirements of the solution. It is usual for simplicity to maintain a constant step length with these methods if $k > 2$.

Each application of a corrector method improves the accuracy of the method by one order, up to a maximum given by the order of accuracy of the corrector. Therefore, if the corrector is not to be iterated, it is usual to use a predictor with an order of accuracy one less than that of the corrector. The predictor is thus not essential, as the value at the previous step may be used as a first crude estimate, but the number of iterations of the corrector may be large.

While, for accuracy, there is a fixed number of relevant iterations, it is desirable for stability purposes to iterate to some predetermined level of convergence. The characteristic root (z_1) of a predictor or corrector when applied to the single variable problem:

$$py = \lambda y \quad \text{with } y(0) = y_0 \tag{I.3.3}$$

may be found from:

$$\sum_{i=0}^{k} (\alpha_i - h\lambda\beta_i)z^{(k-i)} = 0 \tag{I.3.4}$$

When applying a corrector to the problem defined by equation (I.3.3) and rearranging equation (I.3.2) to give:

$$y_{n+1} = \frac{-\sum_{i=1}^{k}(\alpha_i - h\lambda\beta_i)y_{n-i+1}}{(\alpha_0 - h\lambda\beta_0)} \tag{I.3.5}$$

the solution to the problem becomes direct. The predictor is now not necessary as the solution only requires information of y at the previous steps, i.e. at $y_{n=i+1}$ for $i = 1, 2, \ldots k$.

Where the problem contains two variables, one nonintegrable, such that:

$$py = \lambda y + \mu x \quad \text{with } y(0) = y_0 \tag{I.3.6}$$

$$x(0) = x_0$$

$$0 = g(y, x) \tag{I.3.7}$$

then:

$$y_{n+1} = c_{n+1} + m_{n+1} \cdot x_{n+1} \tag{I.3.8}$$

where

$$c_{n+1} = \frac{-\sum_{i=1}^{k}[(\alpha_i - h\lambda\beta_i)y_{n-i+1} - h\mu\beta_i x_{n-i+1}]}{(\alpha_0 - h\lambda\beta_0)} \tag{I.3.9}$$

and

$$m_{n+1} = \frac{-h\mu\beta_0}{(\alpha_0 - h\lambda\beta_0)} \tag{I.3.10}$$

Although c_{n+1} and m_{n+1} are constant at a particular step, the solution is iterative using equations (I.3.7) and (I.3.8). Strictly in this simple case, x_{n+1} in equation (I.3.8) could be removed using equation (I.3.7) but in the general multivariable case this is not so.

The convergence of this method is now a function of the nonlinearity of the system. Provided that the step length is sufficiently small, a simple Jacobi form of iteration gives convergence in only a few iterations. It is equally possible to form a Jacobian matrix and obtain a solution by a Newton iterative process, although the storage necessary is much larger and as before, the step length must be sufficiently small to ensure convergence.

For a multivariable system, equation (I.3.1) is coupled with:

$$0 = G(Y, X) \tag{I.3.11}$$

and the solution of the integrable variables is given by the matrix equation

$$Y_{n+1} = C_{n+1} + M_{n+1} \cdot [Y_{n+1}, X_{n+1}]^t \tag{I.3.12}$$

The elements of the vector C_{n+1} are as given in equation (I.3.9) and the elements of the sparse M_{n+1} matrix are given in equation (I.3.10).

The iterative solution may be started at any point in the loop, if Jacobi iteration is used. Because the number of algebraic variables (X) associated with equations (I.3.1) or (I.3.11) are small, it is most advantageous to extrapolate these algebraic variables and commence with a solution using equation (I.3.11).

The disadvantage of any multi-step method ($k > 2$) is that is not self-starting. After a discontinuity $k - 1$, steps must be performed by some other self-starting method. Unfortunately, it is the period immediately after a step which is most critical as the largest negative eigenvalues are significant. As $k - 1$ is usually small, it is not essential to use an A-stable starting method. Accuracy over this period is of more importance.

1.4. RUNGE−KUTTA METHODS

Runge–Kutta methods are able to achieve high accuracy while remaining single step methods. This is obtained by making further evaluation of the functions within the step. The general form of the equation is:

$$y_{n+1} = y_n + \sum_{i=1}^{v} w_i k_i \tag{I.4.1}$$

where

$$k_i = hf(t_n + c_i h, y_n + \sum_{j=1}^{v} a_{ij} k_j) \quad \text{for } i = 1, 2, \ldots, v \tag{I.4.2}$$

$$\sum_{i=1}^{v} w_i = 1 \tag{I.4.3}$$

Being single-step, these methods are self-starting and the step length need not be constant. If j is restricted so that $j < i$, then the method is explicit and c_1 must be zero. When j is permitted to exceed i, then the method is implict and an iterative solution is necessary.

Also of interest are the forms developed by Merson and Scraton. These are fourth-order methods ($p = 4$) but use five stages ($v = 5$). The extra degree of freedom obtained is used to give an estimate of the local truncation error at that step. This can be used to automatically control the step length.

Although they are accurate, the explicit Runge–Kutta methods are not A-stable. Stability is achieved by ensuring that the step length does not become large compared to any time constant. For a pth order explicit method the characteristic root is:

$$z_1 = 1 + \sum_{i=1}^{p} \frac{1}{i!} h^i \lambda^i + \sum_{i=p+1}^{v} \frac{a_i}{i!} h^i \lambda^i \tag{I.4.4}$$

where the second summation term exists only when $v > p$ and where a_i are constant and dependent on the method.

For some implicit methods the characteristic root is equivalent to a Pade approximant to $e^{h\lambda}$.

The Pade approximant of a function $f(t)$ is given by

$$P_{MN}(f(t)) = \sum_{j=1}^{M} (a_j t^j) / \sum_{j=1}^{N} (b_j t^j) \qquad (I.4.5)$$

and if

$$f(t) = \sum_{j=0}^{\infty} (c_j, t^j) \qquad (I.4.6)$$

then

$$f(t) - P_{MN}(f(t)) = \left(\sum_{j=0}^{\infty} (c_j t^j) \sum_{j=0}^{N} (b_j t^j) - \sum_{j=0}^{M} (a_j t^j) \right) \Big/ \sum_{j=0}^{N} (b_j t^j) \qquad (I.4.7)$$

If the approximant is to have accuracy of order $M + N$ and if $f(0) = P_{MN}(f(0))$ then

$$\sum_{j=0}^{\infty} (c_j t^j) \sum_{j=0}^{N} (b_j t^j) - \sum_{j=0}^{M} (a_j t^j) = \sum_{j=M+N+1}^{\infty} (d_j t^j) \qquad (I.4.8)$$

It has been demonstrated that for approximants of λh where $M = N$, $M = N + 1$ and $M = N + 2$, the modulus is less than unity and thus a method with a characteristic root equivalent to these approximants is A-stable as well as having an order of accuracy of $M + N$.

I.5. REFERENCES

1. L. Lapidus and J. H. Seinfeld, 1971. *Numerical Solution of Ordinary Differential Equations*. Academic Press, New York.
2. G. Dalquist, 1963. 'Stability questions for some numerical methods for ordinary differential equatios', *Proc. Symposia in Applied Mathematics*.
3. V. Zakian, 1975. 'Properties of I_{MN} and J_{MN} approximants and applications to numerical inversion of laplace transforms and initial value problems', *J. of Math. Analysis and Applications*, **50** (1), 191–222.

Appendix II

Power-System Data

The programs described in the text have been applied to the New Zealand Electricity power system. The primary a.c. transmission systems for the North and South Islands are given in Figs. AII.1 and AII.2.

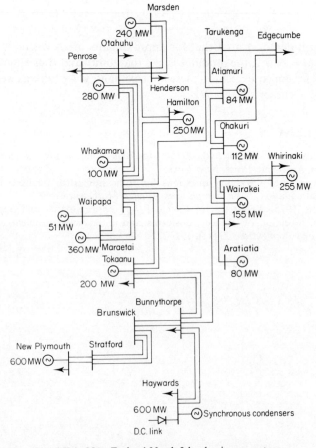

Fig. AII.1. New Zealand North Island primary system

402

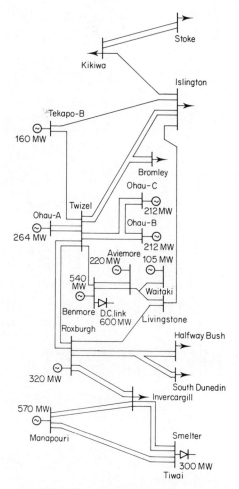

Fig. AII.2. New Zealand South Island primary
system

The data presented here represents the input and output of a transient stability study of the two primary a.c. systems plus the asynchronous h.v.d.c. interconnection. The a.c. fault is in the South Island (rectifier) system and the h.v.d.c. link is assumed controllable throughout the study.

The loading data in the input is the solution from a load-flow program which may be an integral part of the transient stability program or, as in this case, be run separately. Also note that output of the AVR and speed governor information has been restricted to two synchronous machines each.

Examples of the data as required for a short-circuit analysis program and for various load-flow programs are given in the relevant chapters.

ELECTRICAL POWER-SYSTEM TRANSIENT-STABILITY STUDY
DEPARTMENT OF ELECTRICAL ENGINEERING - UNIVERSITY OF CANTERBURY
CHRISTCHURCH, NEW ZEALAND.
26/03/81   10:14 AM

NEW ZEALAND 220KV SYSTEM NO 90
NORTH AND SOUTH ISLANDS WITH HVDC LINK

SYSTEM FREQUENCY =   50.0 HERTZ
M.V.A. BASE   = 100.0 M.V.A.

NO INDUCTION MOTOR DATA INPUT

STEADY-STATE SYSTEM DATA

BUSBAR DATA INPUT

BUSBAR	VOLTAGE	ANGLE	GEN MW	GEN MVAR	LOAD MW	LOAD MVAR
AVIEMORE-11	1.04000	12.6670	220.0000	-1.7000	0.0000	0.0000
AVIEMORE-220	1.04200	7.6930	0.0000	0.0000	0.0000	0.0000
BENMORE-16	1.04200	8.1430	540.0000	95.6000	0.0000	0.0000
BENMORE-220	1.04500	7.4920	0.0000	0.0000	0.0000	0.0000
DROXLET-220	1.00100	-9.8940	0.0000	0.0000	129.0000	38.0000
HALFWAYB-220	1.00200	-7.7430	0.0000	0.0000	183.0000	0.0000
INVERCAR-220	1.00800	-8.1630	0.0000	157.0000	594.0000	164.0000
AIKI4-220	1.00500	-24.0380	0.0000	0.0000	0.0000	0.0000
LIVINGSTN220	1.03800	-9.6240	405.0000	9.1000	0.0000	0.0000
MANAPOURI-11	1.04300	-6.9870	175.0000	11.7000	0.0000	0.0000
OHAU-A---220	1.04500	6.0900	214.0000	36.2000	0.0000	0.0000
OHAU-B---220	1.04500	9.6520	175.0000	0.0000	0.0000	0.0000
OHAU-C---220	1.04300	5.6070	175.0000	102.0000	0.0000	0.0000
OMAKAU---220	1.02350	6.8530	82.0000	36.0000	0.0000	0.0000
ROXBURGH-11	1.04300	-1.3530	160.0000	0.0000	0.0000	-20.0000
SOUTHDN-220	1.02700	-3.3570	0.0000	0.0000	33.0000	0.0000
TIWAI----220	1.04200	2.4020	0.0000	2.2000	0.0000	0.0000
TEKAPO---220	1.04300	29.8200	0.0000	0.0000	0.0000	0.0000
ISLINGTON220	1.00000	8.8600	0.0000	0.0000	420.0000	157.7000
HAYWARDS-220	1.05300	-5.7500	70.0000	0.0000	0.0000	0.0000
HAYWARDS-11	1.04200	-7.5500	70.0000	0.0000	0.0000	0.0000
HAYWARDS-14	1.03200	-29.8230	0.0000	0.0000	0.0000	0.0000
AMATIATIA220	1.02200	30.7880	50.0000	13.0000	0.0000	0.0000
AMATIATIA-11	1.01700	30.1720	0.0000	0.0000	0.0000	0.0000
WANGANUI-220	1.03000	7.7500	0.0000	0.0000	0.0000	0.0000
BUNNYTHPE220	1.02500	2.6260	0.0000	0.0000	66.0000	0.0000
EDGECUMBE220	1.06400	21.6920	0.0000	0.0000	135.5000	11.7000
CARRILL--220	1.05500	18.0250	157.0000	183.0000	416.0000	73.0000
HAYWRD-220	1.04400	-1.5580	0.0000	0.0000	0.0000	0.0000
MAYWARDS-220	0.98000	-4.2440	0.0000	0.0000	0.0000	0.0000

BRANCH DATA INPUT

SENDING BUSBAR	RECEIVING BUSBAR	RESISTANCE P.U.	REACTANCE P.U.	SUSCEPTANCE P.U.	TAP P.C.
BENMORE---1D	BENMORE---DC	0.00001	0.00001	0.00000	0.00
HAYWARDS-DC	HAYWARDS-220	0.00001	-0.99201	0.00000	0.00
HAYWARDS-DC	°	0.00325	-1.04250	0.02304	0.00
AVIEMORE-220	BENMORE---220	0.00325			

SHUNT IMPEDANCE DENOTED BY - *

SYNCHRONOUS MACHINE DATA

M/C PARAMETERS

M/C LOADINGS

M/C USING ALTERNATIVE REF. M/C DENOTED BY - A

AUTOMATIC VOLTAGE REGULATOR DATA

BUSBAR NAME	MC NO	AVR TYPE	FILTER T.CS.	REG. GAIN	REG. T.CS. TA	REG. T.CS. TB	REG. LIMITS MAX MIN	REG. RATE MAX MIN	EXCITER GAIN T.C.	F/BK GAIN	F/BK T.CS. TF	T.CS. TD	EXCITER LIMITS MAX MIN	SAT(NSE) .75MAX MAX

AVR'S CONTROL LOCAL BUSBAR EXCEPT WHERE OTHERWISE SPECIFIED

SPEED GOVERNOR AND TURBINE PARAMETERS

BUSBAR NAME	M/C TYPE NO.	SPEED REGULATION	CONTROL SYSTEM T.CS. T1 T2	CNTRL SYSTEM T.CS. T3	CNTRL LMTS.	TURBINE T.CS. T4	TURBINE LIMIT(MW)	VALVE/GATE OPN TIME (S)

H.V.D.C. LINK DATA

LINK PARAMETERS

1ST AC BUS NAME	2ND AC BUS NAME	RESISTANCE (PU)	(--- NOMINAL VOLTAGES (KV) ---) D.C. 1ST AC BUS 2ND AC BUS		CURRENT LMTS(PU) MAX MIN	
BENMORE---DC	MAYWARDS---DC	0.12780	100.000	105.000 105.000	1.500	0.200

LINK LOADING

AC BUS NAME	TYPE	LINK POWER INPUT MW MVAR	COMMUTATION REACTANCE(PU)	C MEG SLOPE	NO.OF BRDGS	CONTROLS SETTING MARGIN		ALPHA MIN /DELTA CM
BENMORE---DC	RECT	500.43000 244.20000	0.08972	25.00	7	CMNT 0.1200	CMNI(PU)	3.00
MAYWARDS--DC	INV	-476.59000 236.45000	0.07020	25.00	7	0.1200		18.00

SHUNT LOADS

BUSBAR NAME	MEGAWATTS	MEGAVARS	NON-IMPEDANCE TYPE LOAD (PC) VARYING WITH V P FIXED Q
BROMLEY--220	125.60000	38.30000	0 0 0 0
HALFWAYBU220	175.30000	40.30000	0 0 0 0
INVERCARG220	133.20000	42.30000	0 0 0 0
ISLINGTO220	500.10000	12.30000	0 0 0 0
SOUTHDUN-220	34.20000	-12.90000	0 0 0 0
STOKE----220	-53.20000	-25.20000	0 0 0 0
DUNI.TOPP220	460.90000	-56.10000	0 0 0 0
EDGECUMBE220	120.50000	11.70000	0 0 0 0
HAMILTON-220	135.50000	50.30000	0 0 0 0
HAYWARDS-220	100.00000	74.30000	0 0 0 0
HENDERSON220	206.00000	406.00000	0 0 0 0
NEWPLYMTH220	500.00000	30.00000	0 0 0 0
OTAHUHU--220	300.00000	40.30000	0 0 0 0
ROXBURG--220	15.70000	5.30000	0 0 0 0
WAIMAKA-I-220	200.00000	50.00000	0 0 0 0
WHIRINAKI220			

SYSTEM NO. 90   CASE NO. 1

REFERENCE M/C = 1ROXBURGH--11 NO.1
INITIAL STEP LENGTH(S) = 0.0000
STUDY DURATION (SEC) = 0.0000
PRINT INTERVAL (SEC) = 0.0000
MAX POLE PAIR SLIP LIMIT = 0.1330

PRINT INTERVAL CHANGES TO 0.200 SECS AT 0.60 SECS.
ROTOR BACK SWING INCLUDED ON FAULT APPLICATIONS

FAULT APPLICATIONS

FAULT TIMES ON OFF	LOCATION	FAULT TYPE	FAULT IMPEDANCE P.U. RESISTANCE	REACTANCE	EXACT LOCATION	RESISTANCE P.U.	REACTANCE P.U.	SUSCEPTANCE P.U.	TAP P.U.
0.1000 0.2400	TWIZEL---220	3 PHASE	0.00000	0.00000	BUSBAR	-	-	-	-

NO BRANCH SWITCHING OPERATIONS

INITIAL CONDITIONS

SYNCHRONOUS MACHINES

BUSBAR NAME	M/C NO.	ROTOR ANGLE (DEG)	REF. MAIN /ALT	POLE PRS SLPD	ROTOR SLIP P.C.	MECH. POWER MW	ACTIVE POWER OUTPUT MW	REACTIVE MVAR	TERM. VOLTAGE P.U.	TERM. CURRENT P.U.	FIELD VOLTAGE P.U.	FIELD CURRENT P.U.

REFERENCE MACHINE DENOTED BY - *   THE ANGLE OF THIS MACHINE IS RELATIVE TO SYSTEM ZERO

BUSBAR NAME	M/C NO.	CONTROLLED BUS VOLTS	I/P FILTER SIGNAL	AMPLIFIER SIGNAL	FEEDBACK SIGNAL	AUXILIARY SIGNALS SOURCE INPUT OUTPUT
AVIEMORE---18	1	1.040	1.040	1.216	1.049	0.000
BENMORE----16	1	1.045	1.025	1.482	1.025	0.000

AUTO VOLT REGULATORS -

SPEED GOVS + TURBINES -

BUSBAR NAME	M/C NO.	GOVERNOR SIGS(PU)	PUSN(PU) VALVE	HP VALVE PUSN(PU) PWR(MW)	I/C VALVE PUSN(PU) PWR(MW)	TURBINE POWERS (MW) HP	LP
AVIEMORE---11	1	0.000	0.000	0.907	220.814		
BENMORE---16	1	0.000	0.000	0.967	540.943		

H.V.D.C.LINK

AC BUSBAR NAME	TYPE	VOLTAGES(PU) AC DC	CURRENTS(PU) AC DC	POWER INPUT MW MVAR	LINK MW LOSS	FIRING ANGLES (DEG) ALPHA DELTA GAMMA	TYPE	REGULATOR SETTING
BENMORE--DC	RECT	1.465 5.264	5.404	590.228 244.325	23.436	16.1 18.0	CKN1	1.010 (PU)
MAYWARDS--DC	INV	1.700 4.979	5.404	-476.792 236.593		17.1 14.8		

BUSBAR VOLTAGES

BUSBAR	VOLTAGE	ANGLE		BUSBAR	VOLTAGE	ANGLE		BUSBAR	VOLTAGE	ANGLE		BUSBAR	VOLTAGE	ANGLE

MAXIMUM ERROR IN INITIAL CONDITIONS IS 0.601 MVA

SYSTEM NO. 90   CASE NO. 1   TIME = 0.100 S.   STEP LENGTH = 0.0100 S.   STEP NO.= 10   MAX.ITS/STEP = 3   ITS/PRINT = 22

SYNCHRONOUS MACHINES

BUSBAR NAME	M/C NO.	MOTOR FREQ.	REF. VALT/VALT	POLE SLPD	ROTOR SLIP.	MECH. POWER MW	ACTIVE POWER OUTPUT MW	REACTIVE MVAR	TERM. VOLTAGE P.U.	TERM. CURRENT P.U.	FIELD VOLTAGE P.U.	FIELD CURRENT P.U.

AUTO VOLT REGULATORS -

BUSBAR NAME	M/C NO.	CONTROLLED BUS VOLTS	I/P FILTER SIGNAL	AMPLIFIER SIGNAL	FEEDBACK SIGNAL	AUXILIARY SIGNALS SOURCE INPUT OUTPUT
AVIEMORE---11	1	1.040	1.040	1.219	0.000	

BENMORE---16	1	1.025	1.025	1.000	-0.002

SPEED GOVS + TURBINES -

BUSBAR NAME	M/C NO.	GOVERNOR SIGS(PU)		HP VALVE POSN(PU)	PWR(MW)	I/C VALVE POSN(PU)	PWR(MW)	TURBINE POWERS (MW) HP	IP	LP
AVIEMORE--11	1	0.002	0.000	0.907	220.815					
BENMORE---16	1	0.001	0.000	0.967	540.943					

H.V.D.C.LINK

AC BUSBAR NAME	TYPE	VOLTAGES(PU) AC	DC	CURRENTS(PU) AC	DC	POWER INPUT MW	MVAR	LINK MW LOSS	FIRING ANGLES (DEG) ALPHA	DELTA	GAMMA	REGULATOR TYPE	SETTING
BENMORE---DC	RECT	1.025	5.224	5.404	0.958	590.216	244.267	23.436	16.1		17.2	CRNT	1.010 (PU)
HAYWARDS--DC	INV	0.980	4.979	5.404		-476.781	236.589			18.0	14.8		

FAULT AT 'TWIZEL---220' OCCURRED AT 0.1000 S.

SYSTEM NO. 90 CASE NO. 1 TIME = 0.100 S. STEP NO.= 10 ITS TO RECONVERGE = 5

SYNCHRONOUS MACHINES

BUSBAR NAME	M/C NO.		ROTOR ANGLE (DEG)	REF. MAIN /ALT	POLE PRS SLPD	ROTOR SLIP P.C.	MECH. POWER MW	POWER OUTPUT ACTIVE MW	REACTIVE MVAR	TERM. VOLTAGE P.U.	TERM. CURRENT P.U.	FIELD VOLTAGE P.U.	FIELD CURRENT P.U.
AVIEMORE--11	1		27.96	M	0	-0.06	220.814	116.805	351.085	0.725	5.103	1.218	2.132
BENMORE---16	1		31.79	M	0	-0.04	540.942	396.079	773.779	0.754	11.535	1.444	2.718
ISLINGTON220	1		-27.01	M	0	-0.00	1.183	19.819	97.015	0.259	3.825	1.537	1.537
MANAPOURI-14	1		10.24	M	0	-0.01	400.749	359.279	422.333	0.977	5.673	1.321	1.711
OHAU-A----11	1		30.94	M	0	-0.12	214.625	35.988	591.777	0.471	12.677	1.383	4.882
OHAU-B----11	1		29.73	M	0	-0.12	175.447	26.134	503.655	0.470	10.725	1.357	4.900
OHAU-C----11	1		29.72	M	0	-0.12	175.446	28.426	507.892	0.484	10.504	1.363	4.833
ROXBURGH--11	1	*	16.22	*M	0	-0.06	82.925	108.023	442.596	0.773	5.692	1.262	3.383
TEKAPO-B--11	1		27.01	M	0	-0.12	160.258	40.023	353.927	0.430	8.284	1.312	4.031
WAITAKI---11	1		10.37	M	0	-0.03	70.253	44.737	166.051	0.705	2.440	1.150	1.898
ARATIATIA-11	1		55.25	A	0	0.00	70.350	69.790	3.360	1.030	0.679	1.235	1.241
ATIAMURI--11	1		54.69	A	0	0.00	50.170	49.875	14.151	1.048	0.495	1.244	1.251
HAMILTON--11	1		35.91	A	0	0.00	150.339	149.337	6.627	1.039	1.436	1.251	1.255
HAYWARDS-220	1	*	-4.25	*A	0	-0.02	0.209	73.950	292.267	0.949	3.177	1.623	2.013
MARAETAI--11	1		25.71	A	0	0.00	213.742	211.231	78.066	1.064	2.117	1.375	1.384
MARSDEN---11	1		23.27	A	0	0.00	150.112	149.593	67.760	1.070	1.535	1.532	1.533
NEWPLYMTH-11	1		52.99	A	0	0.00	400.297	402.873	119.194	1.062	3.958	1.647	1.682
OHAKURI---11	1		63.79	A	0	0.00	171.062	169.202	8.114	1.036	1.634	1.528	1.563
OTAHUHU---11	1		18.90	A	0	0.00	200.477	200.474	106.553	1.025	2.225	1.537	1.543
TOKAANU---11	1		33.22	A	0	0.00	180.595	180.928	31.599	1.057	1.737	1.276	1.288
WAIPA-A---11	1		30.33	A	0	0.01	40.275	40.089	9.124	1.060	0.388	1.335	1.340
WAIRAKEI--11	1		72.37	A	0	0.00	140.924	139.579	4.306	1.028	1.356	1.312	1.320
WHIRINAKI-11	1		65.54	A	0	0.00	161.223	159.774	23.092	1.029	1.569	1.352	1.358
WHAKAMARU-11	1		36.59	A	0	0.01	91.591	91.497	14.768	1.054	0.880	1.385	1.391

AUTO VOLT REGULATORS -

BUSBAR NAME	M/C NO.	CONTROLLED BUS VOLTS	I/P FILTER SIGNAL	AMPLIFIER SIGNAL	FEEDBACK SIGNAL	AUXILIARY SIGNALS SOURCE	INPUT	OUTPUT
AVIEMORE--11	1	0.745	0.725	1.219	0.000			
BENMORE---16	1	0.754	0.754	1.000	-0.002			

SPEED GOVS + TURBINES -

BUSBAR NAME	M/C NO.	GOVERNOR SIGS(PU)		HP VALVE POSN(PU)	PWR(MW)	I/C VALVE POSN(PU)	PWR(MW)	TURBINE POWERS (MW) HP	IP	LP
AVIEMORE--11	1	0.002	0.000	0.907	220.815					
BENMORE---16	1	0.001	0.000	0.967	540.943					

H.V.D.C.LINK

AC BUSBAR NAME	TYPE	VOLTAGES(PU) AC	DC	CURRENTS(PU) AC	DC	POWER INPUT MW	MVAR	LINK MW LOSS	FIRING ANGLES (DEG) ALPHA	DELTA	GAMMA	REGULATOR TYPE	SETTING
BENMORE---DC	RECT	0.754	3.917	5.243	0.929	363.910	158.827	22.057	3.0		30.4		
HAYWARDS--DC	INV	0.749	3.680	5.243		-341.853	364.999			42.4	8.7	CRNT	1.010 (PU)

SYSTEM NO. 90   CASE NO. 1   TIME = 0.200 S.   STEP LENGTH = 0.0100 S.   STEP NO.= 21   MAX.ITS/STEP = 7   ITS/PRINT = 53

SYNCHRONOUS MACHINES

BUSBAR NAME	M/C NO.	ROTOR ANGLE (DEG)	REF. POLE SLIP	ROTOR SLIP P.U.	MECH. POWER	ACTIVE POWER OUTPUT MW/MVAR	TERM. VOLTAGE P.U.	TERM. CURRENT P.U.	FIELD VOLTAGE P.U.	FIELD CURRENT P.U.

AUTO VOLT REGULATORS   BUSBAR NAME   M/C NO.   CONTROLLED BUS VOLTS   I/P FILTER SIGNAL   AMPLIFIER SIGNAL   FEEDBACK SIGNAL   AUXILIARY SIGNALS SOURCE INPUT OUTPUT

SPEED GOVS + TURBINES   BUSBAR NAME   M/C NO.   GOVERNOR SIGS(PU)   HP VALVE PUSH(PU) PSH(PU)   I/C VALVE PUSH(PU) PSH(PU)   TURBINE POWERS (MW) HP LP

H.V.D.C.LINK

AC BUSBAR NAME   TYPE   VOLTAGES(PU) AC DC   CURRENTS(PU) AC DC   POWER INPUT MW MVAR   LINK MW LOSS   FIRING ANGLES (DEG) ALPHA DELTA GAMMA   REGULATOR TYPE SETTING

SYSTEM NO. 90   CASE NO. 1   TIME = 0.240 S.   STEP LENGTH = 0.0100 S.   STEP NO.= 25   MAX.ITS/STEP = 4   ITS/PRINT = 16

SYNCHRONOUS MACHINES

BUSBAR NAME	M/C NO.	ROTOR ANGLE (DEG)	REF. MAIN PHS SLIP	ROTOR SLIP P.U.	MECH. POWER	ACTIVE POWER OUTPUT MW	REACTIVE MVAR	TERM. VOLTAGE P.U.	FIELD VOLTAGE P.U.	FIELD CURRENT P.U.

AUTO VOLT REGULATORS

BUSBAR NAME	M/C NO.	CONTROLLED BUS VOLTS	I/P FILTER SIGNAL	AMPLIFIER SIGNAL	FEEDBACK SIGNAL	AUXILIARY SIGNALS SOURCE INPUT OUTPUT
AVIEMORE---11	1	0.596	0.596	1.000	0.010	
BENMORE---16	1	0.596	0.596	1.000	0.004	

SPEED GOVS + TURBINES

BUSBAR NAME	M/C NO.	GOVERNOR SIGS(PU)	HP VALVE POSN(PU) PWR(MW)	I/C VALVE POSN(PU) PWR(MW)	TURBINE POWERS (MW) HP IP LP
AVIEMORE---11	1	-0.998 -0.006	0.902 619.440		
BENMORE---16	1	-2.305 -0.002	0.996 549.053		

H.V.D.C.LINK

AC BUSBAR NAME	TYPE	VOLTAGES(PU) AC DC	CURRENTS(PU) AC DC	POWER INPUT MW MVAR	LINK MW LOSS	FIRING ANGLES (DEG) ALPHA DELTA GAMMA	REGULATOR TYPE SETTING
BENMORE--DC RECT		0.396 3.029	5.193 5.193	278.688 138.200	21.037	3.0 52.2	CRNT 1.010 (PU)
HAYWRDS--DC INV		0.490 2.794		-257.051 300.555		34.5 8.1	CRNT 1.010 (PU)

FAULT AT 'TWIZELL---220' REMOVED AT 0.240U S.

SYSTEM NO. 70   CASE NO. 1   TIME = 0.240 S.   STEP NO.= 25   ITS TO RECONVERGE = 5

SYNCHRONOUS MACHINES

BUSBAR NAME	M/C NO.	ROTOR ANGLE (DEG)	REF. MAIN FAULT	POLE PMS SLFD	SLIP P.C.	MECH. POWER MW	POWER OUTPUT ACTIVE REACTIVE MW MVAR	TERM. VOLTAGE P.U.	TERM. CURRENT P.U.	FIELD VOLTAGE P.U.	FIELD CURRENT P.U.

AUTO VOLT REGULATORS

BUSBAR NAME	M/C NO.	CONTROLLED BUS VOLTS	I/P FILTER SIGNAL	AMPLIFIER SIGNAL	FEEDBACK SIGNAL	AUXILIARY SIGNALS SOURCE INPUT OUTPUT
AVIEMORE---11	1	0.793	0.793	1.600	0.010	

415

SYNCHRONOUS MACHINES

BUSBAR NAME	M/C NO.	ROTOR ANGLE (DEG.)	REF. MAIN/ALT.	POLE PKG. SLPD	ROTOR SLIC.	MECH. POWER MW	POWER OUTPUT ACTIVE MW	REACTIVE MVAR	TERM. VOLTAGE P.U.	TERM. CURRENT P.U.	FIELD VOLTAGE P.U.	FIELD CURRENT P.U.

AUTO VOLT REGULATORS -

BUSBAR NAME	M/C NO.	CONTROLLED BUS VOLTS	AMPLIFIER SIGNAL	I/P FILTER SIGNAL	FEEDBACK SIGNAL	AUXILIARY SIGNALS SOURCE INPUT OUTPUT
AVIEMORE---11	1	0.996	1.600	0.996	0.009	
BENMORE---16	1	0.996	1.000	0.996	-0.007	

SPEED GOVS + TURBINES -

BUSBAR NAME	M/C NO.	GOVERNOR SIGS(PU)	MP VALVE POSN(PU PWR(MW)	HP VALVE POSN(PU PWR(MW)	TURBINE POWERS (MW) HP IP LP
AVIEMORE---16	1	-3.090 -0.013	0.894 611.946		
BENMORE---16	1	-3.156 -0.006	0.961 531.754		

H.V.U.C.LINK

AC BUSBAR NAME	TYPE	VOLTAGES(PU) AU UC	CURRENTS(PU) AC DC	POWER INPUT MW MVAR	LINK MW LOSS MVAR	FIRING ANGLES (DEG) ALPHA DELTA GAMMA	REGULATOR TYPE SETTING
BENMORE--DC RECT	0.976 5.194	5.409 0.958	492.970 190.720	23.475	6.2 18.0	24.1 CRNT 1.010 (PU)
HAYWARDS--DC INV	0.965 4.899	5.409	-409.495 234.151		13.0	15.0

SYSTEM NO. 90   CASE NO. 1   TIME = 0.500 S.   STEP LENGTH = 0.0100 S.   STEP NO.= 52   MAX.ITS/STEP = 5   ITS/PRINT = 42

SYNCHRONOUS MACHINES

BUSBAR NAME	M/C NO.	ROTOR ANGLE (DEG.)	REF. MAIN/ALT.	POLE PKG. SLPD	ROTOR SLIC.	MECH. POWER MW	POWER OUTPUT ACTIVE MW	REACTIVE MVAR	TERM. VOLTAGE P.U.	TERM. CURRENT P.U.	FIELD VOLTAGE P.U.	FIELD CURRENT P.U.

417

SYSTEM NO. 90   CASE NO. 1   TIME = 0.600 S.   STEP LENGTH = 0.0100 S.   STEP NO.= 62   MAX.ITS/STEP = 5   ITS/PRINT = 41

H.V.D.C.-LINK

AC BUSBAR NAME	TYPE	VOLTAGES(PU) AC DC	CURRENTS(PU) DC	POWER INPUT MW MVAR	LINK MW LOSS	FIRING ANGLES (DEG) ALPHA DELTA GAMMA	TYPE	REGULATION SETTING

SYSTEM NO. 70   CASE NO. 1   TIME = 0.800 S.   STEP LENGTH = 0.0100 S.   STEP NO.= 80   MAX.ITS/STEP = 4   ITS/PRINT = 80

SYNCHRONOUS MACHINES

SPEED GOVS + TURBINES

H.V.D.C.-LINK

SYSTEM NO. 70   CASE NO. 1   TIME = 1.000 S.   STEP LENGTH = 0.0100 S.   STEP NO.= 97   MAX.ITS/STEP = 11   ITS/PRINT = 82

SYNCHRONOUS MACHINES

419

AUTO VOLT
REGULATORS

BUSBAR NAME	M/C NO.	CONTROLLED BUS VOLTS	I/P FILTER SIGNAL	AMPLIFIER SIGNAL	FEEDBACK SIGNAL	AUXILIARY SIGNALS INPUT OUTPUT
AVIEMORE---16	1	1.092	1.092	1.249	-0.006	
BENMORE---16	1	0.998	0.998	1.000	-0.007	

SPEED GOVS
+ TURBINES

BUSBAR NAME	M/C NO.	GOVERNOR SIGS(PU)	HP VALVE POSN(PU) PWR(MW)	IPC VALVE POSN(PU) PWR(MW)	TURBINE POWERS (MW) HP LP
AVIEMORE---11	1	-1.985 -0.021	0.887 645.775		
BENMORE---16	1	-1.764 -0.018	0.949 530.695		

H.V.D.C.LINK

AC BUSBAR NAME	TYPE	VOLTAGES(PU) AC DC	CURRENTS(PU) AC DC	POWER INPUT MW MVAR	LINK MW LOSS	FIRING ANGLES (DEG) ALPHA DELTA GAMMA	REGULATOR TYPE SETTING
BENMORE---DC RECT		0.778 5.294	5.405 5.405	478.447 213.252	23.445	10.5 18.0	CKNT 1.010 (PU)
HAYWARDS--DC INV		0.576 4.900	5.405 5.405	-475.001 235.594		20.7 14.9	

END OF CASE

NO. OF STEP ITERATIONS 420

END OF SYSTEM NO. 90

END OF PROGRAMME

Subject Index

420